Cellular and Biomolecular Recognition

Synthetic and Non-Biological Molecules

Edited by
Raz Jelinek

WILEY-VCH Verlag GmbH & Co. KGaA

The Editor

Prof. Raz Jelinek
Ben Gurion Univiversity of the Negev
Department of Chemistry
Staedler Minerva Center
84105 Beer Sheva
Israel

All books published by Wiley-VCH are carefully produced. Nevertheless, authors, editors, and publisher do not warrant the information contained in these books, including this book, to be free of errors. Readers are advised to keep in mind that statements, data, illustrations, procedural details or other items may inadvertently be inaccurate.

Library of Congress Card No.: applied for

British Library Cataloguing-in-Publication Data
A catalogue record for this book is available from the British Library.

Bibliographic information published by the Deutsche Nationalbibliothek
The Deutsche Nationalbibliothek lists this publication in the Deutsche Nationalbibliografie; detailed bibliographic data are available on the Internet at http://dnb.d-nb.de

© 2009 WILEY-VCH Verlag GmbH & Co. KGaA, Weinheim

All rights reserved (including those of translation into other languages). No part of this book may be reproduced in any form – by photoprinting, microfilm, or any other means – nor transmitted or translated into a machine language without written permission from the publishers. Registered names, trademarks, etc. used in this book, even when not specifically marked as such, are not to be considered unprotected by law.

Printed in the Federal Republic of Germany
Printed on acid-free paper

Cover design Adam-Design, Weinheim
Typesetting SNP Best-set Typesetter Ltd., Hong Kong
Printing betz-druck GmbH, Darmstadt
Bookbinding Litges & Dopf Buchbinderei GmbH, Heppenheim

ISBN: 978-3-527-32265-7

Cellular and Biomolecular Recognition

Edited by
Raz Jelinek

Further Reading

L.W. Miller (Ed.)

Probes and Tags to Study Biomolecular Function

for Proteins, RNA, and Membranes

2008
ISBN: 978-3-527-31566-6

B. Groner (Ed.)

Peptides as Drugs

Discovery and Development

2009
ISBN: 978-3-527-32205-3

P.J. Walla

Modern Biophysical Chemistry

Detection and Analysis of Biomolecules

2009
ISBN: 978-3-527-32360-9

S. Lutz, U.T. Bornscheuer (Eds.)

Protein Engineering Handbook

2 Volumes

2009
ISBN: 978-3-527-31850-6

R.A. Meyers (Ed.)

Proteins

From Analytics to Structural Genomics

2 Volumes

2006
ISBN: 978-3-527-31608-3

S. Dübel (Ed.)

Handbook of Therapeutic Antibodies

3 Volumes

2007
ISBN: 978-3-527-31453-9

C.A. Mirkin, C.M. Niemeyer (Eds.)

Nanobiotechnology II

More Concepts and Applications

2007
ISBN: 978-3-527-31673-1

Contents

Preface *XIII*

List of Contributors *XV*

1	**Development of Functional Materials from Rod-Like Viruses** *1*	
	Zhongwei Niu, Jianhua Rong, L. Andrew Lee, and Qian Wang	
1.1	Introduction *1*	
1.2	Overview *2*	
1.2.1	TMV *2*	
1.2.2	M13 Bacteriophage *3*	
1.3	Programmable Protein Shells *4*	
1.3.1	Chemical Modifications *5*	
1.3.2	Genetic Modifications *5*	
1.3.2.1	Genetic Modification of TMV *7*	
1.3.2.2	M13 Genetic Modification *7*	
1.3.3	Chemical Modification in Combination with Genetic Mutation *8*	
1.4	Templated Syntheses of Composite Materials *9*	
1.4.1	Synthesis of Inorganic Materials Using TMV as the Template *9*	
1.4.2	Bacteriophage M13 as the Template *11*	
1.5	Self-Assembly of Rod-Like Viruses *12*	
1.5.1	Controlled 1D Assembly *14*	
1.5.1.1	TMV Head-to-Tail Assembly *14*	
1.5.1.2	Conductive 1D TMV Composite Fibers *16*	
1.5.1.3	Weaving M13 Bacteriophage into Robust Fibers *16*	
1.5.1.4	Nanoring Structure *17*	
1.5.2	Fabrication of Thin Films by 2D Self-Assembly *17*	
1.5.3	Controlling the 3D Assembly of TMV and M13 *19*	
1.6	Virus-Based Device and Applications *20*	
1.7	Outlook *22*	
	References *23*	

Cellular and Biomolecular Recognition: Synthetic and Non-Biological Molecules. Edited by Raz Jelinek
Copyright © 2009 WILEY-VCH Verlag GmbH & Co. KGaA, Weinheim
ISBN: 978-3-527-32265-7

2 Biomimetic Nanoparticles Providing Molecularly Defined Binding Sites – Protein-Featuring Structures versus Molecularly Imprinted Polymers *31*
Kirsten Borchers, Sandra Genov, Carmen Gruber-Traub, Klaus Niedergall, Jolafin Plankalayil, Daniela Pufky-Heinrich, Jürgen Riegler, Tino Schreiber, Günter E.M. Tovar, Achim Weber, and Daria Wojciukiewicz

2.1 Introduction *31*
2.2 Core Materials and Functionalities *34*
2.2.1 Inorganic Core Materials *34*
2.2.1.1 Inorganic Crystalline Nanoparticles *34*
2.2.1.2 Particles with Silica Cores *35*
2.2.1.3 Metals and Metal Oxides *36*
2.2.2 Organic Core Materials *37*
2.2.2.1 Polymers, Lipids and Fullerenes *37*
2.3 Functional Shells *42*
2.3.1 Organic Shells *42*
2.3.2 MIPs *44*
2.3.2.1 Tools for MIP Development *44*
2.3.2.2 Bulk MIP and Proteins *45*
2.3.2.3 Nanospheric MIPs in General *47*
2.3.2.4 Nanospheric MIPs and Proteins *48*
2.4 Applications *49*
2.4.1 Biopurification *49*
2.4.1.1 Magnetic Nanoparticles *49*
2.4.1.2 MIPs with Magnetizable Cores *50*
2.4.2 Drug Delivery and Drug Targeting *51*
2.4.2.1 Nanoparticle Systems for Drug Delivery *51*
2.4.2.2 Ligands on Nanoparticle Surfaces *53*
2.4.2.3 Targeting of Specific Cells *53*
2.5 Products *55*
2.5.1 MIPs – Applications and Products *55*
2.5.2 Luminex Assay *55*
2.6 Conclusions *56*
References *57*

3 Interaction Between Silica Particles and Human Epithelial Cells: Atomic Force Microscopy and Fluorescence Study *69*
Igor Sokolov

3.1 Interaction of Silica with Biological Cells: Background *69*
3.2 Interaction of a Silica Particle with the Cell Surface: How It Is Seen with AFM *70*
3.2.1 AFM *70*
3.2.2 AFM on Cells *72*
3.2.2.1 Cell Culture *72*
3.2.2.2 AFM *72*

3.2.3	AFM Probe Preparations 73	
3.2.4	Models to Analyze the Cell Surface: Need for a Two-Layer Model 74	
3.2.5	Experimental Data 78	
3.2.5.1	Surface Brush on Cancer and Normal Cells 78	
3.2.5.2	Measurement of Adhesion: Silica Particle–Cell Interaction 80	
3.2.5.3	Can the Difference in Adhesion Be Used to Detect Cancer Cells? 83	
3.3	Ultra-Bright Fluorescent Silica Particles to Be Used to Study Interaction with Cells 84	
3.4	Ultra-Bright Fluorescent Silica Particles to Distinguish Between Cancer and Normal Cells 85	
3.4.1	Methods and Materials 86	
3.4.1.1	Spectrofluorometric and Optical Measurements of the Particles Attached to Cells 86	
3.4.1.2	Detection of Affinity of Fluorescent Silica Particles to Cells 87	
3.4.2	Experimental Results: Spectrofluorometric and Image Analysis of Cancer and Normal Cervix Cells 87	
3.5	Conclusions 89	
	References 89	

4   Chiral Molecular Imprinting as a Tool for Drug Sensing 97
Sharon Marx

4.1	Introduction 97	
4.2	Electrochemical Drug Sensors 101	
4.3	Optical Drug Sensors 103	
4.4	Mass Drug Sensors 106	
4.5	Conclusions and Summary 107	
	References 107	

5   Catalytic Antibodies for Selective Cancer Chemotherapy 111
Roy Weinstain and Doron Shabat

5.1	Introduction 111	
5.2	Catalytic Antibodies Designed for Prodrug Activation 111	
5.3	Catalytic Antibody 38C2 and Cancer Therapy 114	
5.3.1	General Approach for Prodrug Activation with Antibody 38C2 114	
5.3.2	Bifunctional Antibodies for Targeted Chemotherapy 116	
5.3.3	*In Vitro* and *In Vivo* Evaluations of Antibody 38C2-Catalyzed Prodrug Activation 118	
5.3.4	Polymer Directed Enzyme Prodrug Therapy: An Approach to Target Antibody 38C2 to a Tumor Site 120	
5.3.5	Chemical Adaptor Concept 123	
5.3.6	Self-Immolative Dendrimers Concept 125	
5.3.7	Prodrugs of Dynemicin and Doxorubicin Analogs 126	
5.4	Chemically Programmed Antibodies 128	
5.5	Outlook 134	
	References 134	

6	**Natural and Synthetic Stimulators of the Immune Response** *137*	
	Marine C. Raman and Dominic J. Campopiano	
6.1	Introduction *137*	
6.2	Lipopolysaccharide Endotoxin – A Potent Immunostimulatory Molecule *137*	
6.3	LPS Recognition *138*	
6.4	Septic Shock *139*	
6.5	LPS Biosynthesis *140*	
6.6	Minimal, Modified Lipid A *140*	
6.7	Isolation of "Natural" Kdo_2-Lipid A *143*	
6.8	Synthetic LPS, Lipid A and Their Uses *144*	
6.9	Structural Studies *147*	
6.9.1	LBP, CD14 and FhuA *147*	
6.9.2	MD-2/Lipid IVa Complex *149*	
6.9.3	TLR4–MD-2 Complex *151*	
6.10	Bacterial LPS-Binding Proteins *153*	
6.11	Sphingolipids – Essential Membrane Components of Mammals, Plants, Fungi, Yeast and Bacteria *153*	
6.12	Natural Killer T Cells and GSLs *158*	
6.13	GSL Antigens *159*	
6.14	Structure–Activity Relationships *161*	
6.15	Saccharide Modification *161*	
6.16	Ceramide Modification *163*	
6.17	High-Resolution Structural Analysis of Receptor–GSL Complexes *163*	
6.18	Conclusions *168*	
	References *169*	
7	**Supramolecular Assemblies of Polydiacetylenes for Biomolecular Sensing: Colorimetric and "Turn-On" Fluorescence Approaches** *177*	
	Guangyu Ma and Quan Jason Cheng	
7.1	Introduction *177*	
7.2	Vesicular PDA Sensors for Colorimetric Signaling of Bacterial Pore-Forming Toxin *179*	
7.3	Fabrication of "Turn-On" Fluorescence Vesicle Sensors with PDAs *181*	
7.4	"Mix-and-Detect" Type of "Turn-On" Fluorescence Sensor for Bacterial Toxin *186*	
7.5	Conclusions *189*	
	References *190*	
8	**Multivalent Synthetic Receptors for Proteins** *193*	
	Jolanta Polkowska, Peter Talbiersky, and Thomas Schrader	
8.1	Aminopyrazoles for β-Sheet Capping in Protein Misfolding Events *193*	
8.2	New Mechanisms of Enzyme Inhibition by Molecular Clips and Tweezers *202*	

8.3	Protein Surface Recognition by Tailor-Made Polymers	*210*
8.4	Conclusions and Outlook	*214*
	References	*215*

9 Analysis of Biological Interactions and Recognitions by Surface Plasmon Resonance *219*
Sang Jun Sim and Cuong Cao

9.1	Introduction to Surface Plasmon Resonance Technology	*219*
9.2	Working Principle of SPR	*220*
9.3	Sensor Surface Chemistry and Its Fabrications	*222*
9.4	Important Factors Impacting on the Performance of SPR-Based Analyses of Biological Interactions on the Nonbiological Transducer Surface	*227*
9.4.1	Nonspecific Interactions	*227*
9.4.2	Recognition Elements	*230*
9.4.3	Detection Formats	*230*
9.4.3.1	Direct Binding Assays	*230*
9.4.3.2	Sandwich Detection Assays	*232*
9.4.3.3	Competitive Detection Assays	*232*
9.4.3.4	Inhibition Detection Assays	*232*
9.4.4	Several Approaches for Sensitivity Enhancements of the SPR Bioassays	*233*
9.5	Localized SPR of Inorganic Nanoparticles for Analyses of Biological Interaction	*236*
	References	*239*

10 Membrane-Active Natural and Synthetic Peptides and Peptidomimetics *247*
Regine Willumeit

10.1	Introduction	*247*
10.1.1	Structure of Cell Membranes	*247*
10.1.2	Biophysical Properties of Phospholipids	*248*
10.2	Mode of Action of Membrane-Active Peptides	*249*
10.2.1	Carpet Model	*249*
10.2.2	Barrel Stave Model	*250*
10.2.3	Toroidal or Wormhole Model	*250*
10.2.4	Two-State Model	*251*
10.2.5	"Detergent-Like" Model	*251*
10.2.6	Molecular Mechanism of Membrane Disruption	*251*
10.3	Natural Peptides	*252*
10.3.1	α-Helical Peptides	*253*
10.3.2	β-Sheet Peptides	*254*
10.3.3	Cyclic Peptides	*254*
10.4	Synthetic Peptides	*254*
10.5	Peptidomimetics	*255*

10.6	Conclusions *256*
	References *258*

11 Luminescent Quantum Dot Fluorescence Resonance Energy Transfer-Based Probes in Cellular and Biological Assays *265*

Lifang Shi, Nitsa Rosenzweig, and Zeev Rosenzweig

11.1	Introduction *265*
11.2	Luminescent QDs *266*
11.3	FRET *268*
11.4	QD FRET-Based Protease Probes *269*
11.5	Summary and Conclusions *273*
	References *274*

12 New Proteins for New Sensing Methodologies: The Case of the Protein-Binding Family *281*

Vincenzo Aurilia, Maria Staiano, Mosè Rossi, and Sabato D'Auria

12.1	Introduction *281*
12.2	Galactose/Glucose-Binding Protein from *Escherichia coli* *282*
12.3	Glutamine-Binding Protein from *Escherichia coli* *285*
12.4	Trehalose/Maltose-Binding Protein from the Hyperthermophilic Archaeon *Thermococcus litoralis* *287*
12.5	Lipocalins and Odorant-Binding Protein *289*
12.5.1	Structural Characterization of Pig OBP *291*
12.5.2	Functional Characterization and Biotechnological Application of pOBP *292*
12.6	Conclusions *294*
	References *294*

13 Methods of Analysis for Imaging and Detecting Ions and Molecules *299*

Sung Bae Kim, Hiroaki Tao, and Yoshio Umezawa

13.1	Fluorescent and Luminescent Proteins *299*
13.1.1	GFP and Its Variants *299*
13.1.2	Luciferases *308*
13.2	Functional Peptides *312*
13.3	Representative Technologies for Molecular Imaging *313*
13.3.1	Classical Methods for Sensing Bioactive Small Molecules *313*
13.3.2	Fluorescence (or Förster) Resonance Energy Transfer *317*
13.3.2.1	Probes for Determining Protein Phosphorylation *319*
13.3.2.2	Probes for Determining Steroid-Activated Protein–Protein Interactions *319*
13.3.2.3	A "Flip-Flop"-Type Indicator *319*
13.3.3	Bioluminescence Resonance Energy Transfer *321*
13.3.4	Protein-Fragment Complementation Assay *323*
13.3.5	Intein-Mediated Protein-Splicing Assay *326*

13.3.6	Circular Permutation of Fluorescent and Luminescent Proteins	*331*
13.4	Conclusions and Perspectives	*332*
13.4.1	Frontier Research in Analysis	*332*
13.4.2	Analytical Achievements Inspired by Nature	*332*
13.4.3	Future Directions	*333*
	References	*334*

Index *339*

Preface

Understanding interactions among biological molecules and between biomolecules and cells is at the core of diverse and intersecting disciplines, including biology, biochemistry, pharmaceutics, biophysics, and related fields. However, studying biomolecular interactions in "real life"–in actual cells, tissues, and the like–has generally been a formidable task due to the complexity of biological systems. The utilization of chemical approaches and in particular the development of innovative *biomimetic platforms* have greatly advanced our knowledge of intimate molecular aspects pertaining to interactions between biological entities, and the significance of such interactions in a larger biological context. This book aims to illuminate novel approaches which provide insight into molecular recognition in biological systems.

The Chapters and contributors to this volume span diverse experimental methodologies and research fields. The systems discussed by the authors, while varied, all underlie the important impact multidisciplinary research has had upon elucidating fundamental biomolecular and cellular factors. Specifically, **Cheng** (Chapter 7) presents the application of unique biomimetic polymer assemblies, exhibiting colorimetric and fluorescent properties, for analysis of biological processes. **Campopiano** (Chapter 6) explores molecular interactions involving synthetic and natural molecules in immune systems, critical events for eliciting immune response. **Willumeit** (Chapter 10) focuses on the significance of *membrane interactions* of short peptides in varied biological processes.

Novel and innovative chemical approaches have contributed to our understanding of diverse biomolecular processes. **Rosenzweig** (Chapter 11) discusses the contribution of fluorescent and luminescent quantum dots for analysis of varied biological processes. **Tovar** (Chapter 2) depicts another interesting class of nanoparticles, designed to provide defined binding sites for biological molecules. Elegant synthetic biomimetic systems chemically designed to bind specific proteins are presented by **Schrader** (Chapter 8), while **Wang** (Chapter 1) describes intriguing virus-like particles and their biological functions. **Shabat** (Chapter 5) presents therapeutic applications of *catalytic antibodies*, a promising technology based upon molecular recognition of transient biological species, and **Marx** (Chapter 4) outlines the pharmaceutical and therapeutic potential of the *molecular imprinting* concept, similarly directly related to induced molecular recognition.

Advanced bioanalytical techniques play a significant role in elucidating the molecular basis of diverse biological processes. **Sim** (Chapter 9) summarizes the applications of *surface Plasmon resonance (SPR)*, a powerful methodology for studying biomolecular interactions. **Sokolov** (Chapter 3) discusses application of atomic force microscopy and fluorescence methods for analysis of silica particle interactions with cells. **D'Auria** (Chapter 12) describes the utilization of advanced fluorescence spectroscopy for biosensing applications, and **Umezawa** (Chapter 13) reviews the comprehensive body of work aimed at developing new approaches for studying molecular interactions involving ions and charged molecules.

Raz Jelinek

List of Contributors

Vincenzo Aurilia
CNR
Institute of Protein Biochemistry
Laboratory for Molecular Sensing
Via Pietro Castellino, 111
80131 Naples
Italy

Kirsten Borchers
Fraunhofer Institute for
Interfacial Engineering and
Biotechnology IGB
Nobelstrasse 12
70569 Stuttgart
Germany

Dominic J. Campopiano
University of Edinburgh
School of Chemistry
EaStCHEM
West Mains Road
Edinburgh EH9 3JJ
UK

Cuong Cao
Sungkyunkwan University
Department of Chemical
Engineering
Nano-optics and Biomolecular
Engineering National Laboratory
300 Chunchun dong, Jangan ku
Suwon 440-746
South Korea

Quan Jason Cheng
University of California
Department of Chemistry
501 Big Springs Road
Riverside, CA 92521
USA

Sabato D'Auria
CNR
Institute of Protein Biochemistry
Laboratory for Molecular Sensing
Via Pietro Castellino, 111
80131 Naples
Italy

Sandra Genov
University of Stuttgart
Institute for Interfacial Engineering
Nobelstrasse 12
70569 Stuttgart
Germany

Carmen Gruber-Traub
Fraunhofer Institute for
Interfacial Engineering and
Biotechnology IGB
Nobelstrasse 12
70569 Stuttgart
Germany

Cellular and Biomolecular Recognition: Synthetic and Non-Biological Molecules. Edited by Raz Jelinek
Copyright © 2009 WILEY-VCH Verlag GmbH & Co. KGaA, Weinheim
ISBN: 978-3-527-32265-7

Sung Bae Kim
National Institute of
Advanced Industrial Science and
Technology (AIST)
16-1 Onogawa, Tsukuba
Ibaraki 305-8569
Japan

L. Andrew Lee
University of South Carolina
Department of Chemistry and
Biochemistry and NanoCenter
631 Sumter Street
Columbia, SC 29208
USA

Guangyu Ma
University of California
Department of Chemistry
501 Big Springs Road
Riverside, CA 92521
USA

Sharon Marx
Israel Institute for Biological
Research
Department of Physical
Chemistry
PO Box 19
Ness Ziona 74100
Israel

Klaus Niedergall
University of Stuttgart
Institute for Interfacial
Engineering
Nobelstrasse 12
70569 Stuttgart
Germany

Zhongwei Niu
University of South Carolina
Department of Chemistry and
Biochemistry and NanoCenter
631 Sumter Street
Columbia, SC 29208
USA

Jolafin Plankalayil
University of Stuttgart
Institute for Interfacial Engineering
Nobelstrasse 12
70569 Stuttgart
Germany

Jolanta Polkowska
Universität Duisburg–Essen
Fachbereich Chemie
Institut für Organische Chemie
Universitätsstrasse 5
45117 Essen
Germany

Daniela Pufky-Heinrich
Fraunhofer Institute for
Interfacial Engineering and
Biotechnology IGB
Nobelstrasse 12
70569 Stuttgart
Germany

Marine C. Raman
University of Edinburgh
School of Chemistry
EaStCHEM
West Mains Road
Edinburgh EH9 3JJ
UK

Jürgen Riegler
University of Stuttgart
Institute for Interfacial
Engineering and Fraunhofer
Institute for Interfacial
Engineering and Biotechnology
IGB
Nobelstrasse 12
70569 Stuttgart
Germany

Jianhua Rong
University of South Carolina
Department of Chemistry and
Biochemistry and NanoCenter
631 Sumter Street
Columbia, SC 29208
USA

Nitsa Rosenzweig
University of New Orleans
Department of Chemistry and the
Advanced Materials Research
Institute (AMRI)
2000 Lakeshore Drive
New Orleans, LA 70148
USA

Zeev Rosenzweig
University of New Orleans
Department of Chemistry and the
Advanced Materials Research
Institute (AMRI)
2000 Lakeshore Drive
New Orleans, LA 70148
USA

Mosè Rossi
CNR
Institute of Protein Biochemistry
Laboratory for Molecular Sensing
Via Pietro Castellino, 111
80131 Naples
Italy

Thomas Schrader
Universität Duisburg–Essen
Fachbereich Chemie
Institut für Organische Chemie
Universitätsstrasse 5
45117 Essen
Germany

Tino Schreiber
University of Stuttgart
Institute for Interfacial Engineering
Nobelstrasse 12
70569 Stuttgart
Germany

Doron Shabat
Tel-Aviv University
School of Chemistry
Department of Organic Chemistry
Raymond and Beverly Sackler Faculty
of Exact Sciences
Tel Aviv 69978
Israel

Lifang Shi
University of New Orleans
Department of Chemistry and the
Advanced Materials Research Institute
(AMRI)
2000 Lakeshore Drive
New Orleans, LA 70148
USA

Sang Jun Sim
Sungkyunkwan University
Department of Chemical
Engineering
Nano-optics and Biomolecular
Engineering National Laboratory
300 Chunchun dong, Jangan ku
Suwon 440-746
South Korea

Igor Sokolov
Clarkson University
Department of Physics
Department of Chemical and
Biomolecular Science
NanoBio Laboratory (NABLAB)
8 Clarkson Ave.
Potsdam, NY 13699
USA

Maria Staiano
CNR
Institute of Protein Biochemistry
Laboratory for Molecular Sensing
Via Pietro Castellino, 111
80131 Naples
Italy

Peter Talbiersky
Universität Duisburg–Essen
Fachbereich Chemie
Institut für Organische Chemie
Universitätsstrasse 5
45117 Essen
Germany

Hiroaki Tao
National Institute of
Advanced Industrial Science and
Technology (AIST)
16-1 Onogawa, Tsukuba
Ibaraki 305-8569
Japan

Günter E.M. Tovar
University of Stuttgart
Institute for Interfacial Engineering
and Fraunhofer Institute for Interfacial
Engineering and Biotechnology IGB
Nobelstrasse 12
70569 Stuttgart
Germany

Yoshio Umezawa
University of Tokyo
Department of Chemistry
School of Science
7-3-1 Hongo Bunkyo-ku
Tokyo 113-0033
Japan

Qian Wang
University of South Carolina
Department of Chemistry and
Biochemistry and NanoCenter
631 Sumter Street
Columbia, SC 29208
USA

Achim Weber
University of Stuttgart
Institute for Interfacial Engineering
and Fraunhofer Institute for Interfacial
Engineering and Biotechnology IGB
Nobelstrasse 12
70569 Stuttgart
Germany

Roy Weinstain
Tel-Aviv University
School of Chemistry
Department of Organic Chemistry
Raymond and Beverly Sackler Faculty
of Exact Sciences
Tel Aviv 69978
Israel

Regine Willumeit
GKSS Research Center
Institute for Materials Research
Abt. WFS – Geb. 03
Max-Planck-Strasse 1
21502 Geesthacht
Germany

Daria Wojciukiewicz
University of Stuttgart
Institute for Interfacial Engineering
Nobelstrasse 12
70569 Stuttgart
Germany

1
Development of Functional Materials from Rod-Like Viruses

Zhongwei Niu, Jianhua Rong, L. Andrew Lee, and Qian Wang

1.1
Introduction

Developing functional materials at the nanometer (10^{-9} m) scale with well-defined structures has a great impact on the fabrication of novel biomedical, optical, acoustic and electronic devices [1–6]. In particular, using biological building blocks as templates in materials synthesis is an exciting and emerging area of research [7–12]. Among all available scaffolds, viruses and virus-like particles (VLPs) have attracted much attention due to their well-defined structural features, unique shapes and sizes, genetic programmability, and simple and robust chemistries [12–16]. The viruses provide a wide array of shapes such as rods and spheres, and a variety of sizes spanning from tens to hundreds of nanometers. As naturally available supramolecular systems, viruses are assembled by multiple noncovalent interactions. These protein structures are evolutionary tested, multifaceted systems with highly ordered spatial arrangement, and natural cell targeting and genetic information storing capabilities. The years of dissecting the details of virus infection, replication and assembly pathways have imparted a wealth of information on the stabilities and functionalities of these materials. The structural information provided by X-ray crystallography, nuclear magnetic resonance studies and previous mutagenesis studies lays the firm, primary foundation for viral vector redesigns. Intricately fashioned, multifunctional viruses and VLPs for various applications have spun off from viral vector-based gene therapy and vaccine development founded on the polyvalent antigen display on virus coat proteins. Virus-based assemblies and new materials development is a major hub of current bionanotechnology.

The aim of this chapter is to provide a brief overview of an important subset of viruses, rod-like viruses, as building blocks for the development of novel nanomaterials through genetic and chemical modifications (covalent and noncovalent chemistries), followed by the assembly of such particles into higher ordered structures (Scheme 1.1). Rod-like tobacco mosaic virus (TMV) and M13 bacteriophage are two leading examples as new bio-scaffolds owing to the combined chemical

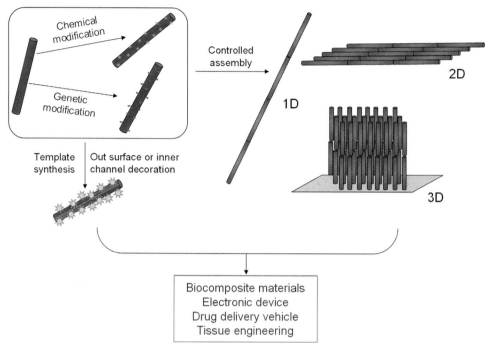

Scheme 1.1 Schematic illustration of self-assembly and preparation of functional materials using rod-like viruses.

functionality of viral coat proteins, monodispersity, liquid crystalline-like organization, and length scales that bridge the gap between top-down and bottom-up strategies for materials development.

1.2
Overview

1.2.1
TMV

TMV, among other rod-like viruses (potato virus X and tomato mosaic virus), shows an unheralded amount of research dating as far back as 1899 as an infectious material [17]. The plant virus persists as a classic example of rod-like plant viruses that has now been studied for over 100 years [18, 19]. Even so, the virus as a supramolecular building block has only been recently envisioned. In this section, we describe a simplified overview of the virus pertinent to defining the benefits of this virus as a supramolecular building block. For an in-depth review of the virus, additional references are listed [20, 21]. The original, wild-type strain (noted as the U1 strain) and many other variants have been extensively studied

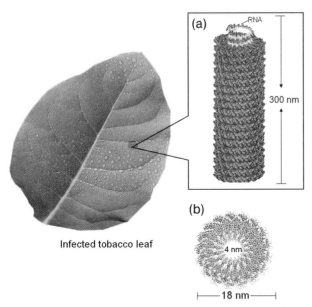

Figure 1.1 (a) Helical organization of TMV coat proteins and (b) cross-section of TMV.

for agricultural purposes, biopharmaceutical applications and recombinant protein expression. The meta-stable structure, which unwinds when inside its host, exhibits some remarkable stability. TMV retains structural integrity even at pH from 3.5 to 9 and up to temperatures of 50 °C. Its diameter measures around 18 nm with a 4-nm cylindrical cavity along the central core that runs through the entire length of the virus (Figure 1.1). The length of native TMV (300 nm) is defined by the encapsulated genomic RNA that stabilizes the 2130 coat proteins that assemble helically around the single-stranded RNA. Interestingly, the coat protein of TMV alone forms three different structural aggregates depending on solvent conditions (i.e., helical, disk and A-protein). These structural variants and a four-layer aggregate of TMV were crystallized and resolved at high resolution, yielding valuable information for understanding the assembly process of TMV [22]. Finally, a great advantage has been that TMV can be economically and easily obtained in large quantities (gram quantities) from infected tobacco plants with a simple purification procedure.

1.2.2
M13 Bacteriophage

M13 is a filamentous bacteriophage composed of circular, single-stranded DNA, which is 6407 nucleotides long encapsulated by approximately 2700 copies of the major coat protein P8, and capped with five copies of four different minor coat proteins (P9, P7, P6 and P3) on the ends (Figure 1.2) [23, 24]. The minor coat protein P3 attaches to the receptor at the tip of the F pilus of the host *Escherichia*

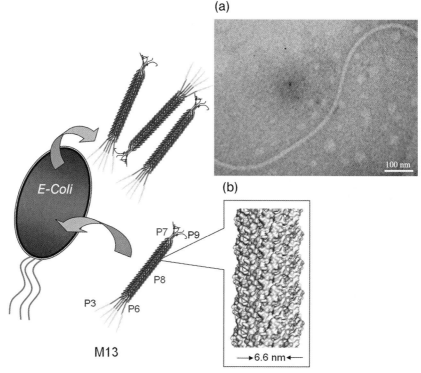

Figure 1.2 (a) TEM image of a M13 particle. (b) Schematic illustration of the helical organization of M13 coat protein P8.

coli. The infection causes turbid plaques in *E. coli*. At other end of the filament are five copies of the surface exposed pIX (P9) and a more buried companion protein, pVII (P7). These two proteins are very small, containing only 33 and 32 amino acids, respectively, although some additional residues can be added to the N-terminal portion of each which are then presented on the outside of the coat. The major coat protein is primarily assembled from a 50 amino acid called pVIII (or P8), which is encoded by gene VIII (or G8) in the phage genome. For a wild-type M13 particle, it takes about approximately 2700 copies of P8 to make the coat about 880 nm long. The diameter of M13 is around 6.6 nm. The coat's dimensions are flexible and the number of P8 copies adjusts to accommodate the size of the single-stranded genome it packages.

1.3
Programmable Protein Shells

Both viruses comprise of an outer protein shell, also known as the virus capsid, which houses the genetic material (RNA in the case of TMV and DNA in M13). These genetic materials possess all the essential information required for the virus to propagate within the host. Hence, by simply altering the genetic codes of the viruses, in particular the gene encoding for the virus capsid, we can redesign the outer shell of the virus with various functionalities. This trait provides an immense advantage for biological systems over synthetic systems. By carefully manipulating the template via genetic and chemical tools, M13 bacteriophage and TMV can be molded with the precise control over the spatial layout of functional groups. Many studies over the past few years highlight this ability of biological templates in materials development. We begin with examples from chemical modification of TMV, followed by molecular cloning in combination with chemistries of both viruses to yield highly uniform templates.

1.3.1
Chemical Modifications

Much of the initial virus chemistries resonate the conventional bioconjugation strategies, targeting endogenous amino acids, such as lysines, glutamic or aspartic acids and cysteines. Less commonly targeted functional groups, such as the phenol ring of tyrosines, have also been incorporated into this chemistry strategy. The systematic characterization on cowpea mosaic virus [25–29], followed by studies with cowpea chlorotic mottle virus, bacteriophage MS2, heat shock protein and turnip yellow mosaic virus, has shed light on the unique chemical reactivities and physical properties of these viruses. It has been reported that the lysines of some filamentous bacteriophages could be addressed using similar bioconjugation strategies [30, 31]. For TMV, only a selected tyrosine residue on the exterior of TMV and glutamates 97 and 106 in the interior appear to be accessible based on the crystal structure (Figure 1.3a). Francis *et al.* showed that Glu97 and Glu106 can be modified by attaching amines through a carbodiimide coupling reaction [33]. For the exterior surface modification, Francis *et al.* have reported the tyrosine residues (Tyr139) of TMV as a viable site for chemical ligation using the electrophilic substitution reaction at the *ortho* position of the phenol ring with diazonium salts [33]. Although this reaction is very efficient, it has two distinct disadvantages: the synthesis of desired starting materials is difficult and the reaction is only suitable for electron-deficient anilines, which dramatically impedes its potential applications. Recently, Wang *et al.* have reported that an alkyne group can be quantitatively attached to tyrosine residues by diazonium coupling and a sequential copper-catalyzed azide–alkyne cycloaddition reaction with azides can efficiently conjugate a wide range of compounds which include florescent dye molecules, small peptides and polymers to the surface of TMV (Figure 1.3) [32].

1.3.2
Genetic Modifications

Many of the viruses in isolated forms are complex macromolecular assemblies of metabolically inert molecules, which can be chemically modified. Alternatively, the viruses can be genetically reprogrammed in their hosts with foreign peptides to express an antigen [34, 35] or to alter the affinity of the recombinant viral particle for

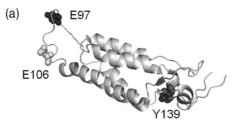

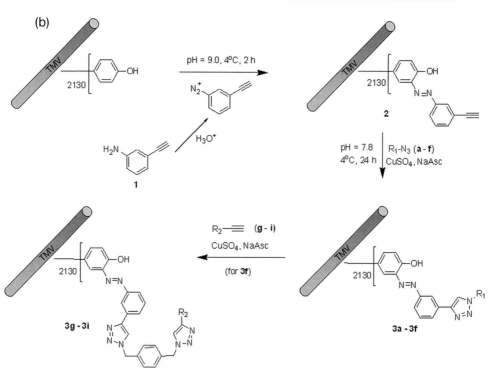

Figure 1.3 (a) Reactive sites for the covalent modification of TMV, i.e., Tyr(Y)139, Glu(E)97 and Glu(E)106 are indicated in a single TMV capsid monomer. (b) Illustration of TMV modification by a two-step sequential reaction. (Part of this figure was adapted with permission from [32]).

different cell surface receptors [36, 37]. Furthermore, single amino acid substitution mutagenesis allows the site-specific incorporation of reactive amino acids, such as lysines or cysteines, or non-natural amino acids [38] on the virus coat protein for regioselective chemical modifications. Using such methods, Schultz *et al.* have incorporated poly(ethylene glycol) [39], alkyne modified amino acids [40] and photoisomerizable amino acids into the proteins [41]. *In vivo* and *in vitro* protein expression systems derived from cell lysates have also been used to drive viral protein synthesis with the non-natural amino acids [40, 42]. Many of the viruses have their X-ray structures resolved at near-atomic resolution, facilitating the generation of the recombinant nanoparticles. These alterations have ranged from single amino acid substitutions to entire protein domain incorporations [43, 44]. To that end, the molecular cloning techniques are well integrated into the generation of hybrid viruses with novel biological, chemical and physical properties.

1.3.2.1 Genetic Modification of TMV

TMV is highly immunogenic in mammalian hosts, which is ideal for adapting TMV-based vectors for antigen display. An initial study by Haynes *et al.* demonstrated the feasibility of incorporating immunogenic peptides as fusion proteins on the plant virus capsids [45]. Short peptide fragments of up to 21 residues had been repeatedly fused to the C-terminus of TMV using a leaky "UAG" stop codon without losing viral replication or assembly [46, 47]. In other studies, a variety of short inserts were made between Ser154 and Gly155 using polymerase chain reaction-based site-directed mutagenesis [48, 49]. Longer peptide sequences of up to 25 amino acids had also been fused to the C-terminus by deleting four to six amino acids at the carboxyl end [50]. Culver *et al.* designed and functionalized the cysteine-mutated TMV particles with fluorescent dyes and the modified TMV particles were then partially disassembled to expose the single-stranded viral RNA. The exposed single-stranded RNA strand was then utilized to hybridize to complementary DNA sequences patterned on surfaces [51, 52]. Francis *et al.* expressed TMV coat protein in a bacterial system to generate cysteine-substituted TMV coat proteins for the incorporation of light-harvesting motifs [53].

1.3.2.2 M13 Genetic Modification

It is known that M13 phage is a high production rate virus, which has a single-strand DNA, approximately 2700 copies of the P8 proteins, five copies each of the P3 and P6 proteins at one end, and five copies each of the P7 and P9 proteins at the other end. The functionalities of these protein groups localized at different locations on a viral particle can be rationally altered independently via genetic engineering [54]. One of the genetic engineering technologies, phage display, has proven to be a powerful technology for selecting polypeptides with the desired biological and physicochemical properties from large molecular libraries. The phage display techniques have been reviewed extensively [55]. Basically, libraries of random DNA sequences can be fused to the genes encoding coat proteins. These are expressed and displayed as fusion peptides on the surface of the phage. The phages are then passed over an immobilized target (such as a biological receptor or inorganic substrate). Nonbinding phages are washed off, and binders are eluted and amplified in *E. coli*. After repeating this process several times, the best binders are sequenced to deduce the amino acid sequences of the displayed peptides. As an example, the type 3 library of M13 viruses is commercially available (New England BioLabs), in which the native M13 genome has been engineered to express around 1×10^9 different random peptide sequences as N-terminal fusions (three to five copies) on the P3 viral protein coat.

In order to synthesize and assemble various inorganic nanowires, Belcher's group genetically modified and expressed the P8 protein by using a phagemid system, which resulted in the fusion of the substrate-specific peptides (e.g., A7, Z8 and J140 peptides for ZnS and CdS, respectively) to the N-terminus of the P8 protein [12, 56]. A type 8 library was constructed by fusing eight random amino acids into the N-terminus of all the P8 proteins with a random population of 10^7–10^8 [57]. This library employs a modified M13KE phage vector by generating restriction sites, *Pst*I and *Bam*HI, in G8 through mutagenesis for the insertion of random codons [G$nm(nnm)_6nn$G], where n can be G, C, A or T and m can T or G. Genome engineering phage employed in the type 8 library can produces 100% expression and monodispersed viral particles [54]. In order to synthesize and assemble various inorganic materials and structures, Belcher's group used a general biopanning technique to select some functional binding motifs on P8 by exposing the type 8 phage library to certain substrate [58]. Weiss also reported variants of the M13 bacteriophage that enable high-copy display of monometic and oligomeric proteins, such as human growth hormone and streptavidin, on major coat protein P8 of the surface of phage particles [59, 60]. Additionally, a type 8-3 phage could be produced with different binding motifs on both P3 and P8 proteins by combining gene 3 (G3) and gene 8 (G8) insertions in a single viral genome after specific binding motifs for targeted substrates were selected from separate P3 and P8 libraries. Therefore, more complicated composite materials could be produced [58]. In addition, P7 and P9 can also be amenable to modification. Janda *et al.* utilized P7 and P9 fusions to display antibody heavy and light chain variable regions [61]. A hexahistidine peptide (AHHHHHH), which binds to Ni(II)-nitrilotriacetic (Ni-NTA) acid complex, was fused to the N-terminus of P9 reported by Belcher's group [62].

1.3.3
Chemical Modification in Combination with Genetic Mutation

For most of viral particles, there are no reactive cysteine residues exposed to exterior surface – a reasonable assumption being that evolution had disfavored particles forming interparticle cross-links via disulfide bonds. This presents the unique opportunity to genetically insert the cysteine residue on strategic locations of viruses and protein shells, after which the sulfhydryl group can be selectively targeted with thiol-selective reagents. Culver *et al.* designed and functionalized the cysteine-substituted TMV particles with fluorescent dyes and the modified TMV particles were then partially disassembled to expose the single-stranded viral RNA [51]. The exposed single-stranded RNA strand was then utilized to hybridize to complementary DNA sequences patterned on surfaces [51, 52]. Francis *et al.* expressed TMV coat protein in a bacterial system to generate cysteine-substituted TMV coat proteins, which were modified with fluorescent chromophores for the purpose of generating a light-harvesting system. By controlling the pH and ionic strength, the proteins self-assembled into long fibrous structures that were capable of positioning the chromophores for efficient energy transfer [53]. These studies highlight an important feature of viruses – chemically reactive groups can be genetically engineered to selectively position drug molecules, imaging agents and biologically relevant molecules on the three-dimensional (3D) template [63].

1.4 Templated Syntheses of Composite Materials

1.4.1 Synthesis of Inorganic Materials Using TMV as the Template

The polar outer and inner surfaces of TMV have been widely exploited as templates to grow metal or metal oxide nanoparticles such as CdS, PbS, gold, nickel, cobalt, silver, copper, iron oxides, CoPt, $FePt_3$ and silica (Table 1.1) [13, 71–75]. From electrophoretic measurements, the isoelectric point (pI) of TMV is around 3.4. At neutral pH, the TMV surface has net negative charge. In order to achieve successful coating, the deposition conditions should be varied in order to match the interaction between the viral surface and the deposition precursor. In the case of silica coating, the reaction pH should be below 3 [64]. As a result, the positively charged TMV surface will have a strong interaction with anionic silicate sols

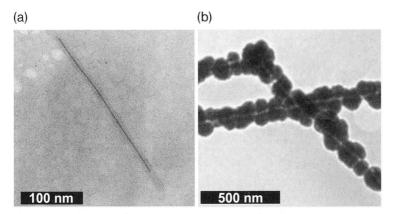

Figure 1.4 TEM images of (a) a single TMV containing a long nickel wire inside the central channel and (b) nickel coated on the outer surface of TMV. (Adapted with permission from [67]).

Table 1.1 Summary of the inorganic nanowires synthesized with native TMV as template.

Composition[a]	Activation[b]	Reducing reagent	pH[c]	Inner/outer[d]
PbS	none	none	5	outer [64]
CdS	none	none	7	outer [64]
Fe(II)/Fe(III) oxide	none	none	9	outer [64]
Silica	none	none	2.5	outer [64]
			8.8	outer [65]
CoPt	none	none	NA	inner [66]
FePt$_3$	none	none	NA	inner [66]
Ni	Pd(II) or Pt(II)	dimethylamine borane (DMAB)	5	inner [67, 68]
	Cu(III)			inner [67, 68]
Co	Pd(II) or Pt(II)	DMAB	5	inner [68]
	Cu(III)			inner and outer [67]
Ag	none	formaldehyde	NA	outer [69]
Au	none	ascorbic acid	6	inner and outer [67]
Ru	none	NaH$_2$PO$_2$ or DMAB	NA	inner and outer [67]
Cu	Pd(II)	DMAB	7.5	inner [70]

a) The final composition of the surface coating on TMV.
b) The predeposition of reducing reagents; "none" = no activation was necessary.
c) NA = pH value not mentioned in the original literature.
d) Indicates the deposition was either inside the inner channel or at the outer surface of TMV.

formed by the hydrolysis of tetraethyl orthosilicate. In comparison, CdS, PbS and iron oxide can be successfully coated on the outer surface at near-neutral pH by specific metal-ion binding with the glutamate and aspartate residues [64]. As for metal deposition, in some case, a suitable activation agent is needed in order to realize successful coating [67]. Pd(II) and Pt(II) are two typical activation agents. The metal deposition can happen either inside the inner channel or at the outer surface of TMV (Figure 1.4) [67]. Furthermore, genetically engineered TMV can improve deposition of metal onto the surface [76, 77]. Basically, native TMV was genetically altered to display multiple metal-binding sites through the insertion of two cysteine residues within the N-terminus of the virus coat protein. *In situ* chemical reductions can successfully deposit the silver, gold and palladium clusters coating onto the genetically modified TMV without any activation agent. In comparison to native TMV, a much higher density of metal coating was observed on the cysteine-inserted TMV.

1.4.2
Bacteriophage M13 as the Template

The M13 viral system is an attractive template for the synthesis and assembly of various materials and structures because of the programmable protein functionalities by genetic engineering and chemical modification, as mentioned previously. In 1992, Stanley Brown pioneered the idea of using bacterial display for the screening and binding of inorganic materials [78]. Belcher et al. extended this phage display technique to decorate M13 virus with different binding peptides with specific recognition for inorganic materials, such as semiconductor materials (e.g., GaAs, InP, ZnS and CdS) and magnetic materials like FePt, CoPt, cobalt and metal gold, or composites of them (Table 1.2) [14, 56, 76, 79, 82]. One of the advantages of this idea is that exact genetic copies of the virus scaffold are easily reproduced in the bacterial host because the protein sequences responsible for these attributes are gene-linked and contained within the capsid of the virus. Another advantage is that the exquisite structure of virus leads to a viable means of synthesizing and

Table 1.2 Summary of the inorganic nanowires synthesized with native M13 as template.

Composition	Location	Binding peptide	Peptide sequence	Crystal phase of materials	References
ZnS	P3, P8	A7	CNNPMHQNC	wurtzite	[12, 14, 56, 79]
ZnS	P3, P8	Z8	VISAHAGSSAAL	zinc blend	[56]
CdS	P3, P8	J140	SLTPLTTSHLRS	wurtzite	[12]
Streptavidin	P3	s1	SWDPYSHLLQHPQ		[58, 80]
Ni	P9	Ni-NTA	AHHHHHH		[62]
CoPt	P3, P8	CP7	CNAGDHANC		[14]
FePt	P3, P8	FP12	HNKHLPSTQPLA	L10	[14, 81]
Au/CaSe	P8/P3	P8#9/s1	VSGSSPDS/SWDPYSHLLQHPQ		[58, 80]
Co_3O_4	P8	E4	EEEE		[16]
Au/Co_3O_4	P8	AuE4	LKAHLPPSRLPS/EEEE		[16]

Single-letter abbreviations for the amino acid residues are as follows: A, Ala; C, Cys; D, Asp; E, Glu; F, Phe; G, Gly; H, His; I, Ile; K, Lys; L, Leu; M, Met; N, Asn; P, Pro; Q, Gln; R, Arg; S, Ser; T, Thr; V, Val; W, Trp; Y, Tyr.

organizing materials on the nanometer scale. In contrast to traditional synthetic methods, this approach allows for the precise control of the crystallinity of the materials by directing nucleation of materials on the nanometer scale at low temperature, which is an exciting new development.

The mineralization of the ZnS systems is a good example to show the merit of M13-based nanowire synthesis [14, 56]. In particular, one phage-bound peptide sequence, named A7, was selected for ZnS, showing a nice control of the size and shape of the final ZnS particles at room temperature under aqueous conditions. The synthesis process involved incubating the viral template with metal salt precursors at low temperature and annealing the mineralized viruses at high temperature. The results showed that the A7 peptide induced the nucleation and separated wurtzite ZnS grown on the virus surface with preferential orientation before annealing [14]. Upon annealing to remove the organic template and minimize of the interfacial energy, the polycrystalline assembled to form single-crystal nanowires still retaining the wurtzite phase and original orientation. The CdS nanowire with the wurtzide phase and CoPt or FePt nanowires with the $L1_0$ phase prepared from genetic modified M13 virus were also prepared using similar method. Figure 1.5 and Table 1.2 show some representative results from Belcher's group.

Furthermore, more complicated nanostructures can be produced by incorporating two or more fusion proteins. The engineering of specific receptors at the ends of the phage enables networks and other assemblies with predictable structures, specific characteristics and defined properties to be created. For example, Belcher *et al.* have used mutated M13 phages with both P8 and P3 engineered as templates to assemble gold and CdSe nanocrystal/hetero-nanocrystal arrays and gold nanowires [58, 83]. In another example, two kinds of materials-specific peptide motifs (AuE4) were engineered together into the major coat P8 protein, which resulted in a hybrid $Au-Co_3O_4$ nanowire [16]. Those works demonstrate that various substrate-specific motifs can be independently selected from type 8 or type 3 libraries and then genetically incorporated into M13 structures to produce versatile hybrid materials with heterofunctionality.

1.5
Self-Assembly of Rod-Like Viruses

Compared to synthetic particles, viruses are truly monodisperse at the nanometer scale, and are thus ideal for the self-assembly study and construction of uniform nanostructured materials. TMV and M13 bacteriophage are widely used as building blocks to construct unique one-dimensional (1D) fibers, two-dimensional (2D) thin films and 3D liquid crystal-like structures.

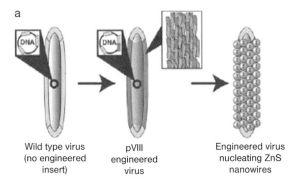

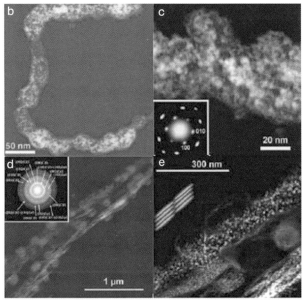

Figure 1.5 (a) Illustration depicting A7 peptide expression on the P8 upon phage amplification and assembly, and then the subsequent nucleation of ZnS nanocrystals. Call-outs depict insertion of the A7 nucleotide sequences, resulting in A7 fusion protein shown as green-shaded areas. Additionally, the call-out of the engineered virus shows detail of the wild-type P8 and the A7-engineered P8 composing the viral coat. (b) Higher magnification scanning TEM (STEM) images of A7–P8-engineered viruses directing ZnS nanocrystal synthesis at 0 °C, showing an individual viral ZnS–virus nanowire. (c) High-angle annular dark field (HAADF) STEM image of a straight region of a viral nanowire at higher magnification showing the close-packed ZnS nanocrystal morphology. (Inset) Electron diffraction pattern, taken from the area shown in (c), shows the hexagonal wurtzite ZnS structure. (d) Images and characterization of ZnS–CdS hybrid nanowires prepared from viruses expressing a stochastic mixture of both the A7–P8 and J140–VIII fusion proteins by using CdS/ZnS nanocrystal synthesis at 25 °C. HAADF STEM image of a viral CdS and ZnS hybrid layered structure. Inset: Electron diffraction pattern of the layered structure showing the coexistence of wurtzite CdS and ZnS phases. (e) HAADF STEM image of the layered structure at higher magnification. Inset: Cartoon illustrating the layered structure composed of viruses and nanocrystals. (Adapted with permission from [12]).

1.5.1
Controlled 1D Assembly

Developing 1D functional structures on nanometer scales defines a new paradigm in the fabrication of novel biomedical, optical, acoustic, electronic and magnetic materials and devices. Numerous methods have been developed for the synthesis of 1D nanostructures, with advances often being sought in terms of structural control and ease of processing. Using biological building blocks as templates in 1D materials synthesis is an exciting and emerging area of research. In addition to using native viral particles as templates to construct inorganic nanowires, a number of efforts have been reported to control the 1D assembly of TMV and M13 particles.

1.5.1.1 TMV Head-to-Tail Assembly
A head-to-tail ordered assembly of wild-type TMV has often been observed as very likely a product of complementary hydrophobic interactions between the dipolar ends of the helical structure [64, 84–87]. In particular, the 1D assembly is dramatically favored in an acidic environment due to the minimization of the repulsion between the carboxylic residues at the assembly interface [88]. The exterior surface of TMV is highly charged and hydrophilic. The particles carry negative charges at neutral pH since the pI of TMV is around 3.4 [89]. Therefore, a monomeric molecule with an amino group (or other positive charged groups), such as aniline, can accumulate on the surface of TMV due to the electrostatic attraction or hydrogen bonding to the negatively charged surface residues of TMV. An *in situ* polymerization should be able to produce a homogenous layer of polymers on the surface of TMV and fix the head-to-tail assembled tube-like structure.

For example, TMV/polyaniline (PANI) composite fibers were produced by the attraction of aniline (and also PANI) to the surface of TMV at neutral pH and a sequential *in situ* polymerization. As shown by transmission electron microscopy (TEM) and atomic force microscopy (AFM), the length of such fibers can reach several micrometers (Figure 1.6). The diameter of fibers increased to 20 nm in comparison to 18 nm of the original TMV measured with TEM (Figure 1.6b). The inner channel could not be detected even after negative staining and no visible gap could be detected from a long composite fiber (Figure 1.6d). This indicates that the head-to-tail protein–protein interaction leads to the formation of fiber-like structures. Such interaction, in principle, is identical to the subunit interactions at any cross-section of the native TMV [85, 90]. In addition, there was no solution PANI formed in the reaction. It is possible that the local concentration of aniline on the TMV surface was much higher than in solution; therefore, *in situ* polymerization was able to produce a thin layer of polymers exclusively on the surface of TMV and fix the head-to-tail assembled tube-like structure. The intrinsic anisotropic morphology of PANI at dilute polymerization conditions further assisted the 1D nanofiber formation.

By combining electron microscopy and AFM, the length and surface morphology of TMV and composite fibers can be readily investigated. However, samples

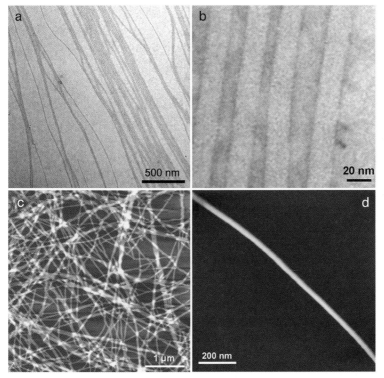

Figure 1.6 TEM images (a and b) and AFM images (c and d) of PANI/TMV nanofibers.

prepared on substrates for electron microscopy and AFM characterization upon drying can potentially alter both their diameter and surface morphology due to the interaction between the surface proteins and substrate [91]. To complement the TEM data, small angle X-ray scattering (SAXS) and *in situ* time-resolved SAXS (TRSAXS) on solution samples were performed to understand the kinetics of PANI/TMV composite nanofiber formation [84]. The difference in cross-sectional structure between wild-type TMV and PANI/TMV was revealed by fitting of the SAXS data using GNOM software [92, 93]. The largest dimension along the cross-section (D) obtained for TMV was around 18 nm, which was consistent with the TMV crystal structure. For the PANI/TMV, the maximum cross-sectional dimension was 30 nm and this increase in length scale could be attributed to the PANI coating of TMV.

Therefore, there are two crucial factors that facilitate the formation of long 1D TMV-composite fibers: (i) accumulation and polymerization of monomers on the surface of TMV, and (ii) prolongation and stabilization of TMV helices. The interaction between monomers and TMV is essential for the long fiber formation. For example, when thiophene was employed as the monomer under a similar polymerization condition, there was no formation of any fiber-like structures, likely due to the much weaker interaction between thiophene and the surface of TMV. To further confirm this, an amino-functionalized thiophene salt was used as the charged monomer, and 1D long fibers were readily observed when a mixture of TMV and amino-functionalized thiophene salt was treated with oxidative regents [94].

1.5.1.2 Conductive 1D TMV Composite Fibers

It is known that the pH of the polymerization reaction has a great influence on the structure and conductivity of PANI. At near-neutral reaction pH, long 1D PANI-coated TMV single nanofibers formed upon treating TMV with a dilute solution of aniline and ammonium persulfate. However, such nanofibers exhibited a homogeneous diameter and high aspect ratio, but no conductivity, likely due to the branched structures of PANI. In order to form conductive PANIs, polymerization reactions were performed under acidic conditions. However, at low reaction pH, only bundle-like structures consisting of parallel arrays of nanofibers were formed. A standard four-probe method was employed to measure the conductivity of PANI/TMV composite nanofibers. At room temperature, the bulk DC conductivities measured were in the range of 0.01–0.1 S cm^{-1} for composite nanofibers synthesized at low reaction pH (2.5 and 4.0). This is comparable to the PANI nanofibers synthesized by other methods [95]. No conductivity was observed for the composite nanofibers formed at a higher reaction pH.

To generate well-dispersed conductive fibers, highly negative charged poly(sulfonated styrene) (PSS) was used both as the dopant acid to enhance the conductivity of PANI and to improve the stability of composite fibers in aqueous solution. The resulted PSS/PANI/TMV composite was formed as predominantly isolated fiber, which could be well-dispersed in a dilute water solution (Figure 1.7). Electronic properties measured using scanning spreading resistance microscopy indicated a conductivity of around $1 \times 10^{-5} \Omega^{-1} cm^{-1}$ [87].

1.5.1.3 Weaving M13 Bacteriophage into Robust Fibers

Silk spiders and silk worms can spin highly engineered continuous fibers by passing aqueous liquid crystalline protein solution through their spinneret. By mimicking this process, Belcher *et al.* used the electrospinning method or wet-spinning process to spin native M13, genetically engineered M13 and quantum dots (QDs) conjugated M13 into robust 1D long fibers with the diameters from tens of nanometers to micrometers [96, 97]. In order to obtain continuous fiber, the virus solution was mixed with polyvinylpyrolidone (PVP) to improve processing ability [96]. The resulting virus-blended PVP fibers were transformed into nonwoven fabrics that retained their ability to infect bacterial hosts. By chemically conjugated amine-terminated cadmium selenide QDs to M13 templates via the carboxylic acid side groups displayed on the P8 proteins, a continuous fiber of micrometer-scale diameter (microfiber) was created through a wet-spinning process while a concentrated QD-conjugated virus solution was spun vertically into glutaraldehyde solution (Figure 1.8) [97]. The M13 fibers containing QDs emitted red light under exposure to ultraviolet (UV) light. These composite fibers have potential applications in optical devices and advanced sensors.

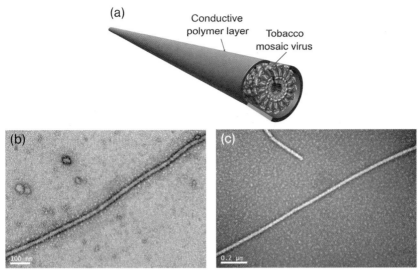

Figure 1.7 (a) Schematic illustration of 1D conductive polymer/TMV composite fiber. (b) TEM image of PSS/PANI/TMV composite long fiber. (c) TEM image of PSS/polypyrrole/TMV composite long fiber.

1.5.1.4 Nanoring Structure

As described earlier, both P3 and P9 proteins that reside at the end of the M13 virus can be used to display peptide, which enable the creation of some other interesting structures, such as rings, squares and other arrays. For example, the bifunctional viruses displayed an anti-streptavidin peptide (binding streptavidin) and hexahistidine peptide (binding to Ni-NTA) at opposite ends of the virus as P3 and P9 fusions, respectively. Stoichiometric addition of the streptavidin–Ni-NTA linker molecule led to the reversible formation of nanorings with circumferences corresponding to the lengths of the DNAs [62].

1.5.2
Fabrication of Thin Films by 2D Self-Assembly

Mirkin's research group reported a direct-write lithographic method that can directly write TMV onto a nanopatterned surface [98]. This method was the so-called dip-pen nanolithography (DPN). TMV nanoarrays were fabricated by initially generating chemical templates of 16-thiohexadecanoic acid (MHA) on a gold thin film by using DPN. By immersing the substrate in an alkanethiol solution, the regions surrounding these features were passivated with a monolayer of 11-thioundecyl-penta (ethylene glycol). The passivation layer can avoid nonspecific binding between TMV and the unpatterned areas. The carboxylic acid groups of MHA were coordinated to Zn^{2+} ions. The metallated substrate was then exposed to TMV solution. Individual TMV particles can be selectively attached to the substrate (Figure 1.9). Using this method, it was possible to isolate and control the orientation of TMV

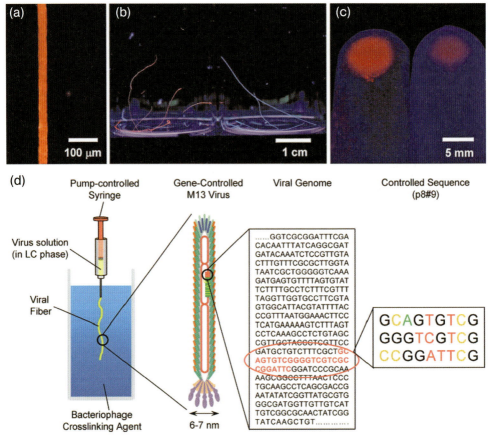

Figure 1.8 Images and schematic design of chemically and engineered functional fibers. (a) Fluorescence microscopy image of a genetically engineered M13 virus fiber conjugated with QDs excited by using UV light. (b) Under exposure to UV light, virus fibers conjugated with QDs emit red light and nonconjugated virus fibers emit blue light. (c) The mutated M13 shows a relatively higher intensity of light emission than the wild-type M13KE virus after QD conjugation. (d) Schematic illustration of electric spinning and genetic modification of M13. LC = liquid crystalline. (Adapted with permission from [97]).

particles in a well-defined manner. Culver's research group reported another method to align genetically modified TMV on the substrate [51]. Briefly, genetically modified TMV nanotemplates were first labeled with fluorescent markers and partially disassembled by alkaline treatment. Results demonstrated that high spatial and sequence specificity are obtained during nanotemplate hybridization, including density control through the modulation of capture DNA concentration.

Recently, Velev *et al.* reported a single-step technique for depositing hierarchically ordered and aligned arrays of TMV particles over macroscopic length scales using convective alignment [99]. Shear-induced alignment is responsible for the

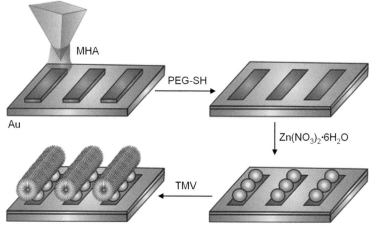

Figure 1.9 Schematic illustration of selective immobilization of a single virus on DPN-generated MHA nanotemplates treated with Zn(NO$_3$)$_2$·6H$_2$O. PEG-SH = poly(ethylene glycol)-thiol. (Adapted with permission from [98]).

long-range organization during the coating process with viscous TMV suspensions. The overall assembled film structure was controlled by the operational parameters, including withdrawal speed and substrate wettability. Gold nanoparticles could be selectively attached onto the fiber surface. This allowed the formation of large uniform coatings with anisotropic conductivity. Using a similar methodology, well-ordered monolayer films were generated using the TMV as a model anisotropic colloid. This convective assembly process rapidly generated films several centimeters in length with all the rods aligned parallel to the direction of assembly [100].

The self-assembly of nanoparticles at fluid interfaces, driven by the reduction in interfacial energy, has been well established. The energetic penalty associated with the formation of an interface is given by the product of the total area of the interface and the interfacial energy [101, 102]. Particles dispersed in one of the phases will segregate to an interface so as to mediate interactions between the fluids. Consequently, the segregation of particles to the interface acts to stabilize the interface. Rod-like TMV can self-assemble at the interface of water and hexane to give a long-range parallel structure.

1.5.3
Controlling the 3D Assembly of TMV and M13

TMV can be arranged in 3D structures [103–105]. Nematic liquid crystals of TMV were used to prepare silica mesostructures and nanoparticles with parallel or radial arrays of linear channels, respectively [106]. The mesostructures were produced as micrometer-size inverse replicas of the nematic phase and had a periodicity of approximately 20 nm. The general stability of TMV liquid crystals suggests that this approach may also be used to prepare a wide range of inorganic oxides, semiconductors and metal-based mesoporous materials.

Just like any rod-like viruses, fd and M13 phages can also form liquid crystals [79, 107, 108]. Belcher's group reported the evidence of chiral smectic C structures of M13 virus-based films – a conformation arising from the helical structure of M13 [79, 109]. The most interesting work was that they had used the viral–inorganic hybrid materials prepared from genetically modified viruses as the basic building block for materials design, which will show great potential for tunable devices. For example, high concentrations of genetically engineered viruses were suspended in ZnS precursor solutions to form viral–ZnS nanocrystal liquid crystalline suspensions with smectic to cholesteric phases at different concentrations of virus. The ZnS nanocrystals and M13 viral systems could form a self-supporting hybrid film material retaining the smectic-like lamellar morphologies when dried in high concentration. Optical characterization revealed that the films were composed of around 72-μm periodic dark and bright band patterns that corresponded to the chiral smectic C structure (Figure 1.10). A periodic length of 895 nm was observed, corresponding to the combination of virus length (860–880 nm) and nanocrystal aggregates (around 20 nm) by scanning electron microscopy. The surface morphology of the viral film exhibited zig-zag chiral smectic O patterns due to the long rod shape of the viruses conjugated to an inorganic head group composed of ZnS nanocrystals. The viral film cast on different substrates could form different liquid crystal phases.

Using the streptavidin-binding M13 mutants, Belcher *et al.* showed a universal approach with which any materials, such as inorganic gold nanoparticles, organic fluorescent dyes (fluorescein) and biological molecules (*R*-phycoerythrin), were able to be aligned in a similar fashion as the ZnS was in viral films [108]. Moreover, because of the ease of genetic engineering of M13, it is possible to assemble a diversity of M13-based composites with more complex structural features and specific functions [16, 97].

1.6
Virus-Based Device and Applications

The practical use of rod-like viruses is to render them with new functionalities, to assemble them into different hierarchical structures and to incorporate them into devices. Balandin *et al.* reported that hybrid virus–inorganic nanostructures, which consist of silica or silicon nanotubes deposited on TMV, had potential application in nanostructure-based nanocircuits [110]. The confined acoustic phonons were found to be redistributed between the nanotube shell and the acoustically soft virus enclosure. As a result, the low-temperature electron mobility in the hybrid virus–silicon nanotube increased by a factor of 4 compared to that of an empty silicon nanotube. Recently, Francis *et al.* reported that TMV could be used for the construction of light-harvesting systems through self-assembly [53]. The building

Figure 1.10 Characterization of mutant M13–ZnS film. (a) Photograph of A7–ZnS viral film. (b) POM (203) birefringent dark and bright band patterns (periodic length 72.8 mm) were observed. These band patterns are optically active, and their patterns reverse depending on the angles between polarizer and analyzer. (c) Schematic structural diagram of the A7–ZnS composite film. (d) AFM image of the free surface. The M13 phage forms parallel aligned herringbone patterns that have almost right angles between the adjacent director (arrows). (Adapted with permission from [79], Copyright by AAAS).

blocks were prepared by attaching fluorescent chromophores to cysteine residues introduced on TMV coat proteins. When placed under the appropriate buffers, these conjugates could be re-assembled into stacks of disks or into rod-like particles that reached hundreds of nanometers in length. By controlling the assembly state, the efficiency of energy transfer could be controlled.

Yang et al. recently reported that TMV, decorated with platinum nanoparticles, could be embedded in a nonconductive polymer to form a sandwich-like structures between two metallic electrodes [111]. It was a prototype of memory device based on conductance switching, which leads to the occurrence of bistable states with an on/off ratio larger than three orders of magnitude. The mechanism of this process was attributed to charge trapping in the nanoparticles for data storage and a tunneling process in the high conductance state. Although many questions concerning the switching mechanism were not answered, this research directed the use of biological objects as basic building blocks in electronic memory devices.

As described above, Co_3O_4 and hybrid Au–Co_3O_4-based composite nanowires can be fabricated using genetically engineered M13 virus as template [16]. These wires had very good specific capacity, and the Au–Co_3O_4 hybrid composite generated higher initial and reversible lithium storage capacity than the pure Co_3O_4 nanowires when tested at the same current rate. Combining the self-assembly of M13 phages on polyelectrolyte multilayers [112] and the nanowire synthesis, 2D ordered monolayers of Co_3O_4 or Au–Co_3O_4 nanowires were produced, which were utilized as electrodes for lithium-ion batteries [16]. The capacity test showed that the assembled monolayer of Co_3O_4 nanowires/Li cells could sustain and deliver 94% of its theoretical capacity at a rate of 1.12 C and 65% at a rate of 5.19 C, demonstrating the capability for a high cycling rate. These results show that basic biological principles can be applied to the rational design and assembly of nanoscale battery components, leading to improved performance in properties such as specific capacity and rate capability. Additionally, the ease of genetic modification of the M13 virus allows for the preparation and assembly of many functional nanomaterials for applications such as light emitting displays, optical detectors, photovoltaic devices, magnetic storage, high-surface-area catalysts, medical diagnostics and supercapacitors.

1.7
Outlook

Nanotechnology revolves around the controlled design, synthesis, and application of particles at the atomic and molecular scales. Viruses decorated with various small molecules to target cells have already demonstrated specific cell-targeting ability. Furthermore, the nanosized probes can be modified with bioimaging agents such as near-infrared fluorescent dyes and magnetic contrast imaging agents at high local concentrations to increase detection sensitivity [113–115]. In particular, rod-like viruses, including TMV and M13, which are extremely difficult to synthesize in the laboratory, are attractive for use as scaffolds for the development of novel functional materials at the nanometer level due to their anisotropic structural features. They have four unique advantages compared to most synthetic particles:

(1) The 3D structures can be characterized at the atomic or near-atomic level.
(2) Potential of controlled self-assembly in a broad length scale.
(3) Genetic control over the composition and surface properties.
(4) Monodispersed particles and economic large-scale production in gram and kilogram quantities.

Therefore, by embracing surface modification to further enhance their physical properties and retaining the biological origins for genetic alterations, rod-like viruses will have great potential in the development of new materials for biomedical and electronic applications.

References

1. Cui, Y. and Lieber, C.M. (2001) Functional nanoscale electronic devices assembled using silicon nanowire building blocks. *Science*, **291**, 851–3.
2. Duan, X., Huang, Y., Cui, Y., Wang, J. and Lieber, C.M. (2001) Indium phosphide nanowires as building blocks for nanoscale electronic and optoelectronic devices. *Nature*, **409**, 66–9.
3. Tao, A., Kim, F., Hess, C., Goldberger, J., He, R., Sun, Y., Xia, Y. and Yang, P. (2003) Langmuir–Blodgett silver nanowire monolayers for molecular sensing using surface-enhanced Raman spectroscopy. *Nano Letters*, **3**, 1229–33.
4. Melosh, N.A., Boukai, A., Diana, F., Gerardot, B., Badolato, A., Petroff, P.M. and Heath, J.R. (2003) Ultrahigh-density nanowire lattices and circuits. *Science*, **300**, 112–15.
5. Busbee, B.D., Obare, S.O. and Murphy, C.J. (2003) An improved synthesis of high-aspect-ratio gold nanorods. *Advanced Materials*, **15**, 414–16.
6. Jana, N.R., Gearheart, L. and Murphy, C.J. (2001) Evidence for seed-mediated nucleation in the formation of gold nanoparticles from gold salts. *Chemistry of Materials*, **13**, 2313–22.
7. Niemeyer, C.M. (2001) Nanoparticles, protein, and nucleic acids: biotechnology meets materials science. *Angewandte Chemie (International Edition in English)*, **40**, 4128–58.
8. Seeman, N.C. (2003) At the crossroads of chemistry, biology, and materials: structural DNA nanotechnology. *Chemistry and Biology*, **10**, 1151–9.
9. Caswell, K.K., Wilson, J.N., Bunz, U.H.F. and Murphy, C.J. (2003) Preferential end-to-end assembly of gold nanorods by biotin-streptavidin connectors. *Journal of the American Chemical Society*, **125**, 13914–15.
10. Dujardin, E. and Mann, S. (2002) Bio-inspired materials chemistry. *Advanced Engineering Materials*, **4**, 461–74.
11. Seeman, N.C. and Belcher, A.M. (2002) Emulating biology: building nanostructures from the bottom up. *Proceedings of the National Academy of Sciences of the United States of America*, **99**, 6451–55.
12. Mao, C., Flynn, C.E., Hayhurst, A., Sweeney, R., Qi, J., Georgiou, G., Iverson, B. and Belcher, A.M. (2003) Viral assembly of oriented quantum dot nanowires. *Proceedings of the National Academy of Sciences of the United States of America*, **100**, 6946–51.
13. Shenton, W., Douglas, T., Young, M., Stubbs, G. and Mann, S. (1999) Inorganic–organic nanotube composites from template mineralization of tobacco mosaic virus. *Advanced Materials*, **11**, 253–6.
14. Mao, C., Solis, D.J., Reiss, B.D., Kottmann, S.T., Sweeney, R.Y., Hayhurst, A., Georgiou, G., Iverson, B. and Belcher, A.M. (2004) Virus-based toolkit for the directed synthesis of magnetic and semiconducting nanowires. *Science*, **303**, 213–17.
15. Lee, S.K., Yun, D.S. and Belcher, A.M. (2006) Cobalt ion mediated self-assembly of genetically engineered bacteriophage for biomimetic Co–Pt hybrid material. *Biomacromolecules*, **7**, 14–17.
16. Nam, K.T., Kim, D.W., Yoo, P.J., Chiang, C.Y., Meethong, N., Hammond, P.T., Chiang, Y.M. and Belcher, A.M. (2006) Virus-enabled synthesis and assembly of nanowires for lithium ion battery electrodes. *Science*, **312**, 885–8.
17. Norrby, E. (2008) Nobel Prizes and the emerging virus concept. *Archives of Virology*, **153**, 1109–23.
18. Beijerinck, M.J. (1898) Ueber ein contagium vivum fluidum als Ursache der Fleckenkrankheit der Tabaksblatter. *Verhandelingen der Koninklijke Akademie van Wetenschappen te Amsterdam*, **65**, 3–21.
19. Kausche, G.A. and Ruska, H. (1939) Adsorption of metallic colloids on protein bodies. I. The reaction of colloidal gold–tobacco mosaic virus. *Kolloid-Zeitschrift*, **89**, 21–6.
20. Lee, L.A. and Wang, Q. (2006) Adaptations of nanoscale viruses and other protein cages for medical applications. *Nanomedicine*, **2**, 137–49.

21 Flynn, C.E., Lee, S.-W., Peelle, B.R. and Belcher, A.M. (2003) Viruses as vehicles for growth, organization and assembly of materials. *Acta Materialia*, **51**, 5867–80.

22 Diaz-Avalos, R. and Caspar, D.L.D. (1998) Structure of the stacked disk aggregate of tobacco mosaic virus protein. *Biophysical Journal*, **74**, 595–603.

23 Opella, S.J., Stewart, P.L. and Valentine, K.G. (1987) Protein structure by solid-state NMR spectroscopy. *Quarterly Reviews of Biophysics*, **19**, 7–49.

24 Opella, S.J. and Stewart, P.L. (1989) Solid-state nuclear magnetic resonance structural studies of proteins. *Methods in Enzymology*, **176**, 242–75.

25 Wang, Q., Lin, T., Johnson, J.E. and Finn, M.G. (2002) Natural supramolecular building blocks: cysteine-added mutants of cowpea mosaic virus. *Chemistry and Biology*, **9**, 813–19.

26 Wang, Q., Kaltgrad, E., Lin, T., Johnson, J.E. and Finn, M.G. (2002) Natural supramolecular building blocks: wild-type cowpea mosaic virus. *Chemistry and Biology*, **9**, 805–11.

27 Wang, Q., Lin, T., Tang, L., Johnson, J.E. and Finn, M.G. (2002) Icosahedral virus particles as addressable nanoscale building blocks. *Angewandte Chemie (International Edition in English)*, **41**, 459–62.

28 Wang, Q., Raja, K.S., Janda, K.D., Lin, T. and Finn, M.G. (2003) Blue fluorescent antibodies as reporters of steric accessibility in virus conjugates. *Bioconjugate Chemistry*, **14**, 38–43.

29 Raja, K.S., Wang, Q. and Finn, M.G. (2003) Icosahedral virus particles as polyvalent carbohydrate display platforms. *ChemBioChem*, **4**, 1348–51.

30 Lettinga, M.P., Barry, E. and Dogic, Z. (2005) Self-diffusion of rod-like viruses in the nematic phase. *Europhysics Letters*, **71**, 692–8.

31 Niu, Z., Bruckman, M.A., Harp, B., Mello, C.M. and Wang, Q. (2008) Bacteriophage M13 as scaffold for preparing conductive polymeric composite fibers. *Nano Research*, **1**, 235–41.

32 Bruckman, M.A., Kaur, G., Lee, L.A., Xie, F., Sepulveda, J., Breitenkamp, R., Zhang, X., Joralemon, M., Russell, T.P., Emrick, T. and Wang, Q. (2008) Surface modification of tobacco mosaic virus with "click" chemistry. *ChemBioChem*, **9**, 519–23.

33 Schlick, T.L., Ding, Z., Kovacs, E.W. and Francis, M.B. (2005) Dual-surface modification of the tobacco mosaic virus. *Journal of the American Chemical Society*, **127**, 3718–23.

34 Uhde, K., Fischer, R. and Commandeur, U. (2005) Expression of multiple foreign epitopes presented as synthetic antigens on the surface of potato virus X particles. *Archives of Virology*, **150**, 327–40.

35 Liu, L., Canizares, M.C., Monger, W., Perrin, Y., Tsakiris, E., Porta, C., Shariat, N., Nicholson, L. and Lomonossoff, G.P. (2005) Cowpea mosaic virus-based systems for the production of antigens and antibodies in plants. *Vaccine*, **23**, 1788–92.

36 Yamada, T., Ueda, M., Seno, M., Kondo, A., Tanizawa, K. and Kuroda, S. (2004) Novel tissue and cell type-specific gene/drug delivery system using surface engineered hepatitis B virus nano-particles. *Current Drug Targets – Infectious Disorders*, **4**, 163–7.

37 Muzyczka, N. and Warrington, K.H., Jr (2005) Custom adeno-associated virus capsids: the next generation of recombinant vectors with novel tropism. *Human Gene Therapy*, **16**, 408–16.

38 Xie, J. and Schultz, P.G. (2005) Adding amino acids to the genetic repertoire. *Current Opinion in Chemical Biology*, **9**, 548–54.

39 Deiters, A., Cropp, T.A., Summerer, D., Mukherji, M. and Schultz, P.G. (2004) Site-specific PEGylation of proteins containing unnatural amino acids. *Bioorganic and Medicinal Chemistry Letters*, **14**, 5743–5.

40 Deiters, A. and Schultz, P.G. (2005) In vivo incorporation of an alkyne into proteins in *Escherichia coli*. *Bioorganic and Medicinal Chemistry Letters*, **15**, 1521–4.

41 Bose, M., Groff, D., Xie, J., Brustad, E. and Schultz, P.G. (2006) The incorporation of a photoisomerizable amino acid into proteins in *E. coli*. *Journal of the American Chemical Society*, **128**, 388–9.

42 Hohsaka, T., Ashizuka, Y., Taira, H., Murakami, H. and Sisido, M. (2001) Incorporation of nonnatural amino acids into proteins by using various four-base codons in an *Escherichia coli in vitro* translation system. *Biochemistry*, **40**, 11060–4.

43 Cruz, S.S., Chapman, S., Roberts, A.G., Roberts, I.M., Prior, D.A. and Oparka, K.J. (1996) Assembly and movement of a plant virus carrying a green fluorescent protein overcoat. *Proceedings of the National Academy of Sciences of the United States of America*, **93**, 6286–90.

44 Toth, R.L., Chapman, S., Carr, F. and Santa Cruz, S. (2001) A novel strategy for the expression of foreign genes from plant virus vectors. *FEBS Letters*, **489**, 215–19.

45 Haynes, J., Cunningham, J., von Seefried, A., Lennick, M., Garvin, R. and Shen, S. (1986) Development of a genetically-engineered, candidate polio vaccine employing the self-assembling properties of tobacco mosaic virus coat protein. *Biotechnology*, **4**, 637–41.

46 Sugiyama, Y., Hamamoto, H., Takemoto, S., Watanabe, Y. and Okada, Y. (1995) Systemic production of foreign peptides on the particle surface of tobacco mosaic virus. *FEBS Letters*, **359**, 247–50.

47 Hamamoto, H., Sugiyama, Y., Nakagawa, N., Hashida, E., Matsunaga, Y., Takemoto, S., Watanabe, Y. and Okada, Y. (1993) A new tobacco mosaic vector and its use for the systemic production of angiotensin-I-converting enzyme inhibitor in transgenic tobacco and tomato. *Biotechnology*, **11**, 930–2.

48 Bendahmane, M., Koo, M., Karrer, E. and Beachy, R.N. (1999) Display of epitopes on the surface of tobacco mosaic virus: impact of charge and isoelectric point of the epitope on virus–host interactions. *Journal of Molecular Biology*, **290**, 9–20.

49 Koo, M., Bendahmane, M., Lettieri, G.A., Paoletti, A.D., Lane, T.E., Fitchen, J.H., Buchmeier, M.J. and Beachy, R.N. (1999) Protective immunity against murine hepatitis virus (MHV) induced by intranasal or subcutaneous administration of hybrids of tobacco mosaic virus that carries an MHV epitope. *Proceedings of the National Academy of Sciences of the United States of America*, **96**, 7774–9.

50 Jiang, L., Li, Q., Li, M., Zhou, Z., Wu, L., Fan, J., Zhang, Q., Zhu, H. and Xu, Z. (2006) A modified TMV-based vector facilitates the expression of longer foreign epitopes in tobacco. *Vaccine*, **24**, 109–15.

51 Yi, H.M., Rubloff, G.W. and Culver, J.N. (2007) TMV microarrays: hybridization-based assembly of DNA-programmed viral nanotemplates. *Langmuir*, **23**, 2663–7.

52 Yi, H.M., Nisar, S., Lee, S., Powers, M.A., Bentley, W.E., Payne, G.F., Ghodssi, R., Rubloff, G.W., Harris, M.T. and Culver, J.N. (2005) Patterned assembly of genetically modified viral nanotemplates via nucleic acid hybridization. *Nano Letters*, **5**, 1931–6.

53 Miller, R.A., Presley, A.D. and Francis, M.B. (2007) Self-assembling light-harvesting systems from synthetically modified tobacco mosaic virus coat proteins. *Journal of the American Chemical Society*, **129**, 3104–9.

54 Armstrong, N., Adey, N.B., McConnell, S.J. and Kay, B.K. (1996) Vectors for phage display, in *Phage Display of Peptides and Proteins: A Laboratory Manual* (eds B.K. Kay, J. Winter and J. McCafferty), Academic Press, San Diego, CA, pp. 35–53.

55 Paschke, M. (2006) Phage display systems and their applications. *Applied Microbiology and Biotechnology*, **70**, 2–11.

56 Flynn, C.E., Mao, C.B., Hayhurst, A., Williams, J.L., Georgiou, G., Iverson, B. and Belcher, A.M. (2003) Synthesis and organization of nanoscale II–VI semiconductor materials using evolved peptide specificity and viral capsid assembly. *Journal of Materials Chemistry*, **13**, 2414–21.

57 Petrenko, V.A., Smith, G.P., Gong, X. and Quinn, T. (1996) A library of organic landscapes on filamentous phage. *Protein Engineering*, **9**, 797–801.

58 Huang, Y., Chiang, C., Lee, S.K., Gao, Y., Hu, E.L., De Yoreo, J. and Belcher, A.M. (2005) Programmable assembly of nanoarchitectures using genetically engineered viruses. *Nano Letters*, **5**, 1429–34.

59 Sidhu, S.S., Weiss, G.A. and Wells, J.A. (2000) High copy display of large proteins on phage for functional selections. *Journal of Molecular Biology*, **296**, 487–95.

60 Weiss, G.A., Wells, J.A. and Sidhu, S.S. (2000) Mutational analysis of the major coat protein of M13 identifies residues that control protein display. *Protein Science*, **9**, 647–54.

61 Gao, C., Mao, S., Kaufmann, G., Wirsching, P., Lerner, R.A. and Janda, K.D. (2002) A method for the generation of combinatorial antibody libraries using pIX phage display. *Proceedings of the National Academy of Sciences of the United States of America*, **99**, 12612–16.

62 Nam, K.T., Peelle, B.R., Lee, S.W. and Belcher, A.M. (2004) Genetically driven assembly of nanorings based on the M13 virus. *Nano Letters*, **4**, 23–7.

63 Endo, M., Fujitsuka, M. and Majima, T. (2007) Porphyrin light-harvesting arrays constructed in the recombinant tobacco mosaic virus scaffold. *Chemistry – A European Journal*, **13**, 8660–6.

64 Shenton, W., Douglas, T., Young, M., Stubbs, G. and Mann, S. (1999) Inorganic–organic nanotube composites from template mineralization of tobacco mosaic virus. *Advanced Materials*, **11**, 253–6.

65 Royston, E., Lee, S.Y., Culver, J.N. and Harris, M.T. (2006) Characterization of silica-coated tobacco mosaic virus. *Journal of Colloid and Interface Science*, **298**, 706–12.

66 Tsukamoto, R., Muraoka, M., Seki, M., Tabata, H. and Yamashita, I. (2007) Synthesis of CoPt and FePt$_3$ nanowires using the central channel of tobacco mosaic virus as a biotemplate. *Chemistry of Materials*, **19**, 2389–91.

67 Knez, M., Sumser, M., Bittner, A.M., Wege, C., Jeske, H., Martin, T.P. and Kern, K. (2004) Spatially selective nucleation of metal clusters on the tobacco mosaic virus. *Advanced Functional Materials*, **14**, 116–24.

68 Knez, M., Bittner, A.M., Boes, F., Wege, C., Jeske, H., Maiss, E. and Kern, K. (2003) Biotemplate synthesis of 3-nm nickel and cobalt nanowires. *Nano Letters*, **3**, 1079–82.

69 Knez, M., Sumser, M., Bittner, A.M., Wege, C., Jeske, H., Kooi, S., Burghard, M. and Kern, K. (2002) Electrochemical modification of individual nano-objects. *Journal of Electroanalytical Chemistry*, **522**, 70–4.

70 Balci, S., Bittner, A.M., Hahn, K., Scheu, C., Knez, M., Kadri, A., Wege, C., Jeske, H. and Kern, K. (2006) Copper nanowires within the central channel of tobacco mosaic virus particles. *Electrochimica Acta*, **51**, 6251–7.

71 Fowler, C.E., Shenton, W., Stubbs, G. and Mann, S. (2001) Tobacco mosaic virus liquid crystals as templates for the interior design of silica mesophases and nanoparticles. *Advanced Materials*, **13**, 1266–69.

72 Fonoberov, V.A. and Balandin, A.A. (2005) Phonon confinement effects in hybrid virus–inorganic nanotubes for nanoelectronic applications. *Nano Letters*, **5**, 1920–3.

73 Royston, E., Lee, S.Y., Culver, J.N. and Harris, M.T. (2006) Characterization of silica-coated tobacco mosaic virus. *Journal of Colloid and Interface Science*, 706–12.

74 Knez, M., Kadri, A., Wege, C., Gosele, U., Jeske, H. and Nielsch, K. (2006) Atomic layer deposition on biological macromolecules: metal oxide coating of tobacco mosaic virus and ferritin. *Nano Letters*, **6**, 1172–7.

75 Knez, M., Sumser, M.P., Bittner, A.M., Wege, C., Jeske, H., Hoffmann, D.M., Kuhnke, K. and Kern, K. (2004) Binding the tobacco mosaic virus to inorganic surfaces. *Langmuir*, **20**, 441–7.

76 Lee, S.Y., Choi, J.W., Royston, E., Janes, D.B., Culver, J.N. and Harris, M.T. (2006) Deposition of platinum clusters on surface-modified tobacco mosaic virus. *Journal of Nanoscience and Nanotechnology*, **6**, 974–81.

77 Lee, S.Y., Royston, E., Culver, J.N. and Harris, M.T. (2005) Improved metal cluster deposition on a genetically engineered tobacco mosaic virus template. *Nanotechnology*, **16**, S435–S441.

78 Brown, S. (1992) Engineered iron oxide-adhesion mutants of the *Escherichia coli* phage receptor. *Proceedings of the National Academy of Sciences of the United States of America*, **89**, 8651–5.

79 Lee, S.W., Mao, C.B., Flynn, C.E. and Belcher, A.M. (2002) Ordering of quantum dots using genetically engineered viruses. *Science*, **296**, 892–5.

80 Lee, S.W., Lee, S.K. and Belcher, A.M. (2003) Virus-based alignment of inorganic, organic, and biological nanosized materials. *Advanced Materials*, **15**, 689–92.

81 Reiss, B.D., Mao, C.B., Solis, D.J., Ryan, K.S., Thomson, T. and Belcher, A.M. (2004) Biological routes to metal alloy ferromagnetic nanostructures. *Nano Letters*, **4**, 1127–32.

82 Whaley, S.R., English, D.S., Hu, E.L., Barbara, P.F. and Belcher, A.M. (2000) Selection of peptides with semiconductor binding specificity for directed nanocrystal assembly. *Nature*, **405**, 665–8.

83 Souza, G.R., Christianson, D.R., Staquicini, F.I., Ozawa, M.G., Snyder, E.Y., Sidman, R.L., Millar, J.H., Arap, W. and Pasqualini, R. (2006) Networks of gold nanoparticles and bacteriophage as biological sensors and cell-targeting agents. *Proceedings of the National Academy of Sciences of the United States of America*, **103**, 1215–20.

84 Niu, Z., Bruckman, M.A., Li, S., Lee, L.A., Lee, B., Pingali, S.V., Thiyagarajan, P. and Wang, Q. (2007) Assembly of tobacco mosaic virus into fibrous and macroscopic bundled arrays mediated by surface aniline polymerization. *Langmuir*, **23**, 6719–24.

85 Klug, A. (1999) The tobacco mosaic virus particle: structure and assembly. *Philosophical Transactions: Biological Sciences*, **354**, 531–5.

86 Niu, Z., Bruckman, M., Kotakadi, V.S., He, J., Emrick, T., Russell, T.P., Yang, L. and Wang, Q. (2006) Study and characterization of tobacco mosaic virus head-to-tail assembly assisted by aniline polymerization. *Chemical Communications*, 3019–21.

87 Niu, Z., Liu, J., Lee, L.A., Bruckman, M.A., Zhao, D., Koley, G. and Wang, Q. (2007) Biological templated synthesis of water-soluble conductive polymeric nanowires. *Nano Letters*, **7**, 3729–33.

88 Lu, B., Stubbs, G. and Culver, J.N. (1996) Carboxylate interactions involved in the disassembly of tobacco mosaic tobamovirus. *Virology*, **225**, 11–20.

89 Wadu-Mesthrige, K., Pati, B., McClain, W.M. and Liu, G.-Y. (1996) Disaggregation of tobacco mosaic virus by bovine serum albumin. *Langmuir*, **12**, 3511–15.

90 Butler, P.J. (1999) Self-assembly of tobacco mosaic virus: the role of an intermediate aggregate in generating both specificity and speed. *Philosophical Transactions: Biological Sciences*, **354**, 537–50.

91 Lee, B., Lo, C.T., Thiyagarajan, P., Winans, R.E., Li, X., Niu, Z. and Wang, Q. (2007) Effect of interfacial interaction on the cross-sectional morphology of tobacco mosaic virus using GISAXS. *Langmuir*, **23**, 11157–63.

92 Svergun, D.I. (1991) Mathematical methods in small-angle scattering data analysis. *Journal of Applied Crystallography*, **24**, 485–92.

93 Svergun, D.I. (1992) Determination of the regularization parameter in indirect-transform methods using perceptual criteria. *Journal of Applied Crystallography*, **25**, 495–503.

94 Bruckman, M.A., Niu, Z., Li, S., Lee, L.A., Varazo, K., Nelson, T., Lavigne, J.J. and Wang, Q. (2007) Development of nanobiocomposite fibers by controlled-

assembly of rod-like tobacco mosaic virus. *NanoBiotechnology*, **3**, 31–9.

95 Zhang, X. and Manohar, S.K. (2005) Narrow pore-diameter polypyrrole nanotubes. *Journal of the American Chemical Society*, **127**, 14156–7.

96 Lee, S.W. and Belcher, A.M. (2004) Virus-based fabrication of micro- and nanofibers using electrospinning. *Nano Letters*, **4**, 387–90.

97 Chiang, C.Y., Mello, C.M., Gu, J.J., Silva, E.C.C.M., Vliet, K.J. and Belcher, A.M. (2007) Weaving genetically engineered functionality into mechanically robust virus fibers. *Advanced Materials*, **19**, 826–32.

98 Vega, R.A., Maspoch, D., Salaita, K. and Mirkin, C.A. (2005) Nanoarrays of single virus particles, *Angewandte Chemie (International Edition in English)*, **44**, 6013–15.

99 Kuncicky, D.M., Naik, R.R. and Velev, O.D. (2006) Rapid deposition and long-range alignment of nanocoatings and arrays of electrically conductive wires from tobacco mosaic virus. *Small*, **2**, 1462–6.

100 Wargacki, S.P., Pate, B. and Vaia, R.A. (2008) Fabrication of 2D ordered films of tobacco mosaic virus (TMV): processing morphology correlations for convective assembly. *Langmuir*, **24**, 5439–44.

101 Lin, Y., Skaff, H., Emrick, T., Dinsmore, A.D. and Russell, T.P. (2003) Nanoparticle assembly and transport at liquid–liquid interfaces. *Science*, **299**, 226–9.

102 Russell, J.T., Lin, Y., Boker, A., Long, S., Carl, P., Zettl, H., He, J., Sill, K., Tangiraia, R., Emrick, T., Littrell, K., Thiyagarajan, P., Cookson, D., Fery, A., Wang, Q. and Russell, T.P. (2005) Self-assembly and cross-linking of bionanoparticles at liquid–liquid interfaces. *Angewandte Chemie (International Edition in English)*, **44**, 2420–6.

103 Urakami, N., Imai, M., Sano, Y. and Takasu, M. (2000) The effects of chondroitin sulfate on the tobacco mosaic virus configuration. *Progress of Theoretical Physics Supplement*, 390–1.

104 Imai, M., Urakami, N., Nakamura, A., Takada, R., Oikawa, R. and Sano, Y. (2002) Polysaccharides induced crystallization of tobacco mosaic virus particles. *Langmuir*, **18**, 9918–23.

105 Nedoluzhko, A. and Douglas, T. (2001) Ordered association of tobacco mosaic virus in the presence of divalent metal ions. *Journal of Inorganic Biochemistry*, **84**, 233–40.

106 Dujardin, E., Peet, C., Stubbs, G., Culver, J.N. and Mann, S. (2003) Organization of metallic nanoparticles using tobacco mosaic virus templates. *Nano Letters*, **3**, 413–17.

107 Dogic, Z. and Fraden, S. (2000) Cholesteric phase in virus suspensions. *Langmuir*, **16**, 7820–4.

108 Filpula, D., Rollence, M., Essig, N., Nagle, J., Achari, A. and Lee, T. (1995) Engineering of immunoglobulin Fc and single-chain Fv proteins in *Escherichia coli*. *Antibody Expression and Engineering*, **604**, 70–85.

109 Lee, S.W., Wood, B.M. and Belcher, A.M. (2003) Chiral smectic C structures of virus-based films. *Langmuir*, **19**, 1592–8.

110 Kalinin, S.V., Jesse, S., Liu, W.L. and Balandin, A.A. (2006) Evidence for possible flexoelectricity in tobacco mosaic viruses used as nanotemplates. *Applied Physics Letters*, **88**, 153902–1.3.

111 Tseng, R.J., Tsai, C.L., Ma, L.P. and Ouyang, J.Y. (2006) Digital memory device based on tobacco mosaic virus conjugated with nanoparticles. *Nature Nanotechnology*, **1**, 72–7.

112 Yoo, P.J., Nam, K.T., Qi, J.F., Lee, S.K., Park, J., Belcher, A.M. and Hammond, P.T. (2006) Spontaneous assembly of viruses on multilayered polymer surfaces. *Nature Materials*, **5**, 234–40.

113 Soto, C.M., Blum, A.S., Vora, G.J., Lebedev, N., Meador, C.E., Won, A.P., Chatterji, A., Johnson, J.E. and Ratna, B.R. (2006) Fluorescent signal amplification of carbocyanine dyes using engineered viral nanoparticles. *Journal of the American Chemical Society*, **128**, 5184–9.

114 Barnhill, H.N., Claudel-Gillet, S., Ziessel, R., Charbonniere, L.J. and Wang, Q. (2007) Prototype protein assembly as scaffold for time-resolved fluoroimmuno assays. *Journal of the American Chemical Society*, **129**, 7799–806.

115 Prasuhn, D.E., Singh, P., Strable, E., Brown, S., Manchester, M. and Finn, M.G. (2008) Plasma clearance of bacteriophage Q beta particles as a function of surface charge. *Journal of the American Chemical Society*, **130**, 1328–34.

2
Biomimetic Nanoparticles Providing Molecularly Defined Binding Sites – Protein-Featuring Structures versus Molecularly Imprinted Polymers

Kirsten Borchers, Sandra Genov, Carmen Gruber-Traub, Klaus Niedergall, Jolafin Plankalayil, Daniela Pufky-Heinrich, Jürgen Riegler, Tino Schreiber, Günter E.M. Tovar, Achim Weber, and Daria Wojciukiewicz

2.1
Introduction

Molecular recognition takes place at the interface between specific moieties, such as cells or individual molecules. Today, chemical nanotechnology enables the generation of artificial nanoparticles with tailor-made properties. Nanoparticles can be generated that act as highly specific partners in molecular recognition reactions. Nanoparticles consist of solid matter and expose an extremely large surface to their surroundings, which makes them principally ideal bearers of molecular recognition sites. Thus, a key question is the following: "How to convert their surface to a molecular recognition element?". This chapter will focus on two alternative routes, either by supramolecular immobilization of biological recognition elements (i.e., proteins such as antibodies or enzymes) or by the supramolecular design of the surface of polymer particles by a specific molecular interaction during the polymerization process (so-called molecular imprinting of polymers).

To enable molecular recognition, a nanoparticle surface must provide molecularly defined binding sites. These consist of organic functions that are positioned at the particle surface such that they can attractively interact with specific molecules by a concerted action. The surface of the nanoparticle must be of organic origin. This can be achieved by building an organic shell around a core material that consists of organic or inorganic material. Additionally, the core material can provide specific physical properties such as a magnetic moment or a specific optical absorbance. Thus, the nanoparticle core could enable a specific interaction of the particle with the external surrounding (e.g., an acceleration of the particle in a magnetic field or a resonance phenomenon at a certain wavelength of an electromagnetic radiation). A variety of different core/shell architectures with a molecularly recognizing surface of the shell and popular core materials will be highlighted in this chapter.

Cellular and Biomolecular Recognition: Synthetic and Non-Biological Molecules. Edited by Raz Jelinek
Copyright © 2009 WILEY-VCH Verlag GmbH & Co. KGaA, Weinheim
ISBN: 978-3-527-32265-7

Recently, inorganic nanoparticles have gained increasing respect. Owing to their size-dependent mesoscopic properties, they are discussed as a valuable material in many life sciences applications. Next to the well-known protocol of gold staining of proteins, cadmium–chalcogenide nanocrystals were used for fluorescence imaging of both extracellular and intracellular matrices. Magnetic nanoparticles built from metals or metal oxides were used for the first time in magnetic resonance imaging and for thermotherapy of cancer. Nevertheless, these particles are not biomimetic as such, but can be used as a functional core material. Silica and organically modified silica are widely used core materials for the preparation of biofunctional nanoparticles. Their relatively simple chemistry – predominantly defined by hydrolysis and condensation reactions – allows for versatile fine-tuning of their composition and excellent control of their surface functions. Additionally, chemical modification of the primary particles is easily achieved using organofunctional silanes – providing that the surface presents free silanol groups, which is almost always the case. At the same time, functional components like fluorescent dyes, quantum dots (QDs), magnetite nanoparticles, drugs or combinations thereof can be encapsulated into silica cores. This encapsulation possibility makes them extremely versatile core elements for a vast variety of desired properties to enable specific applications. Nanoparticulate amorphous silica cores are mostly fabricated via sol-gel reactions. Adding surfactants to the reaction as structure-directing agents results in microporous or mesoporous silica systems. The well-defined pore sizes of such a material offer additional possibilities such as targeted hosting of specific molecules in the pores.

Organic core materials have been widely used for the preparation of nanoparticle cores. In particular, biomedical applications such as drug delivery can take advantage of organic cores. The organic material may be designed to be highly biocompatible and allows for dissolution or encapsulation of organic drugs within its organic matrix. Organic cores are often chosen for envisaged applications such as gene delivery systems or biomarkers – or even biosensors. In this chapter, three different organic systems are highlighted due to their widespread use and interest – polymeric nanoparticles, lipids and fullerenes.

The organic shell that covers the core particle bears the biomimetic feature of molecular recognition. Additionally, these organic shells may perform further functions such as protection of the particle core against oxidation or degradation. Also, the shell may adjust the surface properties of the core material or it may stabilize the particle suspension in order to prevent agglomeration. The shell may also contain fluorescent dyes or be loaded with drugs. In any case it will provide the molecular recognition moiety (i) by stabilizing and directing a biomolecular receptor such as an antibody or (ii) by being molecularly imprinted and thus containing the molecular binding site in an artificial polymer. In the first case, the shell must arbitrate between the material needs of the biomolecule whose complex structure may easily be disturbed and the nanoparticle material that constitutes an artificial, and thus primarily a disturbing, surrounding for the biomolecule.

In the latter case, the shell must be composed such that it can be structurally modified by template molecules during its preparation by a so-called molecular imprinting reaction.

The development of the technique of molecular imprinting has considerably accelerated over recent decades. Molecularly imprinted polymers (MIPs) are usually prepared by radical copolymerization of a functional monomer and a cross-linker in the presence of a template molecule. Thus, a polymer monolith is obtained that may be converted to particles by mechanical processing like grinding. Micro-sized particles with jagged morphology have been produced by this route for many years. In the last few years, a variety of alternative methods for the preparation of particulate and nanoparticulate MIPs have been developed and will be described here.

The highly attractive quest for molecular imprinting of peptides and proteins has been tackled since the early 1990s. Different groups reported successful imprinting of small peptides in organic solvents and also on the preparation of MIPs in water for the recognition of large proteins. A sequence of a protein, a so-called epitope, was used for imprinting; the resulting MIPs were able to recognize proteins containing this sequence under appropriate conditions. Nanoparticulate MIPs prepared via miniemulsion polymerization exhibit a defined size and shape and a high specific surface area. It could be shown that such nanoparticulate MIPs are able to recognize low-molecular-weight compounds up to biomacromolecules such as proteins.

Biomimetic nanoparticles with a magnetic core may be very helpful for separation processes. An ideal separation process optimizes the yield of the product, is easily to perform and maintains low manufacturing costs. In particular, protein purification typically involves several consecutive steps to achieve the desired purity. One possibility to improve and to facilitate the purification of proteins could be the application of MIP beads with a magnetizable core. This kind of MIP bead can bind the protein or the impurities and can be separated very easily from the bulk solution by applying a magnetic field.

The main benefits of nanoparticles used for drug delivery or drug targeting are the controlled release times of the included agents and the possibility to transport drugs to a specific target in the human body. For drug delivery, the most commonly used biomaterials are liposomes, polymer nanoparticles, dendrimers, vesicles and micelles. By conjugating their surface to specific ligands, nanoparticles can be directed to a tissue, an organ or specific cell types, such as macrophages, dendritic cells and tumor cells.

The following sections give an overview of biomimetic nanoparticles that provide molecularly defined binding sites by highlighting nanoparticle systems containing protein-based binding sites and molecularly imprinted binding sites.

2.2
Core Materials and Functionalities

2.2.1
Inorganic Core Materials

2.2.1.1 Inorganic Crystalline Nanoparticles

In this section we introduce some aspects of inorganic crystalline nanoparticles and their use in modern biology/medicine. Related to their size, these particles show some unique chemical and physical properties, which may be useful in a wide range of life sciences applications like imaging, diagnosis or therapeutics. Unfortunately, in respect of this section the research field of inorganic nanoparticles and their use in biology and medicine is too wide to discuss in detail. Additionally, these kinds of particles are not really biomimetic in the full meaning of the word. This section will thus give a rough overview of this research field.

Recently, crystalline inorganic nanoparticles have gained increasing respect in both industry and academia. Owing to their size they show many special properties that are collected in the literature as "mesoscopic properties". Common to all particles at the nanometer scale is the high surface:volume ratio that directly scales in the range of $3/r$ with the radius of the particles. Therefore, crystalline nanoparticles show high surface energies and this leads to many crystallographic defects on the surface. Both effects have been known for a long time and are technically used already in the field of catalysis, for instance [1]. Some other mesoscopic properties of nanoparticles are attributed to the crystal structure. For instance, a permanent electrical dipole moment, which also scales with the diameter of the crystal, allows the potential application of nanoparticles in electrical data storage devices.

In the past few years semiconductor nanocrystals, often referred as QDs, have gained increased interest especially because of their potential application in electropotic devices like light-emitting devices or solar cells. Additionally, QDs are widely discussed as a valuable add-on to imaging application in biology and diagnosis [2, 3]. The wavelength emission itself is very narrow and symmetric. Therefore, theoretically up to eight different antibodies could be labeled with different sized QDs, brought to the target and visualized by one excitation source next to each other simultaneously [4, 5].

Metal and metal oxide nanoparticles are also a focus of interest. Next to the well-known concept of gold staining [6], metal particles are already used or discussed for many applications in modern biology and medicine. Silver particles, for instance, show some antimicrobial features [7]. Gold particles are used as detection labels by reflection or absorption. Other nanoparticles like nickel or magnetite particles are known for their superparamagnetic property, which also occurs at the nanoscale scale of the crystals. With increasing interest in fabricating nanodevices with nano-sized blocks, much attention has been focused on the formulation of a general route to control the size and morphology of nanoparticles

[8, 9]. Next to simple precipitation, a number of physical and chemical routes have been applied to produce nanoscale crystalline materials, including mechanical grinding, sonochemistry, organometallic precursor pyrolysis, metal melt reduction in the micelle phase, electrochemical deposition, etc. [10]. More details about synthesis method for highly crystalline nanoparticles with superior mesoscopic properties, drawing on the example of silica and magnetite particles, will be given in the following sections.

2.2.1.2 Particles with Silica Cores

Silica nanoparticles are widely used for generating surface-functionalized nanoparticles. This can be attributed to the fact that chemical modification of silicon oxide (SiO_x) is easily achieved using (organo-) functional silanes. Silica nanoparticles can be generated by different procedures. Commercially available SiO_x nanopowders produced by high-temperature aerosol processes consist to a great extent of aggregates, and are normally used as filler and reinforcement material.

For use in biotechnology applications, nanoparticles with well-defined spherical shapes and diameters are desirable. Amorphous silica particles that meet these requirements are synthesized via sol-gel reactions. Stöber et al. established a protocol for the synthesis of monodisperse, spherical silica nanoparticles in the range of 5–2000 nm in 1968 [11], which is widely used today [12–14]. The so-called Stöber process is as follows: in an alkaline hydrolysis reaction, tetraethyl orthosilicate (TEOS or differently substituted orthosilicates) is converted into a singly hydrolyzed TEOS monomer:

$$Si(OR)_4 + H_2O \rightarrow (OR)_3 Si(OH) + ROH$$

with R = C_2H_5 for TEOS. Subsequently, this intermediate reaction product condenses to form silica spheres:

$$(OR)_3 Si(OH) + H_2O \rightarrow SiO_2 \downarrow + 3ROH$$

This reaction scheme is a simplification of the versatile condensation processes that lead to the formation of the silica particles. The full range of possibly occurring reaction mechanisms is described by Brinker et al. [15].

The size of the nanoparticles can be controlled by various parameters. It was observed that for particular concentrations of TEOS and ethanol the particle size follows a parabolic behavior for both ammonia and water. Moreover, the particle size is influenced by the reaction temperature [14], the water:TEOS ratio, the monomer used, the ammonia concentration and the solvent [15]. Basically, particle size decreases monotonically as temperature increases in the range of 10–55 °C. When different silica alkoxides (tetramethoxy orthosilicate, TEOS, tetraprothoxy orthosilicate or tetrabuthoxy orthosilicate) are used for particle synthesis in propanol, the particle size follows a parabolic behavior according to the change in size of the carbon chain. The maximum particle size was observed when TEOS was used as precursor monomer [15]. TMOS did not result in spherical, but irregularly shaped nanoparticles. When different solvents are used, the average particle size increases from 30 to 800 nm as the alcohol chain length increases from methanol to butanol [15].

The maximum size of particles achieved by using the described process was 800 nm. In order to prepare larger particles, seed suspensions are used. Silica seed particles are synthesized as described above, and after the reaction has stopped TEOS and water are repeatedly added to the suspension until the particles have grown to the desired size. For reaction conditions that result in monodisperse size distributions, the standard deviation in particle size expressed as percentage of the average particle diameter was found to decrease with increasing particle size from 10–15% for particle sizes below 100 nm to less then 5% for bigger particles [14].

The Stöber process was further enhanced and addition of poly(ethylene glycol) (PEG) to the reaction for particle core synthesis resulted in hybrid particles that could nicely be redispersed after having been stored as a dry solid [16].

Alternative methods of silica particle synthesis like spray-drying and sol-gel emulsion have been applied with the objective of incorporating functional components into the particle cores. For biomedical applications, fluorescent dyes [17], QDs [18], magnetite nanoparticles, drugs [19] and combinations thereof [20–22] have been encapsulated into silica particles.

Apart from solid silica spheres, microporous and mesoporous silica particles with pore sizes in the range 1–2 and 2–50 nm, respectively, can be generated by adding surfactants to the reaction as structure-directing agents [23, 24]. The feasibility to obtain different pore size and geometries offers a wide range of possibilities for targeted hosting of molecules. Such particles can be applied in a broad range of topics like separation and purification processes, catalysis, and controlled drug release.

2.2.1.3 Metals and Metal Oxides

Magnetic materials have shown a tremendous growth of interest for several applications. The intrinsic interaction of the magnetic nanoparticle with an applied magnetic field makes these particles attractive for magnetic field-assisted transport, and separation and analysis [25, 26]. Furthermore, the superparamagnetic attributes of these materials seems to be particularly desirable for biomedical use (e.g., as contrast agents for magnet resonance imaging and as a therapeutic agent for cancer by heating by an external magnetic field) [27]. Much effort has therefore been spent on developing and understanding synthetic aspects, size, and magnetic and chemical behavior.

Apart from iron oxide, pure metals (iron and cobalt), spinel-type ferrimagnets ($MgFe_2O_4$ and $MnFe_2O_4$) as well as alloys ($CoPt_3$ and FePt) have been used as the magnetic phase. In the nanoparticle, the surrounding shell serves not only for stabilization of the magnetic core during the synthesis process and against oxidation, but gives rise to further functionalization depending on the desired application.

Superparamagnetism is dependent on the particle size of the magnetic materials (i.e., ferro-, antiferro- or ferrimagnetic materials). Concerning the fact that the reduced critical volume of the particle develops its superparamagnetic character, it is a special matter whether the particles are built up in an energy-efficient single domain or multidomain. The characteristic diameter of these particles is called the critical particle size D_c. Some examples of critical diameters of different materials are listed in Table 2.1.

Table 2.1 Estimated single-domain size for different spherical nanoparticles [28].

Material	D_c (nm)
Fe	14
Ni	55
Fe_3O_4	128
γ-Fe_2O_3	166

Particles exhibit a superparamagnetic behavior not only if critical particle size has been reached, but also above a certain temperature (blocking temperature) and coercivity near zero. These kinds of particles do not show remnant magnetization, which can be visualized by superconducting quantum interference device (SQUID) measurements as illustrated in Figure 2.1.

Four concepts of magnetic nanoparticles are described in the literature [25]. The most popular concept is coprecipitation. Coated magnetite nanoparticles can be obtained from aqueous Fe(II) and Fe(III) salt solutions within a relatively narrow particle size range.

Another concept describes the thermal decomposition of organometallic compounds in high-boiling organic solvents [29, 30]. Nanoparticles produced by this method can be obtained in high yields possessing a small particle size distribution. Iron nanoparticles have been described that were obtained by decomposition of iron pentacarbonyl [8]. In addition, magnetic nanoparticles are accessible by hydrothermal synthesis. This concept has been established by Wang et al. [31] and has been reviewed recently by Yoshimura et al. [32]. Microemulsions could also be used for the synthesis of magnetic nanoparticles [33]. Microemulsions will become an alternative choice for the synthesis of magnetic nanoparticles especially when they enable for narrower size distributions and when scaled-up microemulsion processes will be available [25].

2.2.2
Organic Core Materials

2.2.2.1 Polymers, Lipids and Fullerenes

Polymer nanoparticles of synthetic origin have been ranked and explored in various fields of research, and go beyond biomedical applications, including their applicability as drug and gene delivery systems [34], biosensor [35, 36], biomarkers, molecular diagnosis systems and nanoelectronics devices [37].

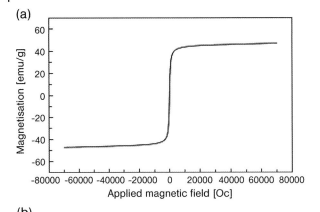
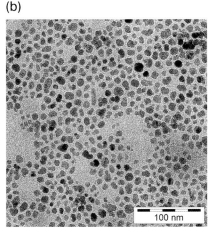

Figure 2.1 (a) Pattern of a SQUID measurement of oleic acid coated magnetite (Fe$_3$O$_4$) nanoparticles which show superparamagnetic behavior. (b) Transmission electron microscopy image of oleic acid-coated magnetite nanoparticles with a diameter from 5 to 12 nm. (Source: Institute for Interfacial Engineering (IGVT), University of Stuttgart).

Preparation methods and their potential field of application in pharmacology are highlighted in detail for three main classes of nanoparticles possessing an organic core–polymeric nanoparticles, lipids and fullerenes. These materials have gained considerable popularity towards bioconjugation and have been well described in several review papers [25, 27, 34, 38].

Polymeric Nanoparticles Polymer particles in the nanometer site range are classified into nanospheres nanocapsules and micelles according to the process used for the preparation of the particle (Figure 2.2) [27].

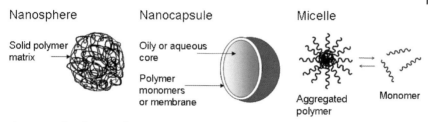

Figure 2.2 Classification of nano-sized polymer particles according to the process used for the preparation of the particle: nanosphere, nanocapsule and micelle, respectively.

Drug loading of the nanoparticles can be realized by two methods: the drug is either incorporated at the time the nanoparticle is produced or the drug is adsorbed by incubating the preformed nanoparticles in the drug solution.

Nanospheres consist of a solid polymer matrix core in which – in the case of a drug delivery system – the drug is dispersed throughout the particle. Nanocapsules are considered as a "reservoir" system. Here the drug is embedded into an oily or aqueous core that is surrounded by a polymer monomer or membrane. For the discussion of both phospholipids and block copolymers, micelles should necessarily be mentioned and should be considered as amphiphilic colloids rather than solid nanoparticles. Micelles are defined as aggregates of amphiphilic molecules dispersed in a liquid colloid. They are in dynamic equilibrium with their monomers, and formed at a specific and narrow concentration range – the critical micelle concentration (CMC).

A variety of synthetic methods for the preparation of polymeric nanoparticles are available. Several techniques have been established that can be divided into two classes. In the bottom-up technique the nanoparticles are synthesized from monomers with methods such as interfacial polymerization and emulsion or microemulsion polymerization. Top-down techniques employ preformed polymers, which are dispersed in a colloidal solution in order to form nano-sized particles. Common methods are emulsion evaporation and diffusion, solvent displacement as well as salting-out techniques.

Nanoparticle formation via electrospraying is an effective process in pharmaceutical industries and is well established. It is advantageous over liquid-based methods since a wide variety of materials can be applied. This topic, including preparation methods and applications, has been review recently by Jaworek [39].

Nanospheres can be prepared by emulsion polymerization, one of the fastest methods and easily scalable. Based on the methodology (oil-in-water or water-in-oil), both biodegradable and nondegradable nanoparticles have been synthesized [40–42]. However, the most common technique for the preparation of nanospheres is solvent displacement, referred to as nanoprecipitation utilizing preformed polymers. Poly(lactic acid) nanoparticles were obtained by dissolution of the polymer in an organic water-miscible solvent. Upon addition of the polymer solution to an aqueous solution, spontaneous emulsification occurs leading to precipitation of the polymer and, thus, formation of nanoparticles [43].

Methods for the preparation of poly(alkylcyanoacrylate) nanocapsules are based on interfacial polymerization [41, 43]. The polymer shell is formed *in situ* after contact of the cyanoacrylate monomer with initiating ions. This technique leads to an efficient encapsulation of active agents (e.g., insulin) into the liquid core, making it advantageous over other preparation methods.

Polymeric micelles have attracted significant attention for drug delivery systems due to their high drug loading capacity as well as their unique disposition behavior in the body [44]. AB-type block copolymers are the most common species. These micelles are assemblies of nanoscale size (10–100 nm in diameter) possessing a core/shell structure in which an inner core is surrounded by a shell of hydrophobic polymers segregated from the aqueous phase by the hydrophobic block of the copolymer.

PEG-based block copolymers are favored candidates, since they are known to be biocompatible, nontoxic, soluble and highly hydrated [45]. A wide variety of different hydrophobic polymer blocks covalently bound to PEG gives rise to numerous types of nanoparticles [38, 46]. Biodegradable PEG–polylactide micelles can be obtained by a dialysis method after anionic ring-opening polymerization of ethylene oxide and subsequent lactide addition. Another method for the formation of polymer micelles is the direct dissolution method [38]. Drug-loaded micelles of pluronic block copolymers were formed by simply adding the polymer to an aqueous solution at concentrations above the CMC [47]. Film casting [48] and the oil-in water emulsion procedure are methods that can be applied to hydrophobic block copolymers by dissolving them in volatile, organic solvents [49].

Biodegradability and biocompatibility play a tremendous role in drug delivery systems (see Figure 2.3). First, rapidly biodegradable nanoparticles were made of poly(methylcyanoacrylate) and poly(ethylcyanoacrylate) as potential lysomotropic carrier systems [50]. Over the years poly(lactic acid) and its copolymer poly(lactic-*co*-glycolide] have become important polymers for drug delivery purposes. Due to their insolubility in aqueous media, nanoparticle syntheses have been challenging, but are now well established. Examples of active agents that have been entrapped include bovine serum albumin (BSA) [51], estrogen, paclitaxel [52] and the fluorescence marker 6-coumarin [51, 53].

Another important structural pattern of polymeric nanoparticles is stimuli sensitivity. It provides a unique opportunity to control the triggered release of the therapeutic agents at the diseased site [54]. *N*-isopropyl acrylamide is a temperature-sensitive hydrophilic moiety that has been studied intensively over the past decades [55]. Thus, thermo-sensitive amphiphilic polymers are generated with suitable hydrophobic segments. Poly(*N*-isopropylacrylamide-*co*-*N*,*N*'-dimethyleamide) micelles were synthesized showing a lower critical solution temperature of 37.7 and 38.2 °C, respectively [56].

Intrinsic properties of nanoparticles can be changed upon their association with active molecules. Fluorescence particles have become widely used as tracers applied in immuno- and genetic fluorescence detection, and cell labeling *in vitro*

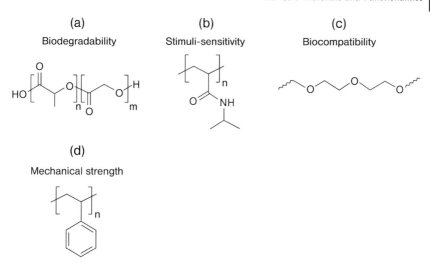

Figure 2.3 Properties of synthetic polymers: structural and functional pattern of various custom-sized synthetic macromolecules: (a) polylactide-*co*-glycolide, (b) poly-*N*-isopropylacrylamide, (c) polyethylene glycole PEG, (d) polystyrene moiety.

and *in vivo* [57–59]. Encapsulation of fluorescence markers into nanoparticles often leads to enhanced performance of the dye compared to the simple dye conjugate. Within the field of polymeric materials as matrices to prepare fluorescence particles, polystyrene and poly(cyanoacrylate) nanoparticles have attracted significant attention [60].

Lipids Nanoparticular lipid-based drug delivery systems include liposomes and solid lipid particles [61]. Phospholipid-based liposomes have been utilized for a variety of applications ranging from drug targeting, drug and gene delivery, diagnosis imaging to biosensors. They can be loaded with a variety of molecules, such as small drug molecules, proteins, nucleotides and even viruses. Stealth liposomes have been reviewed by Moghimi and Szebeni [62]. This sterically protected liposomes are ignored by the immune system and thus exhibit particular prolonged circulation times. Poloaxmer F127 (a diblock copolymer of polyoxyethylene and polyoxypropylene), containing liposomes of phospholipid derivatives, was introduced as temperature-sensitive species. The study demonstrated the possible use of these nanoparticles as drug delivery systems *in vivo* [63].

Solid lipid nanoparticles have attracted increasing attention as an efficient and nontoxic drug carrier. They are available from physiological lipids or synthetic lipid molecules and are prepared either by high-pressure homogenization or microemulsation [61]. The pharmacokinetics and tissue distribution of doxorubicin and tobramycin-loaded stealth solid phospholipid nanoparticles were studied, suggesting their potential use as drug delivery systems to the brain, but not capable of crossing the blood–brain barrier [64, 65].

Fullerenes Fullerenes (also known as bucky balls) have been the subject of great interest because of their potential use in biomedicine [66, 67]. The structure is hollow, and might be used to carry and deliver active agents. However, very few atoms are known to form stable endohedral complexes (e.g., lanthanum, yttrium, scandium and a few noble gases). Many compounds are inclined to break up because the chemical bonds cannot be aligned at the correct angle. A proposed application of these endohedral materials is to shield radioactive tracers inside the fullerene cages and then inject the material into human blood to monitor blood flow. Endohedral materials have also been subject to speculation concerning drug injections, with dissolving atoms caged inside [66].

2.3
Functional Shells

2.3.1
Organic Shells

Nanoparticle cores can be equipped with several functional shells to protect the particle core against oxidation or opsonization, to adjust the surface properties (e.g., in order to render hydrophobic particles water soluble), to prevent agglomeration, or to couple fluorescent dyes or biomolecules.

One of the main features of biomimetic nanoparticles is their capability of molecular recognition reactions. These are mediated by specific molecular structures at the particle's surfaces. These structures may be either composed of biologically derived macromolecules or fully synthetic receptors [13]. If biological building blocks are immobilized onto nanoparticle surfaces in order to fulfill specific tasks, e.g., targeting of drug-loaded nanocarriers or specific labeling of certain analyte molecules, anchor groups must be provided at the particle surface. Basically there is a relatively small set of organic functional groups, which are commonly used as anchors for the conjugation of biomolecules. These conjugation techniques are based on the availability of reactive functional groups that are usually present in biomolecules (i.e., amino, carboxyl and thiol functionalities). An exhaustive overview of the state of the art of bioconjugate techniques including the underlying chemical mechanisms, protocols and a great variety of zero-length, homobifunctional and heterobifunctional as well as trifunctional cross-linkers is given by Hermanson [68].

The most straightforward route for immobilization of peptides and proteins relies on the formation of amide bonds. Frequently this involves the reaction of active esters (e.g., *N*-hydroxysuccinimide ester), with nucleophilic groups at the protein such as amino functionalities present in lysine or histidine residues. Thus, the integration of active ester functions into the nanoparticle shells is one way of providing anchor groups for further biofunctionalization. An alternative method to form amide linkages is the activation of carboxyl moieties via carbodiimide

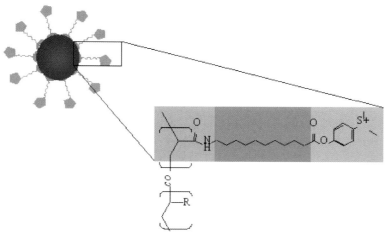

Figure 2.4 P-(11(Acrylamido)undecanoyloxy)phenyl dimethylsulfonium methylsulfate (AUPDS) surfmer: polymerizable surfactant consisting of a polymerizable group, a spacer and an active ester end group for stabilization and immobilization. (Developed at the Institute for Interfacial Engineering (IGVT), University of Stuttgart [69]).

activation for subsequent coupling to amino functions. The most commonly used carbodiimide is 1-ethyl-3-(3-dimethylaminopropyl) carbodiimide hydrochloride (EDC) because of its water solubility. Carboxyl-modified silica nanoparticles have been prepared by covalent conjugation reactions for high-density immobilization of proteins mediated by EDC [12]. An alternative approach utilizes the coupling of surface-bound aldehyde functions with amino functions via Schiff base formation, followed by a reduction reaction.

Coupling of proteins via thiol groups typically involves maleinimide functionalization of the substrate surface. In some rare cases unpaired cysteine residues are present in wild-type proteins. Additional cysteine moieties can be inserted into the amino acid sequence using modern protein engineering techniques.

We have prepared thiol-active, maleinimide-functionalized silica particles for the site-directed coupling of cysteine-histidine-tagged tumor necrosis factor (TNF)-α [14]. Such TNF-α-functionalized nanoparticles acted as cell-mimetic particles by activating TNF receptor TNFR2: this receptor cannot be triggered by soluble TNF molecules, but only by cell membrane-bound TNF and – as was demonstrated – by nanoparticle-bound TNF (see also Section 2.4.2.3).

Apart from post-functionalization of the nanoparticles, the particle interfaces may also be functionalized during particle synthesis.

By using functional surfmers, nanoparticles with activated ester surfaces were prepared in a one-step synthesis (Figure 2.4) [69]. Activated esters readily react with amino-functionalized molecules including proteins and fluorescent dyes. The coupling of native proteins was demonstrated by detecting preserved enzyme activity of particle-bound peroxidase [70].

Another approach to generating functional organic surfaces capable of molecular recognition is the synthesis of MIP nanoparticles (nanoMIPs). This approach is described in the next section.

2.3.2
MIPs

Molecular recognition plays a crucial role in nature and is fundamental for many techniques, wherein detection and separation need to be addressed. Synthetic polymers with molecular binding and recognition sites represent an interesting alternative to commonly used binding matrices that are based on biological molecules. Recently, there has been increasing interest in the use of MIPs for the design of artificial receptor-like binding sites in otherwise nonselective polymeric materials [71–74]. Therein, MIPs overcome drawbacks like the relatively instability of antibodies and enzymes to several organic solvents and extreme pH. Additionally, the use of MIPs has some economic and ecological advantages. In comparison to monoclonal antibodies they could be produced with less expense and are recyclable in general [75].

Roughly, molecular imprinting in polymers can be described as follows. A molecular template is bonded to a monomer to form an imprint precursor. This precursor can be formed by ionic, covalent or coordinative binding. The template–monomer complex now is used in a polymerization step. After polymerization the removal of the template leaves a specific binding site for the template inside the polymeric network. In this way a variety of active agents ranging from low-molecular-weight drugs to peptides, proteins and biomacromolecules might be imprinted [76]. Technically, the imprinted polymeric monoliths are often ground and sieved. Therefore, the surface of the polymer and therein the active binding sites and the rebinding capacity are increased. Nevertheless this process has some disadvantages. The resulting micro- or nanoparticles are not monodisperse, the increase of the binding site is only statistical and the active binding site could possibly be destroyed by grinding. Recently, the approach has been to directly produce MIPs at the mirco- or nanoscale and to ensure the imprinting on the surface of the particles. The so-called nanoMIPs can be produced by several techniques like precipitation polymerization or miniemulsion polymerization. MIPs and nanoMIPs are valuable materials for many scientific and technical applications in sensoric, diagnosis, separation and downstream processing [77–79].

The following section discusses general concepts for producing both bulk and nanoMIPs. Tools for the development of specific MIPs will be discussed. A special focus will be on MIPs that are imprinted against peptides and proteins. Section 2.4.1 shows the exemplary use of MIPs in a technical application.

2.3.2.1 Tools for MIP Development

The accelerated development of new MIPs with better performance can be achieved with new tools established in the past few years. One of these tools is called design of experiments (DoE). This chemometric tool is particularly appropriate to optimize the main factors (amount of functional monomer, amount of cross-linker, porogenic solvent, etc.) that affect the molecular recognition properties. Usually optimizations, rely on monitoring the effect of changing one factor at a time on a response. This method of optimization is time consuming and can produce erroneous optima in experiments where interactions between factors are involved. In comparison, DoE follows a mathematical framework for changing selected factors simultaneously to predict the optimum conditions.

The use of a computational method in combination with molecular modeling software is a promising tool in developing molecular imprinted polymers. Rossi et al. showed how to find the best functional monomer against a target molecule by using a virtual library of functional monomers, modeling software and DoE [80].

Another helpful tool to accelerate the development of MIPs is experimental high-throughput screening. Variations of monomers, templates, solvents and initiators were reacted inside a 96-well plate, and analyses of template rebinding were performed by the evaluation of the supernatants with a multifunctional ultraviolet plate reader. This technique allows the synthesis up to 80 polymers in 24 h and analyzing the binding properties of the MIPs in nearly 1 week [81, 82].

In recent years new techniques have been applied to obtain more data for a better understanding of the mechanisms of molecular imprinting and to investigate the thermodynamics of binding to MIPs. One of these techniques is isothermal titration calorimetry (ITC). ITC is a technique for thermodynamically monitoring chemical reactions initiated by the addition of a component to a reaction mixture. The application of ITC for biological and biomolecular recognition reactions is described in several studies [83, 84].

The first important step of noncovalent molecular imprinting is the preorganization of functional monomer and template. This step is essential for successful molecular imprinting. A useful technique for the characterization of the prepolymerization complex between the functional monomer and the template is nuclear magnetic resonance (NMR) spectroscopy. ^{1}H-NMR spectroscopy is one of the most powerful tools to investigate the chemical shift of H-atoms by virtue of hydrogen bonding or by changing the chemical circumstances [85]. Therefore, the association constant and the stoichiometry of the prepolymerization complex could be calculated [86]. Several groups successfully showed [87, 88] the application of NMR spectroscopy for the rational development of MIPs.

2.3.2.2 Bulk MIP and Proteins

Although molecular imprinting in organic polymers was already developed by Wulff's group in the early 1970s [89], the imprinting of peptides and proteins was only realized for first time in the 1990s. The implementation of this challenge could be reasoned in the properties of proteins. First, proteins have a flexible structure and conformation, which can easily be affected by changes in the environment like temperature, pH and ionic strength. Second, proteins are complex molecules with a lot of potential recognition sites. However, it is generally accepted that MIPs exhibit the greatest specificity and selectivity with few strong interactions between polymer and template. In contrast, there are a lot of weak interactions between the proteins and functional monomer resulting in rather nonspecific binding. Third, most proteins are water soluble and this property is less compatible with mainstream MIP technology, which is based on MIP preparation in organic solvents. This fact limits the choice of functional monomers and cross-linkers because most of them are not water soluble. Moreover, water competes for hydrogen bonding between the template molecule and the monomer [90].

The first success in molecular imprinting of amino acid derivatives and small peptides was achieved by Mosbach's group in the 1990s. They reported on the successful imprinting of a number of chiral compounds, including N-protected amino acid derivatives as template molecules [91]. The prepared MIPs showed efficient enantiomeric resolution of a racemate of the imprint molecule and substrate selectivity for the imprint molecule in a mixture of molecules with similar structure. In the following years the preparation of MIPs for recognition of small peptides and proteins gained increasing respect [92, 93]. However, the used solvent was always an organic one, such as chloroform, tetrahydrofuran or acetonitrile.

As mentioned earlier, the main challenge is linked with the preparation of MIPs in water because most peptides and proteins are water soluble and many natural recognition events such as antigen–antibody binding occur in aqueous media. The first investigations in the field of protein imprinting in water were reported at the end of the 1990s. Bossi *et al.* proposed an approach to the preparation of affinity matrixes based on the surface coating of polystyrene microtiter plates with a thin layer of a stable conjugated polymer polymerized in the presence of various proteins [94]. They used 3-aminophenylboronic acid as the functional monomer, which was polymerized in the presence of microperoxidase, lactoperoxidase, horseradish peroxidase and hemoglobin via oxidation of the monomer by ammonium persulfate. Hjertén *et al.* reported on acrylamide-N,N'-methylenbisacrylamide gel particles, which were polymerized in the presence of cytochrome *c*, transferrin and hemoglobin [95]. After the polymerization and removal of protein templates, the gel particles were packed into a chromatographic column. These chromatographic columns were highly selective (e.g., a "horse myoglobin column" adsorbed horse myoglobin but not whale myoglobin even though these two proteins have similar amino acid sequences and three-dimensional structures). In the following years they modified the polymerization conditions and used absorption spectra for rapid screening of monomers for the design of selective MIPs. To obtain information about the types of bond occurring between the protein molecule and the monomer, they carried out docking experiments indicating that selective adsorption of, for example, hemoglobin by the polyacrylamide matrix is based primarily on hydrogen bonding and dipole–dipole interactions [96].

A completely new approach, the so-called epitope approach, was chosen by Rachkov and Minoura to prepare MIPs by mimicking nature [97]. When recognizing an antigen, an antibody interacts only with a small part of the antigen, the so-called epitope. Thus, it is theoretically possible to use an epitope as a template molecule for imprinting and the resulting MIP must be able to retain the whole protein molecule.

It was shown that by using a tetrapeptide Tyr–Pro–Leu–Gly-NH_2 as template molecule, the MIP can recognize not only the template but also some other peptides possessing the Pro–Leu–Gly-NH_2 sequence, including the peptide hormone oxytocin. Six years later, Nishino et al. used the same approach to prepare MIPs for recognition of three selected proteins: cytochrome c, alcohol dehydrogenase and BSA [98]. The chosen epitopes were the nonapeptides AYLKKATNE for cytochrome c, GRYVVDTSK for alcohol dehydrogenase and VVSTQTALA for BSA. The imprints were carried out on polymer films produced from acrylamide, N,N'-ethylene-bis-acrylamide and PEG200-diacrylate. In rebinding experiments, in each case, the best adsorption was observed for the protein corresponding to the former epitope. Concerning sequence sensitivity, a mismatched sequence that differed only at one position from the original sequence of BSA showed much less rebinding of BSA.

Despite the great advances made in the development of MIPs for the recognition of biomolecules such peptides and proteins, there is still a lot of work that should be done in this field. By now it is known that hydrogen bonds, and hydrophobic and electrostatic interactions are the main mechanisms for template receptor recognition in aqueous media, but the exact recognition mechanism remains unknown. Moreover, there are still many fundamental properties of MIPs prepared for recognition of peptides and proteins that have not yet been systematically investigated. These include temperature, ionic strength, pH, solvent, buffer composition, and the dependence of MIP affinity on cross-linker concentration and functional monomer concentration.

2.3.2.3 Nanospheric MIPs in General

Miniemulsion polymerization offers a general possibility to prepare polymeric nanoparticles [99]. This method enables the generation of MIP nanoparticles in a one-step process with a defined morphology. These nanoparticles exhibit a larger surface:volume ratio compared with the well-established bulk materials [100].

In the first step, nanoparticle nanodroplets are produced by through large shearing forces (e.g., sonication) in a range of 50–500 nm. These nanodroplets are stabilized by an additive against Oswald ripening during the miniemulsion polymerization. Using the direct miniemulsion polymerization process, the reactive mix (oil phase), containing the monomer, the cross-linker and the template, is added to the water phase. The reactive mix in the case of the inverse miniemulsion polymerization is solved in the water phase and added to the oil phase. After the polymerization the template is removed by extraction. The resulting polymeric nanoparticles are called nanoMIPs and are exemplarily shown in Figure 2.5. Now specific binding sites are accessible at the particle surface and this kind of particle can subsequently be used for molecular recognition. This is schematically displayed in Figure 2.6.

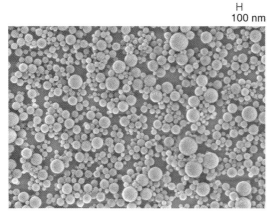

Figure 2.5 Scanning electron micrograph of L-BFA molecularly imprinted nanoparticles consisting of poly(MAA-co-EGDMA) prepared by photo-initiated miniemulsion polymerization. (Adapted from [101]).

New developments approve the production of MIPs in nanometer range via precipitation polymerization [102].

2.3.2.4 Nanospheric MIPs and Proteins

The Tovar's group has researched the development of nanoMIPs via miniemulsion polymerization for the last few years. In 2002, they reported the successful imprinting of L-boc-phenylalanine anilid (L-BFA) [99]. The monomer methacrylic acid (MAA) and the cross-linker ethylene glycol dimethacrylate (EGDMA) were polymerized in the presence of template in a thermally and a photo-initiated way (Figure 2.6) [101, 103].

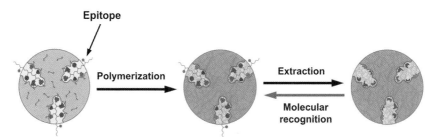

Figure 2.6 General scheme for the preparation of molecularly imprinted nanospheres and their use for molecular recognition. Template molecules induce the formation of binding sites during the miniemulsion polymerization. The templates are extracted from the highly cross-linked particles and are molecularly recognized by the nanosphere-selective binding sites.

Ultrasensitive isothermal titration calorimetry was used to generate thermodynamic data to assess the binding properties of the prepared nanoMIPs. The enthalpy of binding for L-BFA yielded −21.1 ± 1.7 kJ/mol and was significantly higher than that of its stereoisomer (−1.7 ± 1.2 kJ/mol). The nonimprinted polymers yielded enthalpies of only about 1.0 kJ/mol, which means there were no thermodynamically based binding effects on the nonimprinted polymers. The complete results are shown in [103]. The L-boc-nanoMIPs were later deposited as ultrathin layers in composite membranes and used for the selective separation of enantiomers [104].

The miniemulsion polymerization was also used to prepare nanoMIPs for recognition of proteins whereby the inverse miniemulsion was applied. For imprinting of cytochrome c, the epitope approach was used whereby Pal-LKKATNE was selected as the possible epitope [105]. The hydrophobic palmitoyl moiety ensures the effective surface imprinting because of its surface-active properties. For the preparation of suitable particles the ratio of the monomer acrylamide to cross-linker ethylene bisacrylamide and the emulsifier was varied. Arlacel 83 as the emulsifier and the ratio of monomer to cross-linker of 14:1 showed the best result. The imprinted particles bound approximately 9 times more cytochrome c than the nonimprinted particles.

Tan and Tong [106] showed the importance of a suitable emulsifier. They prepared nanoparticles via miniemulsion polymerization with the conventional and modified recipes. The conventional formula included sodium dodecyl sulfate as the surfactant and the modified formula included sodium dodecyl sulfate and polyvinylalcohol as the surfactants. The MIPs prepared with the modified formula showed better results due to recognition effects on the protein ribonuclease A.

The above results have shown the possibility of preparing molecularly imprinted particles for recognition of proteins via miniemulsion polymerization. Nevertheless, there are a lot of aspects that have to be further investigated, such as the exact recognition mechanism between the functional monomer and the template, effects of suitable emulsifier on the recognition mechanism, and much more.

2.4 Applications

2.4.1 Biopurification

2.4.1.1 Magnetic Nanoparticles

Magnetic nanoparticles have become attractive for exploitation in biology and medicine for diagnostic and therapeutic applications. The purified biomolecules include specific proteins [107], antibodies, peptides, nucleic acids [108], enzymes, and even cells [109] and viruses. Magnetic separations eliminate pretreatment, such as centrifugation or filtration. The basic principle of batch magnetic separation is very simple. Magnetic carriers bearing an immobilized affinity for

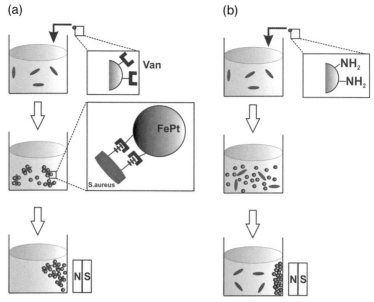

Figure 2.7 Illustration of the capture of bacteria by vancomycin (Van)-conjugated magnetic nanoparticles (a) via a plausible multivalent interaction and the corresponding control experiment (b) [107].

hydrophobic ligand or ion-exchange groups, or magnetic biopolymer particles bearing affinity to the isolated structure, are mixed with a sample containing target compound(s) (Figure 2.7) [107].

Xu et al. [110] have used magnetic FePt nanoparticles to capture and detect vancomycin-resistant enterococci and other Gram-positive bacteria at concentrations of around 10 colony-forming units/ml within 1 h (see Figure 2.8).

Vancomycin can bind to the terminal peptide, D-Ala–D-Ala, on the cell wall of a Gram-positive bacterium via hydrogen bonds. Upon binding of the nanoparticles, the "magnetized" bacteria (with the magnetic nanoparticles attached to the cell surface) are separated from the solution.

2.4.1.2 MIPs with Magnetizable Cores

Molecular imprinting has been widely accepted as a feasible approach for the preparation of tailor-made synthetic materials that are capable of specifically recognizing targeted molecules. Core/shell nanoparticles equipped with the molecular recognition property of MIPs on the shell and a magnetizable core are useful tools in separation and purification processes. The use of these core/shell MIPs as an affinity separation medium can improve the purification process of proteins for medical applications.

If superparamagnetic materials like magnetite can be incorporated into imprinted polymers, magnetic separation can be performed. The advantage of magnetic separation lies in its ease and high efficiency. The cost of separation will be relatively low and it allows fast recovery of the adsorbent. Magnetic separation allows the direct purification of the crude product from a mixture without any pretreatment. Tan and Tong showed in their work how to prepare MIPs with superparamagnetic cores [111]. They incorporated magnetite in a MIP by redox-initiated miniemulsion polymerization. The MIPs consisted of methacrylic acid as functional monomer, ethylene glycol dimethacrylate as cross-linker and ribonuclease A was used as template. The polymerization resulted in spherical polymer beads in the submicron range (700–800 nm) with an average magnetite content of 17.5 wt%. After template extraction the MIPs showed good binding properties in the rebinding tests with ribonuclease A.

2.4.2
Drug Delivery and Drug Targeting

The main benefits of nanoparticles used for drug delivery or drug targeting are the controlled release times of the included agents and the possibility to transport drugs to a specific target in the human body. Applications for drug delivery systems, cancer therapy, diagnosis and immunomodulating activities have recently been reported [27, 112–115]. Future drugs will be increasingly based on proteins as the active agent. If a protein is easily soluble, applications might be based on administering them in a liquid. However, if not, the formulation of the drug must be optimized or completely developed new. Here, the encapsulation of drugs is an efficient tool.

2.4.2.1 Nanoparticle Systems for Drug Delivery

The most commonly used biomaterials for drug delivery are liposomes, polymer nanoparticles and dendrimers. Degradable polymers [e.g., polyesters, poly(lactic acid), poly(glycolic acid), poly(butylcyanoacrylate) and their copolymers] are frequently used for the *in vivo* delivery of drugs to the brain [116–119]. Encapsulation is achieved by using a variety of loading methods, especially the pH gradient method or the ammonium sulfate method [120]. The ideal nanosystems must possess certain properties to allow application in humans:

- Small size for internalization.
- Multivalent attachment for enhanced avidity of interaction with cells.
- Visibility by invasive and noninvasive modalities.
- High drug dose-carrying capacity.
- Control over drug release.
- Safety.

Some of these polymer nanoparticulate systems degrade rather slowly for use in antigen delivery. Furthermore, methods have been developed to trigger their degradation during particular steps all along cell-internalization pathway [121, 122]. Weiss and Lorenz prepared fluorescent polyisoprene nanoparticles and poly(*n*-butylcyanoacrylate) nanoparticles that were synthesized by the miniemulsion

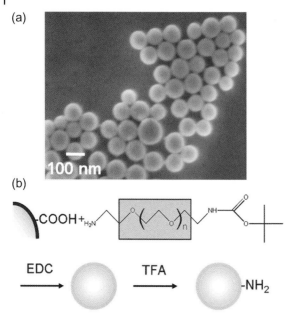

Figure 2.8 (a) Scanning electron microscopy image of core/shell nanoparticles with a silica core and a functional amino PEG amine surface. (Source: Fraunhofer IGB). (b) The modification of carboxylated (particle) surfaces with a functional PEG can be achieved using boc-amino PEG amine ($n = 10$). After coupling via carbodiimide hydrochloride (EDC) and cleavage using trifluoroacetic acid (TFA), a reactive amine anchor group is accessible on the particle surface.

technique as marker particles for cells. The outstanding efficient and fast inclusion of the particles compared to other polymeric particles results in extraordinary accumulation in the cells. Due to their complicated biodegradable properties the particles should be adapted for long-term applications only [123, 124]. Nanoparticulate architecture of dendrimers with an average size of about 10 nm favors the entry in the highly permeable tumor vasculature. The high molecular weight of these structures causes their localization and prevents their escape [125].

However, applications of nanoparticles are limited by their rapid recognition by the reticuloendothelial system. They are removed from the bloodstream within minutes upon intravenous injection, depending on their size and surface characteristics [62, 126]. The rapid uptake of nanoparticles is problematic when the long-term circulation of nanoparticle-loaded drug systems is needed. Therefore, the surface of the nanoparticles can be further coated for long-term circulation with hydrophilic polymers (e.g., PEG) to avoid their rapid recognition by the reticuloendothelial system (Figure 2.8) [127].

The transport of drugs through the blood–brain barrier requires a coating of the nanoparticles consisting of polysorbates, especially polysorbate 80. The coating probably leads to an adsorption of apolipoprotein E from blood plasma onto the nanoparticle surface [128].

2.4.2.2 Ligands on Nanoparticle Surfaces

Nanocarriers offer the possibility of conjugating them to specific ligands, which enable the specific direction to the desired cell type, tissue or organ. Macrophages, dendritic cells and tumor cells are promising cellular targets for the delivery of therapeutic agents. Thus, drugs and peptides of poor stability have been combined with polymers and lipids to obtain very fine, submicron particulate systems that have the ability to interact with the cells and be internalized by them. Tkachenko *et al.* designed multifunctional gold nanoparticle–peptide complexes for nuclear targeting in intact HepG2 cells by entering the cells via receptor-mediated endocytosis [129]. The particles were modified with a shell of BSA conjugated to various cellular targeting peptides and with a 3-maleimido benzoic acid N-hydroxysuccinimide ester linker.

2.4.2.3 Targeting of Specific Cells

Nanoparticulate systems for macrophage-mediated therapies could be applied for bacterial infectious diseases and could also find an application for viral infectious diseases (AIDS therapy) [130], inflammatory diseases like macrophage-mediated neuroinflammatory diseases [131], rheumatoid arthritis [132], atherosclerosis and restenosis [133, 134]. It is well known that macrophages are predominantly involved in the uptake of nanoparticles, leading to their degradation and clearance from the blood stream [135–137]. A nonopsonic attachment to the macrophage surface, via scavenger receptors, was also reported to take place during the phagocytosis of particulate systems [138]. The surface modification of nanoparticles is a crucial issue in terms of their uptake by cells and targeted tissue [139]. Augmented particulate hydrophobicity is known to increase the uptake by forming hydrophobic interactions with the cell surface [140]. A study with modified polystyrene beads showed that the total amount of bound protein increases with increasing hydrophobicity [141].

A restricted number of macrophage phagocytic receptors are involved, like the mannose receptor [142]. This kind of receptor, like other glycoprotein receptors, has the ability to mediate the phagocytosis of saccharide-coated particles as well as the pinocytosis of soluble glycoconjugates [143]. For instance, the incubation of mannose-coated nanoparticles with a mannose receptor-positive mouse macrophage cell line (J-774E) was effective in increasing their uptake [144]. Hoshino *et al.* showed that albumin coupled QD could be endocytosed by mouse macrophages [145, 146].

A possible key to developing the next generation of vaccines is the ability to deliver antigen to dendritic cells more specifically and induce the subsequent activation of T cell immunity [147]. Particles can be functionalized with ligands that bind to receptors and then trigger internalization through endocytosis (i.e., mannose-grafted particles), or through controlling their size they can simply be internalized by macropinocytosis (using nanoparticles in the size range of macromolecules; below 50 nm) [148, 149] or phagocytosis (using microparticles; above 500 nm) [150, 151] without specific recognition [152]. Copland *et al.* showed that mannosylated liposomes 260 nm in size are internalized through receptor-mediated endocytosis to a higher degree and they present antigen to T cells more efficiently than neutral liposomes or free antigen [153].

Medications that can selectively target tumors avoid at the same time the access of the drug to nontarget areas. This helps to overcome limitations relating to chemotherapy using free drug such as poor *in vivo in vitro* correlation and may overcome other possible resistances offered by tumors [154]. For an effective tumor therapy, different nanocarriers like polymeric, inorganic and magnetic nanoparticles and dendrimers are used. A commercial product Abraxane, which uses nanoparticles consisting of albumin and paclitaxel, which is a human protein for breast cancer therapy, is now in a clinical trial [155, 156]. In the trial, the tumor response rate was nearly twice as high for patients who received Abraxane compared to those who received solvent-based paclitaxel.

There are different targeting strategies for tumors cells. Various receptors on the tumor cells, such as transferrin receptors [157], folate receptors [158], epidermal growth factor receptors [159] and lectin receptors [160], can be very useful to provide access of drugs, peptides, etc., into the tumor mass. A range of literature has reviewed that various receptors are overexpressed on the surface of tumor cells to help the tumor mass by providing them with the bioactives required for neovascularization. Molecular diversity and the vasculature of tumors leads to expression of various surface markers and proteins, and these can be utilized as receptors of the ligands via receptor-mediated endocytosis [161]. Diverse applications have been performed such as the supplementation fusiogenic peptides [162, 163], endosomolytic and/or pH-responsive polymers [164, 165] and cell-penetrating peptides [166] that have been shown to facilitate the uptake of various cargoes. Poly(cyanoacrylate) nanoparticles are used for tumor-specific delivery [167] to target transferrin receptors [168]. Nanoparticles with a smaller range of size (1–10 nm) have the capacity to diffuse easily inside the tumor cells. Several anticancer drugs, such as doxorubicin and cisplatin, have already been coupled to transferrin. Compared to uncoupled drugs, they showed higher cytotoxicity [169].

Another way to deliver drugs to any desired target involves the functionalization of the surface of nanoparticles with monoclonal antibodies or ligands to tumor-related receptors, taking advantage of the specific binding ability between an antibody and antigen, or between the ligand and its receptor [170, 171]. There are a wide variety of methods that could be use to attach a tumor marker biomolecule to the magnetic nanoparticle surface. There are several examples that successfully demonstrate an accumulation of superparamagnetic particles applied intravenously, aimed at specific locations by means of external magnets [172]. Visaria *et al.* determined enhancement of tumor thermal therapy using gold nanoparticle-assisted tumor necrosis factor delivery [173].

Tovar *et al.* are using their trademarked Nanocytes technology for the preparation of biofunctional core/shell nanoparticles. These core/shell nanoparticles are particularly suited, for example, to immobilize a specific protein or a protein complex at their shell surface. The protein TNF is mirrored by cell surface membrane receptor proteins [14]. This capability makes it a promising and highly attractive candidate for new strategies in cancer therapy or pathway studies in cells.

TNF is a transmembrane protein [membrane TNF (mTNF)] and initially expressed at the cell surface. From this membrane-bound form the soluble TNF (sTNF) is proteolytically derived by action of the metalloproteinase TNF-α-converting enzyme. Both mTNF and sTNF bind to two different cell membrane receptors, termed TNFR1 and TNFR2. mTNF strongly activates both receptors, whereas sTNF is capable of only stimulating TNFR1. Functionalized particles with a recombinant TNF protein bind to and strongly activate both receptors, thus mimicking the action of mTNF [13]. The biomimetic TNF-Nanocytes mimic the action of the membrane-bound TNF and initiates the mTNF-resembling cellular responses.

2.5 Products

2.5.1 MIPs – Applications and Products

MIPs have been developed for a variety of applications. The most common application is the use as column material for high-performance liquid chromatography [78, 174] and solid-phase extraction [79, 175]. Columns for solid-phase extraction with MIPs as extraction material are commercially available (SupelMIP) for different analyte molecules. Another field of interest for MIP application is sensor technology as biological sensing systems or biosensors (e.g., enzyme/substrate, antibody/antigen) have several disadvantages compared to MIPs. These include sensitivity to extremes of temperature, pressure or pH and many are incompatible with organic solvents. In some instances, there is a lack of a suitable biosensing material for a particular analyte, while in others the cost and time to develop biosensors are excessive. Sensors based on MIP technology have the advantages of low cost and highly versatile procedures for demanding areas concerning the long-lasting serviceability of the sensor device and resistance against chemical and physical stress. Several sensors were developed to a vast range of analytes, such as vapors, micropollutants [176] and complex mixtures [177]. MIPs have been shown to possess binding characteristics and cross-reactivity profiles similar to those of antibodies and biological receptors [178, 179]. For this reason it is possible to apply MIPs in immunoassays to substitute the used antibodies. Another possible application of MIPs is their use as specific adsorbers in protein purification [98] or in waste water treatment [180].

2.5.2 Luminex Assay

Luminex offers a bioassay technology that is based on multifunctional microparticles. The so-called xMAP technology makes use of the particle inherent advantage of combining various core functions with different surface functions. This open architecture technology provides a set of differently color-coded polystyrene microbeads that are equipped with surface-bound anchor functions for capture molecule conjugation. In this way, bioassays may be carried out in suspension: the liquid reaction kinetics give fast and reproducible results, and as up to 100 assays may be performed in the same sample – distinguishable by the color-coded particle cores – smaller sample volumes are required. A collection of publications reporting on assays performed by xMAP technology is provided on the company's website (http://www.luminexcorp.com/01_xMAPTechnology).

2.6
Conclusions

Synthetic inorganic and organic nanoparticles are a the focus of interest in the field of pharmacology and biomedical devices. They find primary applications as carrier systems, including drug delivery systems and drug targeting systems. Intrinsic properties can be enhanced upon association with active molecules. Magnetization, fluorescence or a permanent electric dipole gives rise to various biomedical applications, such as imaging devices, cell labeling or contrast agents for magnet resonance devices.

Surface functionalization of the particle is crucial for biomimetic use. Defined binding sites have to be integrated. Two main strategies exist to establish a chemical functionality on the surface of the nanoparticle: immobilization of biomolecules or molecular imprinting of specific structural elements in a polymer surface.

Specific reaction sites for biomolecule immobilization can be introduced directly during particle synthesis (e.g., via surfmer chemistry). More often, nanoparticles must undergo postsynthetic modifications to render them chemically functional. Biofunctionalization techniques are well established, and relate to the chemical and physical properties of the nanoparticle as well as the type of biomolecule that has to be coupled.

Molecular recognition plays a crucial role in nature and is fundamental for many techniques wherein detection and separation should be addressed. Synthetic polymers with molecular binding and recognition sites represent an interesting alternative to commonly used binding matrices based on biological molecules. Recently, there has been increasing interest in the use of MIPs for the design of artificial receptor-like binding sites in otherwise nonselective polymerics. Therein, MIPs overcome drawbacks like the relatively instability of antibodies and enzymes to several organic solvents and extreme pH. Additionally, the use of MIPs has some economic and ecological advantages. In comparison to monoclonal antibodies they could be produced with less expense and are recyclable in general. MIPs have been developed for a variety of applications. The dominant application area of MIPs is still the field of analytical chemistry, but recent intense efforts have been focused on the exploitation of MIPs in various fields like purification, separation, sensor technology and immunoassays.

The use of nanocarriers for the delivery of therapeutic agents has received considerable attention for medical and pharmaceutical applications. These systems can be tailor-made for specific physicochemical requirements, and they display low toxic and immunogenic effects. The application of these nanoparticles in drug delivery and drug targeting is a seminal opportunity for a highly specific therapy. For drug delivery, the most commonly used biomaterials are liposomes, polymer nanoparticles, dendrimers, vesicles and micelles. The wide choice of chemical composition of the nanoparticles allows the encapsulation of a wide variety of drugs or other compounds.

The generation of fully synthetic binding sites by molecular imprinting of polymers is highly attractive. Enormous advances have been achieved lately and possible applications can be envisaged in various technological fields from downstream processing of valuable molecular products to environmental protection by specifically absorbing unwanted substances. With the progress in this field, the near future may see an increasing use of biomimetic particles and nanoparticles in applications other than in life sciences.

Acknowledgments

The authors thank Marc Herold, Mathias Lehmann, Melanie Dettling, Marion Herz, Jürgen Schmucker, Herwig Brunner, and Thomas Hirth (all Fraunhofer IGB, Stuttgart) for helpful discussions and scientific support, Monika Riedl (Fraunhofer IGB, Stuttgart) for SEM measurements, Ralf Thormann (Freiburg Materials Research Center FMF, Freiburg) for TEM measurements, Udo Welzel (Max-Planck-Institute for Metals Research, Stuttgart) for magnetic measurements, Eike Müller (Fraunhofer IGB, Stuttgart) for the MIP graphics in this chapter and on the book cover, and the Bundesministerium für Bildung und Forschung BMBF, Baden-Württemberg, and the Fraunhofer-Gesellschaft for financial support.

References

1 Sanders, T., Papas, P. and Veser, G. (2008) Supported nanocomposite catalysts for high-temperature partial oxidation of methane. *Chemical Engineering Journal*, **142** (1), 122–32.

2 Bruchez, M., Jr, Moronne, M., Gin, P., Weiss, S. and Alivisatos, A. (1998) Semiconductor nanocrystals as fluorescent biological labels. *Science*, **281** (5385), 2013.

3 Chan, W. and Nie, S. (1998) Quantum dot bioconjugates for ultrasensitive nonisotopic detection. *Science*, **281** (5385), 2016.

4 Riegler, J.T.N. (2004) Application of luminescent nanocrystals as labels for biological molecules. *Analytical and Bioanalytical Chemistry*, **379** (7–8), 913–19.

5 Gao, X., Chan, W.C.W. and Nie, S. (2002) Quantum-dot nanocrystals for ultrasensitive biological labeling and multicolor optical encoding. *Journal of Biomedical Optics*, **7** (4), 532–7.

6 Hermanson, G. (2008) *Bioconjugate Techniques*, Academic Press, New York.

7 Fernandez, E.J., Garcia-Barrasa, J., Laguna, A., Lopez-de-Luzuriaga, J.M., Monge, M. and Torres, C. (2008) The preparation of highly active antimicrobial silver nanoparticles by

an organometallic approach. *Nanotechnology*, **19**, 1–6.
8 Puntes, V., Krishnan, K. and Alivisatos, A. (2001) Colloidal nanocrystal shape and size control: the case of cobalt. *Science*, **291** (5511), 2115–17.
9 Sun, S., Murray, C.B., Weller, D., Folks, L. and Moser, A. (2000) Monodisperse FePt nanoparticles and ferromagnetic FePt nanocrystal superlattices. *Science*, **287** (5460), 1989.
10 Ramesh, S., Prozorov, R. and Gedanken, A. (1997) Ultrasound driven deposition and reactivity of nanophasic amorphous iron clusters with surface silanols of submicrospherical silica. *Chemistry of Materials*, **9**, 2996–3004.
11 Stöber, W., Fink, A. and Bohn, E. (1968) Controlled growth of monodisperse silica spheres in the micron range. *Journal of Colloid and Interface Science*, **26**, 62–9.
12 Schiestel, T., Brunner, H. and Tovar, G.E.M. (2004) Controlled surface functionalization of silica nanospheres by covalent conjugation reactions and preparations of high density streptavidin nanoparticles. *Journal of Nanoscience and Nanotechnology*, **4** (5), 504–11.
13 Weber, A., Gruber-Traub, C., Herold, M., Borchers, K. and Tovar, G.E.M. (2006) Biomimetic nanoparticles. *NanoS*, **2** (6), 20–7.
14 Bryde, S., Grunwald, I., Hammer, A., Krippner-Heidenreich, A., Schiestel, T., Brunner, H., Tovar, G.E.M., Pfizenmaier, K. and Scheurich, P. (2005) Tumor necrosis factor (TNF)-functionalized nanostructured particles for the stimulation of membrane TNF-Specific cell responses. *Bioconjugate Chemistry*, **16**, 1459–67.
15 Brinker, J.C. and Scherer, G.W. (1990) *Sol-Gel Science: The Physics and Chemistry of Sol-Gel Processing*, Academic Press, San Diego, CA.
16 Bogush, T.Z. (1988) Preparation of monodisperse silica particles: control of size and mass fraction. *Journal of Non-Crystalline Solids*, **104**, 95–106.
17 Whan Yoo, S.Y. and Hyung, K. (2006) Influence of reaction parameters on size and shape of silica nanoparticles. *Journal of Nanoscience and Nanotechnology*, **6**, 3343–6.
18 Kopelman, R., Xu, H., Yan, F., Monson, E.E. and Tang, W. (2002) Preparation and characterization of poly (ethylene glycol)-coated Stöber silica nanoparticles for biomedical applications. *Proceedings of SPIE*, **4626**, 383–93.
19 Burns, A., Ow, H. and Wiesner, U. (2006) Fluorescent core–shell silica nanoparticles: towards "Lab on a Particle" architectures for nanobiotechnology. *Chemical Society Reviews*, **35** (11), 1028–42.
20 Guo, J., Yang, W., Wang, C., He, J. and Chen, J. (2006) Poly(N-isopropylacrylamide)-coated luminescent/magnetic silica microspheres: preparation, characterization, and biomedical applications. *Chemistry of Materials*, **18**, 5554–62.
21 Corr, S.A., Rakovich, Y.P. and Gunko, Y.K. (2008) Multifunctional magnetic–fluorescent nanocomposites for biomedical applications. *Nanoscale Research Letters*, **3** (3), 87–104.
22 Law, W.-C., Yong, K.-T., Roy, I., Xu, G., Ding, H., Bergey, E.J., Zeng, H. and Prasad, P.N. (2008) Optically and magnetically doped organically modified silica nanoparticles as efficient magnetically guided biomarkers for two-photon imaging of live cancer cells. *Journal of Physical Chemistry C*, **112** (21), 7972–7.
23 Buchel, G., Grun, M., Unger, K.K., Matsumoto, A. and Tsutsumi, K. (1998) Tailored syntheses of nanostructured silicas: control of particle morphology, particle size and pore size. *Supramolecular Science*, **5** (3–4), 253–9.
24 Yano, K. and Fukushima, Y. (2003) Particle size control of mono-dispersed super-microporous silica spheres. *Journal of Materials Chemistry*, **13**, 2577–81.
25 Lu, A.-H., Salabas, E.L. and Schüth, F. (2007) Magnetic nanoparticles: synthesis, protection, functionalization, and application. *Angewandte Chemie (International Edition in English)*, **46**, 1222–44.
26 Latham, A.H. and Williams, M.E. (2008) Controlling transport and chemical

functionality of magnetic nanoparticles. *Accounts of Chemical Research*, **41** (3), 411–20.

27 Brigger, I., Dubernet, C. and Couvreur, P. (2002) Nanoparticles in cancer therapy and diagnosis. *Advanced Drug Delivery Reviews*, **54**, 631–51.

28 Leslie-Pelecky, D.L. (1996) Magnetic properties of nanostructured materials. *Chemical Materials*, **8**, 1770–83.

29 Ahniyaza, A., Seisenbaevab, G.A., Häggströmc, L., Kamalic, S., Kesslerb, V.G., Nordbladd, P., Johanssone, C. and Bergström, L. (2008) Preparation of iron oxide nanocrystals by surfactant-free or oleic acid-assisted thermal decomposition of a Fe(III) alkoxide. *Journal of Magnetism and Magnetic Materials*, **320**, 781–7.

30 Redl, F.X., Black, C.T., Papaefthymiou, G.C., Sandstrom, R.L., Yin, M., Zeng, H., Murray, C.B. and O'Brien, S.P. (2004) Magnetic, electronic, and structural characterization of nonstoichiometric iron oxides at the nanoscale. *Journal of the American Chemical Society*, **126**, 14583–99.

31 Wang, X., Zhuang, J., Peng, Q. and Li, Y. (2005) A general strategy for nanocrystal synthesis. *Nature*, **437**, 121–4.

32 Yoshimura, M. and Byrappa, K. (2008) Hydrothermal processing of materials: past, present and future. *Journal of Material Science*, **43**, 2085–103.

33 Langevin, D. (1992) Micelles and microemulsions. *Annual Reviews Physics and Chemistry*, **43**, 341–69.

34 Kreuter, J. (2007) Nanoparticles – a historical perspective. *International Journal of Pharmaceutics*, **331**, 1–10.

35 Weber, A., Knecht, S., Brunner, H. and Tovar, G.E.M. (2004) Modular structure of biochips based on microstructured deposition of functional nanoparticles. *Engineering in Life Sciences*, **4** (1), 93–7.

36 Weber, A., Knecht, S., Brunner, H. and Tovar, G.E.M. (2003) Modularer Aufbau von Biochips durch mikrostrukturierte Abscheidung von funktionellen Nanopartikeln. *Chemie Ingenieur Technik*, **75** (4), 437–41.

37 Saunders, B.R. and Turner, M.L. (2008) Nanoparticle–polymer photovoltaic cells. *Advances in Colloid and Interface Science*, **138** (1), 1–23.

38 Letchford, K. and Burt, H. (2007) A review of the formation and classification of amphiphilic block copolymer nanoparticulate structures: micelles, nanospheres, nanocapsules and polymersomes. *European Journal of Pharmaceutics and Biopharmaceutics*, **65**, 259–69.

39 Jaworek, A. (2007) Micro- and nano-particle production electrospraying. *Powder Technology*, **176** (1), 18–35.

40 Pereverzeva, E., Treschalin, I., Bodyagin, D., Maksimenko, O., Kreuter, J. and Gelperina, S. (2008) Intravenous tolerance of a nanoparticle-based formulation of doxorubicin in healthy rats. *Toxicology Letters*, **178** (1), 9–19.

41 Reis, C.P., Neufeld, R.J., Ribeiro, A.J. and Veiga, F.V. (2006) Nanoencapsulation I. Methods of drug-loaded polymeric nanoparticles. *Nanomedicine: Nanotechnology, Biology, and Medicine*, **2**, 8–21.

42 Roney, C., Kulkarni, P., Arora, V., Antich, P., Bonte, F., Wu, A., Mallikarjuna, N.N., Nanohar, S., Liang, H.-F., Kulkarni, A.R., Sung, H.-W., Sairam, M. and Aminabhavi, T.M. (2005) Targeted nanoparticles for drug delivery through the blood-brain barrier for Alzheimer's disease. *Journal of Controlled Release*, **108**, 193–214.

43 Vauthier, C., Dubernet, C., Fattal, E., Pinto-Alphandary, H. and Couvreur, P. (2003) Poly(alkylcyanoacrylates) as biodegradable materials for biomedical applications. *Advanced Drug Delivery Reviews*, **55** (4), 519–48.

44 Kumar, N., Ravikumar, M.N.V. and Domb, A.J. (2001) Biodegradable block copolymers. *Advanced Drug Delivery Reviews*, **53**, 23–44.

45 Bontha, S., Kabanov, A.V. and Bronich, T.K. (2006) Polymer micelles with cross-linked ionic cores for delivery of anticancer drugs. *Journal of Controlled Release*, **114**, 163–74.

46 Otsukaa, H., Nagasaki, Y. and Kataoka, K. (2003) PEGylated nanoparticles for

47 Kabanov, A.V. (2002) Pluronic block copolymers in drug delivery: from micellar nanocontainers to biological response modifiers. *Critical Reviews in Therapeutic Drug Carrier Systems*, **19** (1), 1–72.

48 Burt, H.M., Zhang, X., Toleikis, P., Embree, L. and Hunter, W.L. (1999) Development of copolymers of poly(D,L-lactide) and methoxypoly-ethylene glycol as micellar carriers of paclitaxel. *Colloids and Surfaces B: Biointerfaces*, **16**, 161–71.

49 Sant, V.P., Smith, D. and Leroux, J.-C. (2004) Novel pH-sensitive supra-molecular assemblies for oral delivery of poorly water soluble drugs: preparation and characterization. *Journal of Controlled Release*, **97** (2), 301–12.

50 Couvreur, P., Kante, B., Roland, M., Guiot, P., Bauduin, P. and Speiser, P. (1979) Polycyanoacrylate nanocapsules as potential lysomotropic carriers: preparation, morphological and sorption properties. *Journal of Pharmacy and Pharmacology*, **31** (5), 331–2.

51 Davda, J. and Labhasetwar, V. (2002) Characterization of nanoparticle uptake by endothelial cells. *International Journal of Pharmaceutics*, **233**, 51–9.

52 Astete, C.E. and Sabliov, C.M. (2006) Synthesis and characterization of PLGA nanoparticles. *Journal of Biomaterials Science Polymer Edition*, **17** (3), 247–89.

53 Quaddoumi, M.G., Gukasyan, H.J., Davda, J., Labhasetwar, V., Kim, K.-J. and Lee, V.H.L. (2003) Clathrin and cleoveolin-1 expression in primary pigmented rabbit conjunctival epithelial cells: role in PLGA nanoparticle endocytosis. *Molecular Vision*, **9**, 559–69.

54 Ganta, S., Devalapally, H., Shahiwala, A. and Amiji, M. (2008) A review of stimuli-responsive nanocarriers for drug and gene delivery. *Journal of Controlled Release*, **126**, 197–204.

55 Yoshida, R., Uchida, K., Kaneko, Y., Sakal, K., Kikuchi, A., Sakurai, Y. and Okano, T. (1995) Comb-type grafted hydrogels with rapid deswelling response to temperature changes. *Nature*, **374**, 240–2.

56 Liu, X.-M., Yang, Y.-Y. and Leong, K.W. (2003) Thermally responsive polymeric micellar nanoparticles self-assembled from cholesteryl end-capped random poly(*N*-isopropylacrylamide-*co*-*N*,*N*-dimethylacrylamide): synthesis, temperature-sensitivity, and morphologies. *Journal of Colloid and Interface Science*, **266**, 295–303.

57 West, J.L. and Halas, N.J. (2003) Engineered nanomaterials for biophotonics applications: improving sensing, imaging, and therapeutics. *Annual Review of Biomedical Engineering*, **5**, 285–92.

58 Thurn, K.T., Brown, E.M.B., Wu, A., Vogt, S., Lai, B., Maser, J., Paunesku, T. and Woloschak, G.E. (2007) Nanoparticles for applications in cellular imaging. *Nanoscale Research Letters*, **2** (9), 430–41.

59 Ambade, A.V., Sandanaraj, B.S., Klaikherd, A. and Thayumanavan, S. (2007) Fluorescent polyelectrolytes as protein sensors. *Polymer International*, **56** (4), 474–81.

60 Holzapfel, V., Musyanovych, A., Landfester, K., Lorenz, M.R. and Mailänder, V. (2005) Preparation of fluorescent carboxyl and amino functionalized polystyrene particles by miniemulsion polymerization as markers for cells. *Macromolecular Chemistry and Physics*, **206** (24), 2440–9.

61 Almeida, A.J. and Souto, E. (2007) Solid lipid nanoparticles as drug delivery system for peptides and proteins. *Advanced Drug Delivery Reviews*, **59**, 478–90.

62 Moghimi, S.M. and Szebeni, J. (2003) Stealth liposomes and long circulating nanoparticles: critical issues in pharmacokinetics, opsonization and protein-binding properties. *Progress in Lipid Research*, **42** (6), 463–78.

63 Wells, J., Sen, A. and Hui, S.W. (2003) Localized delivery to CT-26 tumors in mice using thermosensitive liposomes. *International Journal of Pharmaceutics*, **261**, 105–14.

64 Zara, G.P., Cavalli, R., Bargoni, A., Fundarograve, A., Vighetto, D. and Gasco, M.R. (2002) Intravenous administration to rabbits of non-stealth and stealth doxorubicin-loaded solid lipid nanoparticles at increasing concentrations of stealth agent: pharmacokinetics and distribution of doxorubicin in brain and other tissues. *Journal of Drug Targeting*, **10** (4), 327–35.

65 Cavalli, R., Zara, G.P., Caputo, O., Bargoni, A., Fundaro, A. and Gasco, M.R. (2000) Transmucosal transport of tobramycin incorporated in SLN after duodenal administration to rats. Part I – a pharmacokinetic study. *Pharmacological Research*, **42** (6), 541–5.

66 Partha, R., Lackey, M., Hirsch, M., Casscells, S.W. and Conyers, J.L. (2007) Self assembly of amphiphillic C60 fullerene derivatives into nanoscale supramolecular structures. *Journal of Nanobiotechnology*, **5** (6), 1–11.

67 Patni, S. and Bhatia, A.L. (2008) Nanotechnology: a double edge sword. *Asian Journal of Experimental Sciences*, **22** (2), 153–66.

68 Hermanson, G.T. (1995) *Bioconjugate Techniques*, Academic Press, San Diego, CA.

69 Herold, M., Brunner, H. and Tovar, G.E.M. (2003) Polymer nanoparticles with activated-ester surface by using functional surfmers. *Macromolecular Chemistry and Physics*, **204** (5–6), 770–8.

70 Herold, M., Tovar, G.E.M., Håkanson, M. and Brunner, H. (2005) Copolymer nanoparticles with activated ester surface for the facile immobilization of enzymes. *Polymer Preprints*, **46** (2), 1233–4.

71 Wulff, G. (1995) Molecular imprinting in cross-linked materials with the aid of molecular templates – a way towards artificial antibodies. *Angewandte Chemie (International Edition in English)*, **34**, 1812–32.

72 Sellergren, B., Karmalkar, R.N. and Shea, K.J. (2000) Enantioselective ester hydrolysis catalyzed by imprinted polymers. *Journal of Organic Chemistry*, **65** (13), 4009–27.

73 Andersson, L.I. (2000) Molecular imprinting for drug bioanalysis – a review on the application of imprinted polymers to solid-phase extraction and binding assay. *Journal of Chromatography B*, **739** (1), 163–73.

74 Whitcombe, M.J. and Vulfson, E.N. (2001) Imprinted polymers. *Advanced Materials*, **13** (7), 467–78.

75 Alexander, C., Andersson, H.S., Andersson, L.I., Ansell, R.J., Kirsch, N., Nicholls, I.A., O'Mahony, J. and Whitcombe, M.J. (2006) Review: Molekular imprinting science and technology: a survey of the literature for the years up to and including 2003. *Journal of Molecular Recognition*, **19**, 106–80.

76 Vaihinger, D. (2008) Untersuchung zur Darstellung, Charakterisierung und Verarbeitung von mit boc-Phenylanilin-anilid molekular geprägter Mikrogele aus hochvernetzer Copolymere. Dissertationsschrift, Universität Stuttgart.

77 Mosbach, K. and Ramstroem, O. (1996) The emerging technique of molecular imprinting and its future impact on biotechnology. *Biotechnology*, **14**, 163–70.

78 Spivak, D., Gilmore, M.A. and Shea, K.J. (1997) Evaluation of binding and origins of specifity of 9-ethyladenine imprinted polymers. *Journal of the American Chemical Society*, **119** (19), 4388–93.

79 Lanza, F. and Sellergren, B. (2001) The application of molecular imprinting technology to solid phase extraction. *Chromatographia*, **53**, 599–611.

80 Rossi, C. and Haupt, K. (2007) Application of the Doehlert experimental design to molecularly imprinted polymers: surface response optimization of specific template recognition as a function of the type and degree of cross-linking. *Analytical and Bioanalytical Chemistry*, **389**, 455–60.

81 Lanza, F., Hall, A.J., Sellergren, B., Bereczki, A., Horvai, G., Bayoudh, S., Cormack, P.A.G. and Sherrington, D.C. (2001) Development of a semiautomated procedure for the synthesis and evaluation of molecularly imprinted polymers applied to the search for

functional monomers for phenytoin and nifedipine. *Analytica Chimica Acta*, **435**, 91–106.

82 Dirion, B.S.E. and Sellergren, B. (2003) Development of a high throughput synthesis technique for the optimization of MIPs for 17beta-estradiol, in Proceedings of the Material Research Society Symposium, **787**.

83 Ladbury, J.E. and Chowdhry, B.Z. (eds) (1998) *Biocalorimetry: Applications of Calorimetry in the Biological Sciences*, John Wiley & Sons, Ltd, Chichester.

84 Doyle, M.L. (1997) Characterization of binding interactions by isothermal titration calorimetry. *Current Opinion in Biotechnology*, **8**, 31–5.

85 Komiyama, M., Mukawa, T.T. and Asanuma, T.H. (2003) *Molecular Imprinting – From Fundamentals to Applications*, Wiley-VCH Verlag, Weinheim.

86 Matsui, J., Kubo, H. and Takeuchi, T. (1998) Design and preparation of molecularly imprinted atrazine-receptor polymers: investigation of functional monomers and solvents. *Analytical Sciences*, **14** (4), 699–702.

87 Sellergren, B., Lepistoe, M. and Mosbach, K. (1988) Highly enantio-selective and substrate-selective polymers obtained by molecular imprinting utilizing noncovalent interactions. NMR and chromatographic studies on the nature of recognition. *Journal of the American Chemical Society*, **110** (17), 5853–60.

88 O'Mahony, J., Molinelli, A., Nolan, K., Smyth, M.R. and Mizaikoff, B. (2005) Towards the rational development of molecularly imprinted polymers: ^{1}H NMR studies on hydrophobicity and ion-pair interactions as driving forces for selectivity. *Biosensors and Bioelectronics*, **20** (9), 1884–93.

89 Wulff, G. and Sarhan, A. (1972) The use of polymers with enzyme-analogous structure for the resolution of racemates. *Angewandte Chemie (International Edition in English)*, **11**, 341ff.

90 Nicholls, I.A., Ramstroem, O. and Mosbach, K. (1995) Insights into the role of the hydrogen bond and hydrophobic effect on recognition in molecularly imprinted polymer synthetic peptide receptor mimics. *Journal of Chromatography A*, **691**, 349–53.

91 Andersson, L.I. and Mosbach, K. (1990) Enantiomeric resolution on molecularly imprinted polymers prepared with only non-covalent and non-ionic interactions. *Journal of Chromatography*, **516**, 313–22.

92 Kempe, M. and Mosbach, K. (1995) Separation of amino acids, peptides and proteins on molecularly imprinted stationary phases. *Journal of Chromatography A*, **691**, 317–23.

93 Kempe, M. (2000) Oxytocin receptors mimetics prepared by molecular imprinting. *Letters in Peptide Science*, **7**, 27–33.

94 Bossi, A., Piletsky, S.A., Piletska, E.V., Righetti, P.G. and Turner, A.P.F. (2001) Surface-grafted molecularly imprinted polymers for protein recognition. *Analytical Chemistry*, **73** (21), 5281–6.

95 Liao, J.-L., Wang, Y. and Hjerten, S. (1996) A novel support with artificially created recognition for the selective removal of proteins and for affinity chromatography. *Chromatographia*, **5/6**, 259–62.

96 Tong, D., Hetenyi, C., Bikaddi, Z., Gao, J.-P. and Hjerten, S. (2001) Some studies of the chromatographic properties of gels ("artificial antibodies/receptors") for selective adsorption of proteins. *Chromatographia*, **54**, 7–14.

97 Rachkov, A. and Minoura, N. (2000) Recognition of oxytocin and oxytocin-related peptides in aqueous media using a molecularly imprinted polymer synthesized by the epitope approach. *Journal of Chromatography A*, **889** (1–2), 111–18.

98 Nishino, H., Huang, C.-S. and Shea, K.J. (2006) Selective protein capture by epitope imprinting, *Angewandte Chemie (International Edition in English)*, **45**, 2392–6.

99 Vaihinger, D., Landfester, K., Kräuter, I., Brunner, H. and Tovar, G.E.M. (2002) Molecularly imprinted polymer

nanospheres as synthetic affinity receptors obtained by miniemulsions polymerization. *Macromolecular Chemistry and Physics*, **203**, 1965–73.

100 Wulff, G. (1995) Molecular imprinting in cross-linked materials with the aid of molecular templates – a way towards artificials antibodies. *Angewandte Chemie (International Edition in English)*, **34**, 1812–32.

101 Sezgin, S., Weber, A., Herold, M., Gruber-Traub, C., Brunner, H. and Tovar, G.E.M. (2006) Kinetic and thermodynamic behaviour of recognition processes employing nano-spherical L-boc-phenylalanine anilide molecularly imprinted poly(MAA-*co*-EGDMA) monoliths. *Polymer Preprints*, **47** (2), 860–1.

102 Yoshimatsu, K., Reimhult, K., Krozer, A., Mosbach, K., Sode, K. and Ye, L. (2007) Uniform molecularly imprinted microspheres and nanoparticles prepared by precipitation polymerization: the control of particle size suitable for different analytical applications. *Analytica Chimica Acta*, **584** (1), 112–21.

103 Weber, A., Dettling, M., Brunner, H. and Tovar, G.E.M. (2002) Isothermal titration calorimetry of molecularly imprinted polymer nanospheres. *Macromolecular Rapid Communications*, **23** (14), 824–8.

104 Lehmann, M., Brunner, H. and Tovar, G.E.M. (2003) Molekular geprägte Nanopartikel als selektive Phase in Kompositmembranen: Hydrodynamik und Stofftrennung in nanoskaligen Schüttungen. *Chemie Ingenieur Technik*, **75** (1–2), 149–53.

105 Gruber-Traub, C., Weber, A., Dettling, M., Herz, M., Herold, M., Brunner, H. and Tovar, G.E.M. (2006) Nanocytes – inverse miniemulsion polymerization technology for specific protein recognition. *Polymer Preprints*, **47** (2), 901–2.

106 Tan, C.J. and Tong, Y.W. (2007) The effect of protein structural conformation on nanoparticle molecular imprinting of ribonuclease a using miniemulsion polymerization. *Langmuir*, **23**, 2722–30.

107 Chiang, C.-L., Chen, C.-Y. and Chang, L.-W. (2008) Purification of recombinant enhanced green fluorescent protein expressed in *Escherichia coli* with new immobilized metal ion affinity magnetic absorbents. *Journal of Chromatography B*, **864**, 116–22.

108 Willner, I., Cheglakov, Z., Weizmann, Y. and Sharon, E. (2008) Analysis of DNA and single-base mutations using magnetic particles for purification, amplification and DNAzyme detection. *Analyst*, **133**, 923–7.

109 Schwalbe, M., Pachmann, K., Höffken, K. and Clement, J.H. (2006) Improvement of the separation of tumour cells from peripheral blood cells using magnetic nanoparticles. *Journal of Physical Chemistry B: Condensed Matter*, **18**, 2865–76.

110 Gu, H., Ho, P.-L., Tsang, K.W.T., Wang, L. and Xu, B. (2003) Using biofunctional magnetic nanoparticles to capture vancomycin-resistant enterococci and other Gram-positive bacteria at ultralow concentration. *Journal of the American Chemical Society*, **125**, 15702–3.

111 Tan, C.J. and Tong, Y.W. (2007) Preparation of superparamagnetic ribonuclease a surface-imprinted submicrometer particles for protein recognition in aqueous media. *Analytical Chemistry*, **79**, 299–306.

112 Brannon-Peppas, L. and Blanchette, J.O. (2004) Nanoparticle and targeted systems for cancer therapy. *Advanced Drug Delivery Reviews*, **56** (11), 1649–59.

113 Mainardes, R.M. and Silva, L.P. (2004) Drug delivery systems: past, present, and future. *Current Drug Targets*, **5** (5), 449–55.

114 Ravi Kumar, M., Hellermann, G., Lockey, R.F. and Mohapatra, D. (2004) Nanoparticle-mediated gene delivery: state of the art. *Expert Opinion on Biological Therapy*, **4** (4), 1213–24.

115 Rihova, B. (2002) Immunomodulating activities of soluble synthetic polymer-bound drugs. *Advanced Drug Delivery Reviews*, **54** (5), 653–74.

116 Lutolf, M.P. and Hubbell, J.A. (2005) Synthetic biomaterials as instructive extracellular microenvironments for

morphogenesis in tissue engineering. *Nature Biotechnology*, **23** (1), 47–55.

117 Peppas, N.A. and Langer, R. (1994) New challenges in biomaterials. *Science*, **263** (5154), 1715–20.

118 Prieto, M.J.B., Delie, F., Fattal, E., Tartar, A., Puisieux, F., Gulik, A. and Couvreur, P. (1994) Characterization of V3 BRU peptide-loaded small PLGA microspheres prepared by a (w1/o)w2 emulsion solvent evaporation method. *International Journal of Pharmaceutics*, **111** (2), 137–45.

119 Kreuter, J., Alyautdin, R.N., Kharkevich, D.A. and Ivanov, A.A. (1995) Passage of peptides through the blood-brain barrier with colloidal polymer particles (nanoparticles). *Brain Research*, **674** (1), 171–4.

120 Fahmy, T.M., Fong, P.M., Park, J., Constable, T. and Saltzman, W.M. (2007) Nanosystems for simultaneous imaging and drug delivery to T cells. *AAPS Journal*, **9** (2), E171–80.

121 Discher, D.E. and Eisenberg, A. (2002) Polymer vesicles. *Science*, **297** (5583), 967–73.

122 Napoli, A., Valentini, M., Tirelli, N., Muller, M. and Hubbell, J.A. (2004) Oxidation-responsive polymeric vesicles. *Nature Materials*, **3** (3), 183–9.

123 Lorenz, M.R., Kohnle, M.V., Dass, M., Walther, P., Hocherl, A., Ziener, U., Landfester, K. and Mailander, V. (2008) Synthesis of fluorescent polyisoprene nanoparticles and their uptake into various cells. *Macromolecular Bioscience*, **8** (8), 711–27.

124 Weiss, C.K., Lorenz, M.R., Landfester, K. and Mailänder, V. (2007) Cellular uptake behavior of unfunctionalized and functionalized pBCA particles prepared in a miniemulsion. *Macromolecular Bioscience*, **7** (7), 883–96.

125 Shukla, R., Thomas, T.P., Peters, J.L., Desai, A.M., Kukowska-Latallo, J., Patri, A.K., Kotlyar, A. and Baker, J.R., Jr (2006) HER2 specific tumor targeting with dendrimer conjugated anti-HER2 mAb. *Bioconjugate Chemistry*, **17** (5), 1109–15.

126 Verdun, C., Brasseur, F., Vranckx, H., Couvreur, P. and Roland, M. (1990) Tissue distribution of doxorubicin associated with polyisohexylcyanoacrylate nanoparticles. *Cancer Chemotherapy and Pharmacology*, **26** (1), 13–18.

127 Torchilin, V.P. and Trubetskoy, V.S. (1995) Which polymers can make nanoparticulate drug carriers long-circulating? *Advanced Drug Delivery Reviews*, **16** (2–3), 141–55.

128 Kreuter, J. (2001) Nanoparticulate systems for brain delivery of drugs. *Advanced Drug Delivery Reviews*, **47**, 65–81.

129 Tkachenko, A.G., Xie, H., Coleman, D., Glomm, W., Ryan, J., Anderson, M.F., Franzen, S. and Feldheim, D.L. (2003) Multifunctional gold nanoparticle–peptide complexes for nuclear targeting. *Journal of the American Chemical Society*, **125**, 4700–1.

130 Bender, A.R., von Briesen, H., Kreuter, J., Duncan, I.B. and Rubsamen-Waigmann, H. (1996) Efficiency of nanoparticles as a carrier system for antiviral agents in human immunodeficiency virus-infected human monocytes/macrophages *in vitro*. *Antimicrobial Agents and Chemotherapy*, **40** (6), 1467–71.

131 Merodio, M., Irache, J.M., Eclancher, F., Mirshahi, M. and Villarroya, H. (2000) Distribution of albumin nanoparticles in animals induced with the experimental allergic encephalomyelitis. *Journal of Drug Targeting*, **8** (5), 289–303.

132 Rabinovich, G.A. (2000) Apoptosis as a target for gene therapy in rheumatoid arthritis. *Memórias do Instituto Oswaldo Cruz*, **95** (Suppl. 1), 225–33.

133 Libby, P. (2001) Current concepts of the pathogenesis of the acute coronary syndromes. *Circulation*, **104** (3), 365–72.

134 Cooper, A.I. and Holmes, W.P.H. (1999) Synthesis of highly cross-linked polymers in supercritical carbon dioxide by heterogeneous polymerization. *Macromolecules*, **32**, 2156–66.

135 Leu, D., Manthey, B., Kreuter, J., Speiser, P. and DeLuca, P.P. (1984) Distribution and elimination of coated polymethyl [2-^{14}C]methacrylate nanoparticles after intravenous injection in rats. *Journal of Pharmacological Sciences*, **73** (10), 1433–7.

136 Leroux, J.C., De Jaeghere, F., Anner, B., Doelker, E. and Gurny, R. (1995) An investigation on the role of plasma and serum opsonins on the internalization of biodegradable poly(D,L-lactic acid) nanoparticles by human monocytes. *Life Sciences*, **57** (7), 695–703.

137 Peracchia, M.T., Harnisch, S., Pinto-Alphandary, H., Gulik, A., Dedieu, J.C., Desmaele, D., d'Angelo, J., Muller, R.H. and Couvreur, P. (1999) Visualization of *in vitro* protein-rejecting properties of PEGylated stealth polycyanoacrylate nanoparticles. *Biomaterials*, **20** (14), 1269–75.

138 Kobzik, L. (1995) Lung macrophage uptake of unopsonized environmental particulates. Role of scavenger-type receptors. *Journal of Immunology*, **155** (1), 367–76.

139 Labhasetwar, V., Song, C., Humphrey, W., Shebuski, R. and Levy, R.J. (1998) Arterial uptake of biodegradable nanoparticles: effect of surface modifications. *Journal of Pharmacological Sciences*, **87** (10), 1229–34.

140 Tabata, Y. and Ikada, Y. (1988) Effect of the size and surface charge of polymer microspheres on their phagocytosis by macrophage. *Biomaterials*, **9** (4), 356–62.

141 Blunk, T., Hochstrasser, D.F., Sanchez, J.C., Muller, B.W. and Muller, R.H. (1993) Colloidal carriers for intravenous drug targeting: plasma protein adsorption patterns on surface-modified latex particles evaluated by two-dimensional polyacrylamide gel electrophoresis. *Electrophoresis*, **14** (12), 1382–7.

142 Dickert, F.L. and Hayden, O. (2002) Bioimprinting of polymers and sol-gel phases. selective detection of yeasts with imprinted polymers. *Analytical Chemistry*, **74**, 1302–6.

143 Warr, G.A. (1980) A macrophage receptor for (mannose/glucosamine)-glycoproteins of potential importance in phagocytic activity. *Biochemical and Biophysical Research Communications*, **93** (3), 737–45.

144 Cui, D. and Gao, H. (2003) Advance and prospect of bionanomaterials. *Biotechnology Progress*, **19**, 683–92.

145 Henry-Toulme, N., Grouselle, M. and Ramaseilles, C. (1995) Multidrug resistance bypass in cells exposed to doxorubicin-loaded nanospheres. Absence of endocytosis. *Biochemical Pharmacology*, **50** (8), 1135–9.

146 Li, S. and Huang, L. (1997) *In vivo* gene transfer via intravenous administration of cationic lipid-protamine–DNA (LPD) complexes. *Gene Therapy*, **4** (9), 891–900.

147 O'Hagan, D.T. and Valiante, N.M. (2003) Recent advances in the discovery and delivery of vaccine adjuvants. *Nature Reviews Drug Discovery*, **2** (9), 727–35.

148 Norbury, C.C., Hewlett, L.J., Prescott, A.R., Shastri, N. and Watts, C. (1995) Class I MHC presentation of exogenous soluble antigen via macropinocytosis in bone marrow macrophages. *Immunity*, **3** (6), 783–91.

149 Sallusto, F., Cella, M., Danieli, C. and Lanzavecchia, A. (1995) Dendritic cells use macropinocytosis and the mannose receptor to concentrate macromolecules in the major histocompatibility complex class II compartment: downregulation by cytokines and bacterial products. *Journal of Experimental Medicine*, **182** (2), 389–400.

150 Lutsiak, M.E., Robinson, D.R., Coester, C., Kwon, G.S. and Samuel, J. (2002) Analysis of poly(D,L-lactic-*co*-glycolic acid) nanosphere uptake by human dendritic cells and macrophages *in vitro*. *Pharmaceutical Research*, **19** (10), 1480–7.

151 Thiele, L., Merkle, H.P. and Walter, E. (2003) Phagocytosis and phagosomal fate of surface-modified microparticles in dendritic cells and macrophages. *Pharmaceutical Research*, **20** (2), 221–8.

152 Lanzavecchia, A. (1996) Mechanisms of antigen uptake for presentation. *Current Opinion in Immunology*, **8** (3), 348–54.

153 Copland, M.J., Baird, M.A., Rades, T., McKenzie, J.L., Becker, B., Reck, F., Tyler, P.C. and Davies, N.M. (2003) Liposomal delivery of antigen to human dendritic cells. *Vaccine*, **21** (9–10), 883–90.

154 Sinek, J., Frieboes, H., Zheng, X. and Cristini, V. (2004) Two-dimensional chemotherapy simulations demonstrate fundamental transport and tumor

response limitations involving nanoparticles. *Biomedical Microdevices*, **6** (4), 297–309.

155 Wong, J., Brugger, A., Khare, A., Chaubal, M., Papadopoulos, P., Rabinow, B., Kipp, J. and Ning, J. (2008) Suspensions for intravenous (IV) injection: a review of development, preclinical and clinical aspects. *Advanced Drug Delivery Reviews*, **60** (8), 939–54.

156 Nyman, D.W., Campbell, K.J., Hersh, E., Long, K., Richardson, K., Trieu, V., Desai, N., Hawkins, M.J. and Von Hoff, D.D. (2005) Phase I and pharmacokinetics trial of ABI-007, a novel nanoparticle formulation of paclitaxel in patients with advanced nonhematologic malignancies. *Journal of Clinical Oncology*, **23** (31), 7785–93.

157 Pun, S.H., Tack, F., Bellocq, N.C., Cheng, J., Grubbs, B.H., Jensen, G.S., Davis, M.E., Brewster, M., Janicot, M., Janssens, B., Floren, W. and Bakker, A. (2004) Targeted delivery of RNA-cleaving DNA enzyme (DNAzyme) to tumor tissue by transferrin-modified, cyclodextrin-based particles. *Cancer Biology and Therapy*, **3** (7), 641–50.

158 Kukowska-Latallo, J.F., Candido, K.A., Cao, Z., Nigavekar, S.S., Majoros, I.J., Thomas, T.P., Balogh, L.P., Khan, M.K. and Baker, J.R., Jr (2005) Nanoparticle targeting of anticancer drug improves therapeutic response in animal model of human epithelial cancer. *Cancer Research*, **65** (12), 5317–24.

159 Schwechheimer, K., Huang, S. and Cavenee, W.K. (1995) EGFR gene amplification–rearrangement in human glioblastomas. *International Journal of Cancer*, **62** (2), 145–8.

160 Banaszczyk, M.G., Lollo, C.P., Kwoh, D.Y., Phillips, A.T., Amini, A., Wu, D.P., Mullen, P.M., Coffin, C.C., Brostoff, S.W. and Carlo, D.J. (1999) Poly-L-lysine-graft-PEG comb-type polycation copolymers for gene delivery. *Journal of Macromolecular Science – Pure and Applied Chemistry*, **36** (7 and 8), 1061–84.

161 Zurita, A.J., Arap, W. and Pasqualini, R. (2003) Mapping tumor vascular diversity by screening phage display libraries.
Journal of Controlled Release, **91** (1–2), 183–6.

162 Li, W., Nicol, F. and Szoka, F.C., Jr (2004) GALA: a designed synthetic pH-responsive amphipathic peptide with applications in drug and gene delivery. *Advanced Drug Delivery Reviews*, **56** (7), 967–85.

163 Plank, C., Zauner, W. and Wagner, E. (1998) Application of membrane-active peptides for drug and gene delivery across cellular membranes. *Advanced Drug Delivery Reviews*, **34** (1), 21–35.

164 Bieber, T., Meissner, W., Kostin, S., Niemann, A. and Elsasser, H.P. (2002) Intracellular route and transcriptional competence of polyethylenimine–DNA complexes. *Journal of Controlled Release*, **82** (2–3), 441–54.

165 Fattal, E., Couvreur, P. and Dubernet, C. (2004) "Smart" delivery of antisense oligonucleotides by anionic pH-sensitive liposomes. *Advanced Drug Delivery Reviews*, **56** (7), 931–46.

166 Futaki, S. (2005) Membrane-permeable arginine-rich peptides and the translocation mechanisms. *Advanced Drug Delivery Reviews*, **57** (4), 547–58.

167 Xu, Z., Gu, W., Huang, J., Sui, H., Zhou, Z., Yang, Y., Yan, Z. and Li, Y. (2005) *In vitro* and *in vivo* evaluation of actively targetable nanoparticles for paclitaxel delivery. *International Journal of Pharmacology*, **288** (2), 361–8.

168 Reddy, J.A., Dean, D., Kennedy, M.D. and Low, P.S. (1999) Optimization of folate-conjugated liposomal vectors for folate receptor-mediated gene therapy. *Journal of Pharmaceutical Sciences*, **88** (11), 1112–18.

169 Daniels, T.R., Delgado, T., Helguera, G. and Penichet, M.L. (2006) The transferrin receptor part II: targeted delivery of therapeutic agents into cancer cells. *Clinical Immunology*, **121** (2), 159–76.

170 Sinha, R., Kim, G.J., Nie, S. and Shin, D.M. (2006) Nanotechnology in cancer therapeutics: bioconjugated nanoparticles for drug delivery. *Molecular Cancer Therapeutics*, **5** (8), 1909–17.

171 Duguet, E., Vasseur, S., Mornet, S. and Devoisselle, J.-M. (2006) Magnetic

nanoparticles and their applications in medicine. *Nanomedicine*, **1** (2), 157–68.

172 Goya, G.F., Grazú, V.R. and Ibarra, M. (2008) Magnetic nanoparticles for cancer therapy. *Current Nanoscience*, **4**, 1–16.

173 Visaria, R.K., Griffin, R.J., Williams, B.W., Ebbini, E.S., Paciotti, G.F., Song, C.W. and Bischof, J.C. (2006) Enhancement of tumor thermal therapy using gold nanoparticle-assisted tumor necrosis factor-α delivery. *Molecular Cancer Therapeutics*, **5** (4), 1014–20.

174 Ansell, R.J., Kriz, D. and Mosbach, K. (1996) Molecularly imprinted polymers for bioanalysis: chromatography, binding assays and biomimetic sensors. *Current Opinion in Biotechnology*, **7**, 89–94.

175 Masqué, N., Marcé, R.M. and Borrull, F. (2001) Molecularly imprinted polymers: new tailor-made materials for selective solid-phase extraction. *Trends in Analytical Chemistry*, **20** (9), 477–86.

176 Blanco-López, M.C., Lobo-Castañón, M.-J., Miranda-Ordieres, A.J. and Tuñón-Blanco, P. (2003) Voltammetric response of diclofenac-molecularly imprinted film modified carbon electrodes. *Analytical and Bioanalytical Chemistry*, **377**, 257–61.

177 Dickert, F.L. and Thierer, S. (1996) Molecularly imprinted polymers for optochemical sensors. *Advanced Materials*, **8** (12), 987–90.

178 Andersson, L.I., Mueller, R., Vlatakis, G. and Mosbach, K. (1995) Mimics of the binding sites of opioid receptors obtained by molecular imprinting of enkephalin and morphine. *Proceedings of the National Academy of Sciences of the United States of America*, **92**, 4788–92.

179 Ramström, O., Ye, L. and Mosbach, K. (1996) Artificial antibodies to molecular imprinting. *Chemistry and Biology*, **3** (6), 471–7.

180 Randhawa, M., Gartner, I., Becker, C., Student, J., Chai, M. and Mueller, A. (2007) Imprinted polymers for water purification. *Journal of Applied Polymer Science*, **106**, 3321–6.

3
Interaction Between Silica Particles and Human Epithelial Cells: Atomic Force Microscopy and Fluorescence Study

Igor Sokolov

3.1
Interaction of Silica with Biological Cells: Background

Silica is the second most abundant mineral on earth. Interaction between biological cells and silica particles has been intensively studied for more than half a century. Being biologically inert, it can nevertheless cause silicosis, a form of lung disease caused by inhalation of pretty large amount of crystalline silica (quartz) dust. Silica has been broadly used as a substrate for cell growth. In 1949, the use of silica gels made of colloidal silica was reported [1] as a medium for growing microorganisms, for example, as a substitute for an agar medium. Later, silica-coated polymeric beads were used as a substrate for cell culture dishes. The use of nanoscale silica to produce a tissue culture surface was developed later [2, 3].

Colloidal silica has also been broadly used as a media for cell separation [4]. As originally discovered in 1959, colloidal silica was the best of all substances tested, although it was later found that pure silica as well as fused silica (quartz) were toxic to cells, causing hemolysis of red blood cells [5]. However, when polysaccharides were added to stabilize colloidal silica gradients, it inhibited the toxic effects of silica [6–8]. For example, adsorbing polyvinylpyrrolidone polymer to 10–30 nm silica particles allowed researchers to obtain iso-osmotic, pH-neutral and high-density silica solutions, so-called Percoll [9]. Percoll, however, has shown some toxicity effects for some sperm cells. Apart from that, it had a problem of its sterilization in the autoclave [10]. Later-developed silanized colloidal silica particles [11] show neither toxicity nor problems of sterilization.

Thus, one can conclude that the problem of silica toxicity is not completely solved. It is believed that the presence of silanol groups has been well inversely correlated with the toxicity of silica particles [2,3]. We believe that the main problem is in the variability of the silica surface. The number of silanol groups on the virgin silica surface depends on the history of preparation and storage of silica. Secondly, when silica is introduced in cell growth medium, it immediately starts to be coated with proteins and polysaccharides of the medium. This changes the chemistry of the silica surface, and consequently, alters its interaction with the cell membrane.

Cellular and Biomolecular Recognition: Synthetic and Non-Biological Molecules. Edited by Raz Jelinek
Copyright © 2009 WILEY-VCH Verlag GmbH & Co. KGaA, Weinheim
ISBN: 978-3-527-32265-7

If such a silica particle crosses the cell membrane and enters the cell, it will undergo complex biochemical reactions. Finally, silica is slowly biodegradable [12]. In the organism, it is converted to silicon acid, which is easily metabolically removed.

Almost all processes mentioned above are dependent on the size of silica particles. This is particularly important for nano-sized particles. For example, it has recently been shown that permeation of phospholipid membranes by silica nanoparticles is strongly size-dependent [13]. All this makes it difficult to solve the problem of interaction between silica particles and cells.

With the development of scanning probe techniques, in particular, atomic force microscopy (AFM) [14–18], optical tweezers [19, 20] and magnetic beads/tweezers [21–27], it is now possible to study the interaction between individual silica particles and cell surface. Furthermore, recently synthesized ultra-bright fluorescent silica particles [28] open new possibilities for the study of cell–silica particle interaction by using fluorescence microscopy, a widely used technique in biology. AFM and fluorescence can be used in synergy. This is described in this chapter.

3.2
Interaction of a Silica Particle with the Cell Surface: How It Is Seen with AFM

Surface properties of human epithelial cervical cells have recently been studied with AFM [29]. It was shown that the AFM "sees" the cell surface heavily covered with a sort of "brush". The presence of such brush was known before that study, but not taken into account in the AFM research (in the other words, it was believed that the cell surface was flat). Thus, when a particle interacts with a cell, one should consider the interaction with the brush first. To avoid contamination of silica surface by organics in the growth media, we performed all our measurements when viable cells were immersed in Hank's balanced salt solution (HBSS). Here, we describe the technique, how and what exactly was measured.

3.2.1
AFM

AFM is still a relatively novel method [30], particularly in biological applications. AFM is one of the major techniques responsible for the emergence what is nowadays called nanotechnology. This technique is based on the detection of forces acting between a sharp probe, the AFM tip, and a sample surface (Figure 3.1). The tip is attached to a sufficiently flexible cantilever. Any motion of the cantilever is detected by various methods (e.g., an optical detection system of the deflection in which a laser beam is reflected from the cantilever and detected by a photodiode). The tip is brought to a contact, engaged with the surface of interest. When scanning over a surface of interest, the AFM system records the deflections of the cantilever with subnanometer precision. The deflection signal (or any derivatives/harmonics of the deflection) is recorded digitally and can be visualized on a computer as a raster image in real-time.

3.2 Interaction of a Silica Particle with the Cell Surface: How It Is Seen with AFM

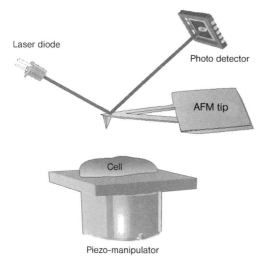

Figure 3.1 Schematic view of the AFM method [31].

The AFM technique has much more capability than just simply microscopy. It can be used to measure forces acting between the AFM tip and the surface of interest at a specific point. It can be done when the cantilever stops scanning in the lateral direction and oscillates in the vertical direction only. A typical force dependence recorded in such a case is shown in Figure 3.2. The force F of bending of the cantilever (vertical axis) is plotted versus the vertical position z of the sample. When the tip is far away from the surface, there is no deflection and consequently the force is equal to zero (under the assumption of no long-range forces). When the z position of the sample increases (the sample is moved up by the scanner, Figure 3.1), the tip–sample distance decreases. At some point, the tip touches the

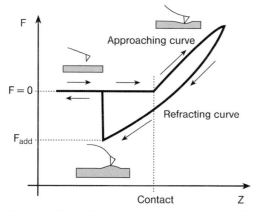

Figure 3.2 Typical force–distance curve recorded by AFM in force mode. Both the approaching and retracting curves are shown [31].

sample (position of contact). After that the tip and sample move up together. This part of the force curve is called the region of constant compliance. At some point, which can be controlled via the AFM software, the sample stops and retracts down. The force curve before this point is called the approaching curve. The force curve recorded when the sample is retracted down is called the retracting curve. During the retraction, the surface may display its nontrivial viscoelastic properties. This typically results in a hysteresis between the approaching and retracting curves shown in Figure 3.2.

An important feature of the retracting curve is the presence of nonzero force required to detach the tip from the surface. This is the so-called adhesion force. It appears due to the presence of strong short-range forces (like van der Waals force and hydrogen bonding) acting between the tip and surface while at contact.

The AFM technique has a number of features that make it extremely valuable in biology, in particular, to study cells [20, 32–79]. It can be used for studying biological objects directly in their natural conditions, particularly in buffer solutions, *in situ* and *in vitro*, if not *in vivo*. There is virtually no sample preparation apart from one requirement–the object of study should be attached to some surface. There are virtually no limitations on the temperature of the solution/sample, chemical composition the type of the medium (either nonaqueous or aqueous liquid). The only one limitation exists for the most popular optical systems of detection–the medium should be transparent for the laser light used to detect the cantilever detection.

3.2.2
AFM on Cells

3.2.2.1 Cell Culture
Primary cultures of human cervical epithelial cells were prepared by a two-stage enzymatic digestion of cervical tissue and cells were maintained in keratinocyte serum-free medium (Invitrogen, Carlsbad, CA). Cervical cancer cells were isolated from primary cervical carcinomas as described in Woodworth *et al.* [80]. All normal and cancer tissue was obtained from the Cooperative Human Tissue Network. Normal cervical cells were used at passages 2–4 when they were actively growing and carcinoma cell lines were used at passages 30–40. All cells were plated in 60-mm tissue culture dishes. Epithelial cells adhered tightly to the bottom of the tissue culture dishes.

3.2.2.2 AFM
Using AFM to study biological cells has its own particular features. The description of the AFM given in the previous section was universal, specific neither to a particular AFM manufacturer nor the sample type. For the results described below we used a particular AFM, a Nanoscope Dimension 3100 (Digital Instruments/Veeco, Santa Barbara, CA). Below we will describe only of the important features of the method that we used, referring to the original papers [29, 31, 81, 82] for more detail.

Compared to nonbiological objects, studying cells has one serious difficulty sometime overseen by material scientists. The cells are intrinsically variable. To make any quantitative conclusion, one has to measure a statistically sound number of cells. For example, to collect sufficient statistics in the methods used to measure mechanical properties of cells [31, 83–86], one needed to measure at least 10–15 cells and a few hundreds of force curves per cell. Fortunately, the later part of the procedure is somewhat automated. For example, the so-called force volume mode allows one to collect sufficient numbers of force curves per region of interest. Apart from the force curves, this mode simultaneously records information about the surface topography. This is important because the models used to quantify the measurements have been developed for a sphere (or another well-defined geometry) over a horizontal plane. For the particular example of human epithelial cells [31, 85, 86], the collected force curves were processed only over relatively flat areas of the cells (below 10–15° of the slope angle). The force curves were collected over areas of several hundred square microns with the vertical ramp size within 4–5 μm. The AFM probe moved up and down during the force collection with a frequency of 2 Hz to avoid viscoelastic effects. While we could not avoid the viscoelastic effects completely, to be consistent we performed all measurements with the same oscillation frequency of 2 Hz. Working with slower speeds is almost impractical because of too long time is required for such a scanning.

The time-dependent adhesion forces between the silica sphere attached to the AFM cantilever (see the next section) and cell surfaces were measured in HBSS solution on viable cells [81]. Using a built-in AFM software option, we kept the AFM probe in contact with the cell surface for a predefined time, from 0 to 60 s. After that the probe was retracted and the adhesion force was recorded.

Another important point is the intrinsic nonlinearities of the AFM scanners. Typically, this effect is ignored when the vertical travel of the scanner is small. As cells are typically rather soft, it requires substantial vertical travel of the scanner to detect the interaction with the cell. Typically this is a size of a few microns. In such a case the scanner nonlinearities are already substantial. To avoid this problem, a special NPoint closed-loop scanner (200 μm × 200 μm × 30 μm, XYZ) was used in this study. A closed loop is paramount because it provides the linearity required for quantitative description of the force curves with such extended scan range. A relatively large vertical Z-range was particularly important because the cell height was around 10–15 μm. Adding the ramp size needed to collect the force curves (at approximately a half of the maximum range), one can easily get a number required for the Z-scan range (ramp size) close to the full extent of the scanner.

3.2.3
AFM Probe Preparations

To study the interaction between a silica particle and a cell, one can attach the silica particle to the AFM cantilever. To be able to quantify such interaction, it is important to have a particle of well-defined geometry (e.g., a sphere). At present, commercial availability of such cantilevers is fairly limited, particularly for use in

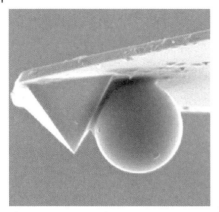

Figure 3.3 Scanning electron microscopy example of gluing of a 5-μm silica sphere to a standard AFM cantilever [86]. One can see quite a large difference between the potential contact area of the regular (pyramidal) tip and the spherical probe.

liquids. Fortunately, such probes can be relatively easy prepared by gluing a silica particle to an AFM cantilever by epoxy resin with a micromanipulator (see, e.g., [85–87]). A scanning electron microscopy image (Figure 3.3) shows an example of gluing of a 5-μm silica sphere to a standard AFM cantilever. One can see quite a large difference between the potential contact area of the regular (pyramidal) tip and the spherical probe. As was described in [85, 86, 88], such a modified probe has numerous advantages in obtaining quantitative data on soft surfaces. (1) The probe has a considerably larger area of contact with the cell compared to a typical commercial probe (apex radius of curvature around 5–50 nm) commonly used for such measurements. The larger area of probe–cell contact results in averaging local variation in rigidity compared to that measured with the regular sharp probe. This leads to lesser variation of the collected data. Consequently, one needs a smaller number of measurements to gather the necessary statistics. This is important because of the limited time (2–3 h) for measurements when the cells were alive in the HBSS solution. (2) Excessively sharp probes may give smaller sensitivity than needed to detect surface molecular brushes. (3) Using the 5-μm probes, we caused less potential disturbance of the cell during the scanning [86]. (4) Using a micron-sized silica ball as the AFM probe, we presumably do not overstress the cell surface (staying in the linear stress–strain regime) as can be in the case of the sharp AFM tip [89], and consequently, we may use the classical Sneddon models to describe deformations of the cell body.

3.2.4
Models to Analyze the Cell Surface: Need for a Two-Layer Model

It is important to note that the cell is not a homogenous medium. However, as was shown in the literature [90–93], the approximation of a homogenous medium,

3.2 Interaction of a Silica Particle with the Cell Surface: How It Is Seen with AFM | 75

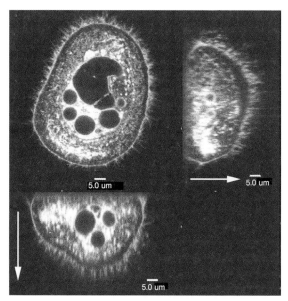

Figure 3.4 High-resolution confocal image of a human epithelial cell showing the brush associated with membrane corrugations. To distinguish the cell surface brush and filopodia developed on the cell culture dish surface, we also show three-dimensional cross-sections of cells. The arrows indicate the vertical direction pointing out of the dish. All scale bars are 5 µm. (Taken from [82]).

which is used in the Hertz–Sneddon model, works quite well in the description of the mechanical properties of cells. To find the Young's modulus of the cell medium, the Hertz–Sneddon model [94, 95] is used because of relatively low adhesion between the AFM probe, a silica sphere, and the cell. In this model, the Young's modulus E is given by:

$$E = \frac{3}{4}\frac{1-v^2}{\sqrt{R}}\frac{dF}{d(p^{3/2})} \tag{3.1}$$

where F is the load force, R is the radius of the silica sphere and p is the probe penetration into the cell. The Poisson ratio v is typically unknown (ranging from 0.3 to 0.5), but the factor $1-v^2$ is simply included in the definition of the Young's modulus.

Nevertheless, when we use not a very sharp probe (the reason for that was described previously), we can start to feel the presence of a "brush" layer on the cell surface. The existence of such a layer is known in biology. For example, Figure 3.4 shows a high-resolution confocal image of a single cell. A fluorescent dye is used to highlight the cellular lipid membrane. One can clearly see that the cell is surrounded by a sort of a brush. This is important – it means that when a silica particle approaches such a cell, it will interact with the brush first. The same

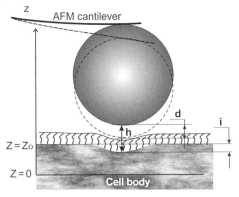

Figure 3.5 Spherical AFM probe deforms both the brush and cell body [29].

happens when the particle is attached to the AFM cantilever and serves as an AFM probe. Coming into contact, the probe will deform both the brush and cell body. Figure 3.5 graphically describes this situation.

The two-layer model suggested in [29] separates the elastic deformation of the cell body from the force dependence of the brush layer. Notations used in Figure 3.5 are as follows. Z is the relative scanner position of the cantilever, d is the cantilever deflection, Z_0 is the nondeformed position of the inner layer of sample, i is the deformation of sample, $Z = 0$ is for the maximum deflection (assigned by the AFM user), and h is the separation between the inner layer and the tip. Looking at the geometry shown in Figure 3.4, one can find the following equation connecting the above parameters:

$$h = Z - Z_0 + i + d \tag{3.2}$$

In these notations, to find the geometry of the brush, one first has to find the force due to the brush [i.e., the cantilever deflection d (load force) as a function of h]. Parameters Z and d are directly measured while collecting the force load curves. The other two parameters (i, Z_0) have to be found. Using the Hertz–Sneddon model, one gets:

$$i = \eta d^{2/3}, \eta = \left[\frac{9}{16}\frac{k}{E}\sqrt{\frac{R_t + R_s}{R_t R_s}}\right]^{2/3} \tag{3.3}$$

where R_t and R_s are the radii of curvature of the tip and sample, respectively (in the case of a flat surface R_s is infinite).

The main assumption here is that the geometry of the interface between the layers is the same as the AFM probe. In such a case, the probe pushing on the top layer and the top layer pushing the second layer, both interfaces have the same the radius of curvature. In some sense, it means that the radius of the tip (sphere) does not change while "acting" through the top layer. This is why R_t in Equation 3.3 is the radius of the AFM probe. For example, this assumption is plausible

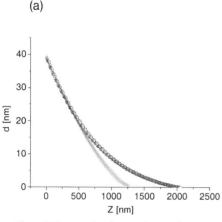
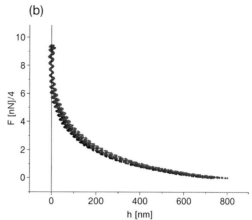

Figure 3.6 Example of using the two-layer model to separate force interaction between the brush layer and the AFM spherical probe from the elastic response of the cell body. (a) Fitting the rigid contact to determine Z_0 and the Young's modulus of the cell body (several lines are shown to demonstrate variability of the fitting taking different amount of points near the maximum contact from 0–200 to 0–450 nm). (b) The force F as a function of h obtained from Equation 3.4. (Taken from supplementary materials of [82]).

if the top layer is a rather rare brush, which is a plausible assumption as one can see from the results.

Combining Equations 3.2 and 3.3, we have:

$$h = Z - Z_0 + \eta d^{2/3} + d \tag{3.4}$$

In order to solve the problem [find $d(h)$], we need to find Z_0. This is a relatively simple to do if at the moment of maximum load, the spherical probe squeezes the first layer so that $h \ll Z_0$. Obviously, this depends on the value of maximum load. We will demonstrate the validity of this assumption below. Thus, one can both find Z_0 and the Young's modulus of the cell body by putting $h = 0$ at the moment of maximum load. Then Equation 3.4 will be given by:

$$Z = Z_0 - \eta d^{2/3} - d \tag{3.5}$$

near the point of maximum load.

Let us demonstrate how the procedure described above works for the example of a representative cervical cell. Figure 3.6a shows a typical $d(Z)$ curve recorded on the surface of a (normal) cervical cell. The force can be found by multiplying the deflection d by the cantilever spring constant k, around 0.25 N/m in this case. Fitting Equation 3.5 near the maximum load ($0 < Z < 200$ nm), one can get $E = 2.6$ kPa and $Z_0 = 1250$ nm. The results are rather stable with respect to the choice of the fitting interval. For example, if one takes the fitting interval

$0 < Z < 450$ nm, then $E = 2.3$ kPa and $Z_0 = 1330$ nm. This uncertainty is still well below the variability between different cells. Finally, the sought $d(h)$ (or the force F) dependence as a function of h can be obtained from Equation 3.4 by plugging in the found values of E and Z_0. This is shown in Figure 3.6b.

To demonstrate self-consistency of the model, let us show that the assumption $h = 0$ (more precisely $h \ll Z_0$) near the point of maximum contract is valid for the cells analyzed here. For the example of Figure 3.6, one can check that when the probe reaches distances $Z < 450$ nm, the cantilever deflection is above 18 nm (Figure 3.6a). This corresponds to $h < 40$ nm (4.5 nN force of Figure 3.6b, [4.5 nN/0.25 N/m] = 18 nm), which is definitely smaller than $Z_0 = 1330$ nm. For the probe distances $Z < 200$ nm, the assumption is satisfied even better ($h \sim 0$ nm $\ll Z_0 = 1250$ nm). This shows the consistency of our model.

Such high "squeezeability" of the first layer is an intrinsic requirement of this model. However, it is clear how to develop a model for a "rigid brush". If the top layer is dense enough it can obviously be squeezed, roughly speaking, until it becomes stiffer than the inner layer, which will then be pushed by the squeezed top layer. In such a case, the model would be applicable for a relatively large h. For smaller h, the distance between the probe and second layer will be decreased faster compared to that predicted by the model when the probe approaches the cell.

3.2.5
Experimental Data

3.2.5.1 Surface Brush on Cancer and Normal Cells

Application of the above models to analyze force curves collected on human cervical epithelial cells showed very interesting results [82]. Both normal and malignant cells were analyzed. The difference in the Young's moduli of normal and cancer cells was found to be statistically insignificant. The striking difference was found for the brush layer on the cell surface. In contrast to normal cells, malignant cells have a heterogeneous brush with two major lengths. This was derived as follows. After processing the force curves as described in the previous section, the collections of the forces acting between the brush and the AFM probe (an example is shown in Figure 3.6b), the graph shown in Figure 3.7 was obtained. One can clearly see the exponential force dependences (straight line in the used logarithmic scale). While the normal cells were characterized by a single exponential dependence, the malignant cells showed at least double-slope straight lines (i.e., the sum of two exponents). A known example of exponential dependence that could also contribute to the probe–cell interaction is the electrical double-layer interaction. However, it is definitely to be excluded from the list of possibilities because the screening range of such force, the Debye length, in the physiological solution is within the range of nanometers, whereas the decay lengths of the curves shown in Figure 3.7 are within hundreds of nanometers. Another example of the exponential dependence is the force measured on molecular brushes. Interaction with a brush gives steric repulsion, which can be

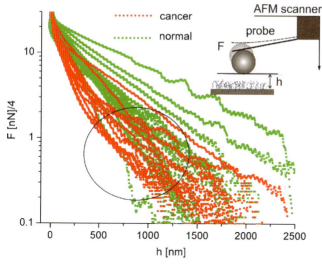

Figure 3.7 Long-range forces of interaction between the AFM probe and the cell brush. The force (shown in logarithmic scale) is plotted versus the probe–cell body distance h. A feature of cancer cells, slower exponential decay in the force dependence at large distance, is highlighted by the circle. (Taken from [82]).

approximated by an exponential force dependence [29, 96]. The Derjaguin approximation gives the force of interaction between a spherical AFM probe and a flat surface [97]:

$$F_{steric} \approx 50 k_B T R N^{3/2} \exp(-2\pi h/L) L \qquad (3.6)$$

where L is the equilibrium thickness of the brush layer, N is the surface density of the brush molecules (grafting density), R is the radius of the AFM probe and T is the temperature of the medium. Equation 3.6 is a good description of a brush for $0.1 < h/L < 0.8$.

The easiest was to explain the two-slope behavior is to assume the force due to two brushes of different lengths, which is described by:

$$F_{steric} \approx 50 k_B T R \left[N_1^{3/2} \exp\left(-\frac{2\pi}{L_1} h\right) L_1 + N_2^{3/2} \exp\left(-\frac{2\pi}{L_2} h\right) L_2 \right] \qquad (3.7)$$

Here, the indexes of N and L correspond to the first and second brushes. Definitely, Equation 3.7 is an approximation. Strictly speaking, it is no longer applicable if the distance h is between the lengths L_1 and L_2, as well as greater than the larger L. However, the exponential functions vanish quickly, leaving rather small corrections. A similar approximation was used previously (e.g., [98]). Secondly, the fitted forces are the averaged ones over the surface of each cell. Therefore, the grafting densities N_1 and N_2 should be treated as the *effective* ones.

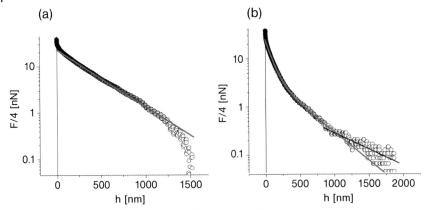

Figure 3.8 Representative force curves on (a) normal and (b) cancer cells. Double-brush behavior of cancer cells is seen. Solid lines are the model, Equation 3.6 for normal cells and Equation 3.7 for cancel cells. Additional straight lines for larger h show that cancer cells demonstrate even a weak third brush. (Taken from [82]).

Figure 3.8 shows representative examples of fitting the force curves with Equations 3.6 and 3.7. One can see a very good fit. In contrast with the normal cell force curves, for large h the forces lie above the fitting line. This means that cancer cells demonstrate even a weak *third* brush. However, it is close to the limit of AFM sensitivity to make any quantitative conclusions.

Comparing normal and cervical cell brushes, one can see a convincing difference between these types of cells. While normal cells have a single-length brush, cancer cells have a double-length brush; the long length is about the size of the normal cell brush (there is no statistically significant difference), but the shorter length is around 5 times smaller (the difference is statistically significant). The grafting density of the long brush is almost twice lower than that of normal cells, but the short brush is around 2 times denser than the brush of normal cells. All graphing densities are statistically different. Figure 3.9 shows a graphical presentation of the brushes with the derived parameters.

3.2.5.2 Measurement of Adhesion: Silica Particle–Cell Interaction

As we have seen in the previous section, the force measurements show a clear difference in the geometry of the cytoplasmic membrane in normal and cancer cells. Therefore, it is plausible to expect a difference in the adhesion between a silica particle, the AFM probe and the surfaces of these types of cells. This was recently analyzed [81]. Processing the retraction force curves, adhesion forces between a silica probe and cell surface were found. The averaged per cell adhesions are plotted as histograms in Figure 3.10. One can indeed see some difference between the adhesions to normal versus cancer cells. As calculated, the average on all cells is 0.39 ± 0.20 nN for normal cells and 0.54 ± 0.16 nN for cancer cells.

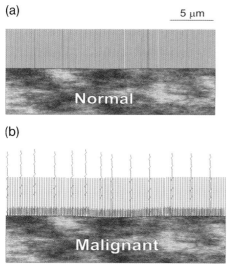

Figure 3.9 Graphical presentation of the brushes based on the derived parameters on normal (a) and cancer (b) cells. The 5-μm bar is shown for comparison. (Adopted from [82]).

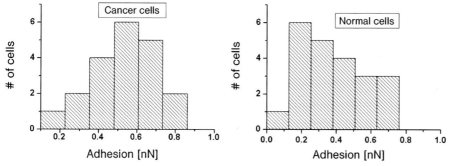

Figure 3.10 Histograms of adhesion of a silica particle to the cells averaged per each cell. (Taken from [81]).

It is interesting to note that the cancer cells show no larger (even a bit less) variability than normal cells, the same tendency as for the "brush" forces shown in Figure 3.7. Thus, we can consider this as further evidence that we are dealing with interaction with the surface brush when measuring the adhesion. The fact that we found slightly higher adhesion to cancer than normal cells is somewhat against a generally accepted paradigm that says cancer cells are less sticky in order to be more invasive.

Studying adhesion further, it was found [81] that the adhesion phenomenon is more complicated; it depends on the duration of silica–cell contact. Increasing the duration when the probe touches cell surface, the reversal of the adhesion was

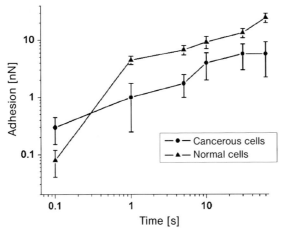

Figure 3.11 Time-dependent adhesion between a silica particle and the cell surface. The AFM silica particle probe stays in contact with the cells for a definite period of time (from 0 to 60 s) before retracting. (Taken from [81]).

observed. Specifically, cancer cells became lesser adhesive then normal cells with the increase of the contact time. Figure 3.11 shows this behavior. The silica probe stays in contact with the cell for a defined period of time (0, 1, 5, 10, 30 and 60 s) and then is pulled back. It should be noted that the adhesion also depends on the force which the AFM probe exerts on the cell. The forces used to obtain data in Figure 3.10 were about 4 times lower than used to obtain Figure 3.11. This was done to avoid possible disturbance of the cells exposed to a long action of the silica probe. In principle, the observed phenomena could potentially be used to detect cancer cells. We will analyze this possibility in the next section.

Can the observed behavior be explained by our proposed brush parameters shown in Figure 3.9? Presumably, the brush has adhesion to silica. Being part of the cell surface, the brush has glycocalyx molecules, polysaccharides, microvilli and/or microridges. All of those have hydrogen (attraction), van der Waals (attraction), steric (repulsion) and weak electrostatic (can be both attractive and repulsive) interactions with silanol groups of the hydrated silica surface of the probe. For example, proteoglycans of the glycocalyx seem to be only weakly negatively charged [82]; a strong competitive charge could prevent molecules from sticking to the negatively charged silica. The time dependence of adhesion could be explained by the "sea grass effect". Owing to thermal motion, the longer brush can envelope the silica surface, spreading over the surface to find the sticky points (active head groups or charge concentrations) more effectively with time than the shorter brush. At the beginning, the silica sphere penetrates more easily through a relatively rare long cancerous brush, while the brush of normal cells, being almost 2 times denser, keeps the silica ball repelled for some time (the repulsion by the brush of normal cells is around 3 times stronger due to the steric repulsion, Equation 3.6). After several seconds, the silica sphere presumably becomes enveloped by the long brush of normal cells, but it can still be repelled by the fairly dense and short second (inner) brush of cancer cells.

3.2.5.3 Can the Difference in Adhesion Be Used to Detect Cancer Cells?

To answer the question of this subsection, a simple experiment was performed [81]. Silica particles were dispersed over the normal and cancer cells. The particles were allowed to precipitate on cells in a culture dish for a definite period of time. After that the particles were removed by a controlled washing out the cells with HBSS buffer (using an orbital shaker; for more details, see Section 3.4 in which a similar procedure using fluorescent particles is described). The force acting on cells from the silica particles during precipitation was definitely much smaller (around 0.8 pN due to gravity) than the one used in the AFM measurements described previously (around 3 nN). Keeping in mind the "sea grass" mechanism described in the previous section, a longer time of contact was used to compensate for the smaller force. It was found that about 120 s waiting time before removing the particles was the optimum to see lower adhesion to normal versus cancer cells. Figure 3.12 shows an example of single cells either surrounded by silica particles (small round circles) in the case of normal cells (Figure 3.12a) or covered by the silica particles adhered in the case of cancer cells (Figure 3.12b). When waiting longer, the particles develop so strong adhesion that it was too hard to remove them by a controllable washing. While a clear difference is seen in Figure 3.12, we need to make a quantitative conclusion on a statistically large number of cells. It is not that easy to do with optically neutral silica particles. One would need to create software capable of distinguishing between the cell, culture dish and silica particles to calculate the amount of particles adhered per unit area of the cell surface. While in principle it could be done, an easier approach is to use

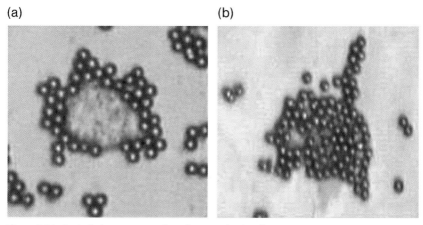

Figure 3.12 Optical phase images of single normal (a) and cancer cervical cells (b) exposed to 5-μm silica spheres (small round circles). (Taken from [81]).

fluorescent silica particles. This was done in [81]. The next section describes briefly the particles used in [81].

3.3
Ultra-Bright Fluorescent Silica Particles to Be Used to Study Interaction with Cells

Fluorescent silica particles have a number of advantages that make them an attractive candidate for fluorescent tagging of cells. Silica surface can be easily functionalized with various chemistry. A large variety of fluorescent dyes can be encapsulated inside silica. Since typically fluorescent dyes are toxic, encapsulation inside silica decreases as the apparent dye toxicity. To prevent leakage of the dyes out of the porous matrix, dyes were covalently bound to the silica matrix [99–103]. While the photostability of such materials was higher than the stability of pure dyes, it did not prevent bleaching substances, including oxygen, from penetrating inside such a composite material. In the case of xerogel, it is rather hard to use it for labeling. Recently, there have been several reports about the incorporation of fluorescent lasing dyes into nanoporous patterned silica films [104] and silica rods [105].

Recently, a new one-step self-assembly of nanoporous silica particles with encapsulated organic dyes has been developed [28, 106], in which the dyes are physically entrapped inside the silica matrix. These particles have a silica matrix in which there are hexagonally packed cylindrical channels. The particles could also be described as arrays of closely packed cylindrical channels. A schematic of such a particle is shown in Figure 3.13.

The main feature of such particles is their fluorescent brightness. It comes from the following effect. Alkane chains of surfactant molecules (shown in Figure 3.13 as zig-zag vertical lines with the head groups adjacent to silica walls) act as separators between the dye molecules, preventing dimerization of the dye molecules in the direction along the channels. In the perpendicular directions, silica walls play the role of separators to prevent dimerization. This results in up to around 10 000 times brighter than the maximum fluorescence from the same dye dissolved in

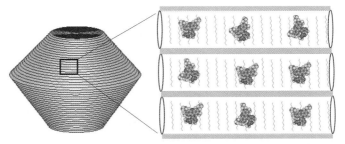

Figure 3.13 Schematics of location of the dyes inside the synthesized shapes. Right side of the image presents a "zoomed" area of the channels with the dye encapsulated inside around 3-nm channels. (Taken from [28]).

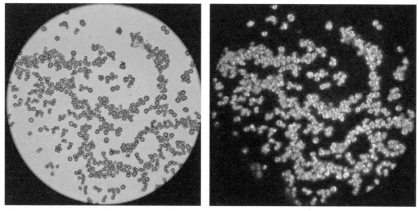

Figure 3.14 Transmitted image (a) and dark-field image (b) of the particles with encapsulated rhodamine 6G dye. (Taken from [28]).

aqueous solution at its maximum concentration before self-quenching (due to dimerization of the fluorescent molecules). Comparing this fluorescence with the brightest micron-size particles assembled from aqueous compatible quantum dots [107] encapsulated in polymeric particles of similar size (around 1.2 μm) reported recently [108], one can find that the particles (scaled to the same size) are about 170–260 times brighter. This makes the particles the brightest tags presently available.

The particles are typically a mixture of discoids (a shape shown in Figure 3.13) and fibers. The fibers are can be filtered out fairly easily. Alternatively, there are syntheses in which only discoids can be synthesized. An optical image is shown in Figure 3.14. A regular transmitted and a dark field image are shown. One can see that the typically whitish dark field image appears as greenish. This is due to a very strong fluorescence coming from the particles, which is comparable to the amount of scattered white light. The typical size of the particles is around 5–6 μm.

3.4
Ultra-Bright Fluorescent Silica Particles to Distinguish Between Cancer and Normal Cells

Using the ultra-bright fluorescent silica particles described in the previous section, we will describe a simple potential method to distinguish between cancer and normal cells *in vitro*. Being silica from outside, these particles interact with the cells as we described in Section 3.2. The goal of this section is to demonstrate how one can use fluorescent silica particles to distinguish between cancer and normal cells by using the different silica–cell interaction described in Section 3.2.

As we discussed, in principle, one could count the number of silica particles adhered to cancer versus normal cells. Such a counting is greatly simplified for the case of fluorescent particles. Instead of counting the number of particles, one can measure the amount of fluorescent light coming from a particular area. Heaving a strong fluorescence signal from each particle, it makes the measurement of fluorescent light rather straightforward. It does not require high sensitivity spectrometers typically used for measuring fluorescence. A fairly simple instrumental setup capable of measuring such fluorescence is described in the next section.

3.4.1
Methods and Materials

Cells used in this section were prepared in the same way as described in Section 3.2.

3.4.1.1 Spectrofluorometric and Optical Measurements of the Particles Attached to Cells

The following assembly was set up to record optical images of cells and record emission intensity of the fluorescent particles (Figure 3.15). An Olympus BH2-UMA microscope was connected to a JVC TK-1280U color video camera. The images of cells were captured using FlashBus MV version 3.91 software. To record the fluorescent emission, an optic fiber (UVIR1000) with a C-type adapter was used to connect the microscope to an Ocean Optics USB 2000 spectrometer. A Cyonics air-cooled argon ion laser was used as the light source. It should be noted that the use of laser light here is not necessary, but is helpful. An advantage in using laser is a fairly strong fluorescence signal and, consequently, very fast measurements can be done. A 500-nm notch filter (Omega Optical) was utilized to filter the laser light. The spectrographs were recorded using OOIBase32 software. To average the emission over the culture dish area, we defocused the laser by rotating the focus knob of the microscope by three full turns in the counter-clockwise direction (to increase the distance between the objective and sample). This procedure was maintained the same for spectroflurometric measurements of all samples.

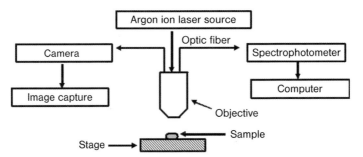

Figure 3.15 Schematic diagram of the optical measurement system.

3.4.1.2 Detection of Affinity of Fluorescent Silica Particles to Cells

The ultra-bright fluorescent silica particles were prepared as described above (see [28, 106] for more detail). After carefully washing the particles to remove any traces of the synthesizing chemistry, the particles were dispersed in HBSS buffer solution to form a colloidal dispersion. The concentration of the dispersion was maintained at around $15\,cm^3/l$. The cells in the 60-mm culture dishes were washed twice with HBSS solution for 2 min each. The cells were then exposed to 1 ml of the colloidal dispersion for 2 min. During this 2-min period, the dishes were subjected to an initial 30 s of shaking on a Boekel Scientific Ocelot 260300F at minimum speed, to ensure uniform distribution of the colloidal solution and to minimize particle agglomeration. The cells were then washed with HBSS solution 2 times thoroughly using the same Boekel shakier, to remove excess particles that had not adhered to the cells. The culture dishes were then dried in ambient air at room temperature.

It is important to note the density of cells in the culture dish. A control experiment of the affinity was carried out for a light density population of cells [81]; we will not describe that experiment. That was done only to verify that a heavy density population (up to confluent density) of cells behaves similar to the light density population. The major measurements were done on the cells which just reached confluent coverage of the culture dish bottom. This allows us to exclude the possible affinity of the particles to the culture dish bottom as well as dealing with different total areas of cells in different culture dishes. In such a case, the processing of the results becomes fairly simple. The amount of the adhered to fluorescent particles is linear proportional to the fluorescent signal.

3.4.2
Experimental Results: Spectrofluorometric and Image Analysis of Cancer and Normal Cervix Cells

The cells exposed to the silica particles were imaged both with optical microscopy and fluorescent microscopy. Inspection of the typical optical images represented in Figure 3.16 indeed corroborated that the fluorescent silica particles adhered more to cancer cells. A fairly smooth background seen in Figure 3.16 is the confluent layer of cells. Excess of silica particles can be clearly seen on the cancer cells (Figure 3.16a). Comparison of the emission spectrographs of the cancer and normal cells revealed a fairly clear-cut distinction between the two types of cells. The total fluorescent emissions from the cancer cells were on an average 2 times stronger than that from normal cells shown in Figure 3.16. This increase in intensity was due to a greater number of silica particles per unit area attached to the cancer cells when compared to the normal cells.

This study was repeated on cells taken from three healthy individuals and three cancer patients. Figure 3.17 shows the averaged total fluorescent emissions calibrated in absolute energy units as well as the standard deviations for each cell line. The deviations were the results of variability of the measured values

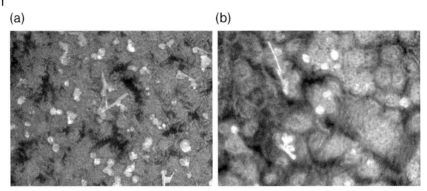

Figure 3.16 Optical reflection images taken in white light of confluent cancer (a) and normal cell (b) layers with the silica particles attached to the cell layers. (Taken from [81]).

over the culture dish (the area of single measurement was approximately 3 mm in diameter).

Although the majority of the data is statistically different (e.g., the difference between CTX2 and HCX30 in Figure 3.17), it is not sufficient for an unambiguous conclusion on malignancy of cells. Obviously, the difference between the latter cell lines could be made stronger if one took a larger area of the culture dish for the signal collection. The goal of this work is to demonstrate the effect of different interactions between silica particles and cancer versus normal cells. For the development of a clinical protocol to distinguish between cancer and normal cells, one would need more human subjects to test.

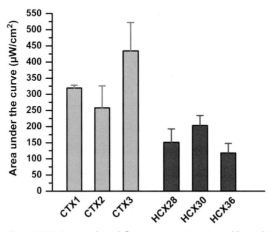

Figure 3.17 Averaged total fluorescent emissions calibrated in absolute energy units as well as the standard deviations for each cell lines taken from cancer patients (CTX lines) and healthy individuals (HCX strains). (Taken from [81]).

3.5
Conclusions

We have described the problem of interaction of silica particles with cell surfaces using an example of human epithelial cervical cells. The study was done using the AFM; silica particle was attached to the AFM cantilever. Analyzing force curves, it was found that the cell surface is not smooth. In the other words, the particles interact with a "brush"-like surface consisting of microvilli and microridges (membrane protrusions). Consequently, the adhesion of particles, particularly silica, was expected to be dependent of the brush structure. Moreover, because both the brush and cell body are sufficiently soft (viscoelastic), it was found that the adhesion depended on time of contact. Statistically different brushes were found for cancer and normal cells. Subsequently, different adhesion was also observed. Analyzing possible intermolecular forces, we can conclude that the interaction between silica particles and the cell surface is mostly a nonspecific mix of hydrogen and van der Waals interactions. The adhesion is proportional to the time of contact, and presumably, the increasing area of the contact.

One of the possible applications of this method can be related to the detection of cancer cells. Here, we described such a method based on counting the silica particles adhered to cell surfaces. To make the method simple, we used fluorescent silica particles. As a result, instead of counting the number of particles, the total fluorescent signal is measured. The method is simple, comparatively inexpensive and fairly unambiguous in interpretation.

Acknowledgments

Partial supports from ARO (grant W911NF-05-1-0339), NSF (0304143), and NanoBio Laboratory (NABLAB) are acknowledged. I am thankful to Prof. Woodworth for useful discussions on the topic of this review.

References

1 Temple, K.L. (1949) A new method for the preparation of silica gel plates. *Journal of Bacteriology*, **57** (3), 383.
2 Barnes, D. and Sato, G. (1980) Methods for growth of cultured cells in serum-free medium. *Analytical Biochemistry*, **102** (2), 255–70.
3 Wolcott, C.C. (1998) Colloidal silica films for cell culture. US Patent 5, 814, 550.
4 Mateyko, G.M. and Kopac, M.J. (1959) Isopyknotic cushioning for density gradient centrifugation. *Experimental Cell Research*, **17** (3), 524–6.
5 Stadler, K. and Stober, W. (1965) Hemolytic activity of suspensions of different silica modifications and inert dusts. *Nature*, **207**, 874–5.
6 Pertoft, H. (1966) Gradient centrifugation in colloidal silica–polysaccharide media. *Biochimica et Biophysica Acta*, **126** (3), 594–6.
7 Pertoft, H., Back, O. and Lindahl-Kiessling, K. (1968) Separation of various blood cells in colloidal silica–polyvinyl-pyrrolidone gradients. *Experimental Cell Research*, **50** (2), 355–68.

8 Pertoft, H., Rubin, K., Kjellen, L., Laurent, T.C. and Klingeborn, B. (1977) The viability of cells grown or centrifuged in a new density gradient medium, Percoll™. *Experimental Cell Research*, **110** (2), 449–57.

9 Pertoft, H., Hiertenstein, M. and Kagedal, L. (1979) Cell separation in a new density gradient medium, Percoll, in *Cell Populations, Methodological Surveys* (ed. D. Reid), Ellis Horwood, Chichester, pp. 67–80.

10 Pertoft, H. (2000) Fractionation of cells and subcellular particles with Percoll. *Journal of Biochemical and Biophysical Methods*, **44** (1–2), 1–30.

11 Centola, G.M., Herko, R., Andolina, E. and Weisensel, S. (1998) Comparison of sperm separation methods: effect on recovery, motility, motion parameters, and hyperactivation. *Fertility and Sterility*, **70** (6), 1173–5.

12 Iler, R.K. (1979) *The Chemistry of Silica: Solubility, Polymerization, Colloid and Surface Properties and Biochemistry of Silica 1979*, John Wiley & Sons, New York.

13 Roiter, Y., Ornatska, M., Rammohan, A.R., Balakrishnan, J., Heine, D.R. and Minko, S. (2008) Interaction of nanoparticles with lipid membrane. *Nano Letters*, **8** (3), 941–4.

14 Habelitz, S., Rodriguez, B.J., Marshall, S.J., Marshall, G.W., Kalinin, S.V. and Gruverman, A. (2007) Peritubular dentin lacks piezoelectricity. *Journal of Dental Research*, **86** (9), 908–11.

15 Kalinin, S.V., Rodriguez, B.J., Jesse, S., Shin, J., Baddorf, A.P., Gupta, P., Jain, H., Williams, D.B. and Gruverman, A. (2006) Vector piezoresponse force microscopy. *Microsc Microanal*, **12** (3), 206–20.

16 Drake, B., Prater, C.B., Weisenhorn, A.L., Gould, S.A., Albrecht, T.R., Quate, C.F., Cannell, D.S., Hansma, H.G. and Hansma, P.K. (1989) Imaging crystals, polymers, and processes in water with the atomic force microscope. *Science*, **243** (4898), 1586–9.

17 Brown, H.G. and Hoh, J.H. (1997) Entropic exclusion by neurofilament sidearms: a mechanism for maintaining interfilament spacing. *Biochemistry*, **36** (49), 15035–40.

18 Kumar, S. and Hoh, J.H. (2004) Modulation of repulsive forces between neurofilaments by sidearm phosphorylation. *Biochemical and Biophysical Research Communications*, **324** (2), 489–96.

19 Cuvelier, D., Derenyi, I., Bassereau, P. and Nassoy, P. (2005) Coalescence of membrane tethers: experiments, theory, and applications. *Biophysical Journal*, **88** (4), 2714–26.

20 Svoboda, K., Schmidt, C.F., Branton, D. and Block, S.M. (1992) Conformation and elasticity of the isolated red blood cell membrane skeleton. *Biophysical Journal*, **63** (3), 784–93.

21 Bausch, A.R., Moller, W. and Sackmann, E. (1999) Measurement of local viscoelasticity and forces in living cells by magnetic tweezers. *Biophysical Journal*, **76** (1 Pt 1), 573–9.

22 Berrios, J.C., Schroeder, M.A. and Hubmayr, R.D. (2001) Mechanical properties of alveolar epithelial cells in culture. *Journal of Applied Physiology*, **91** (1), 65–73.

23 Deng, L., Fairbank, N.J., Cole, D.J., Fredberg, J.J. and Maksym, G.N. (2005) Airway smooth muscle tone modulates mechanically induced cytoskeletal stiffening and remodeling. *Journal of Applied Physiology*, **99** (2), 634–41.

24 Feneberg, W., Aepfelbacher, M. and Sackmann, E. (2004) Microviscoelasticity of the apical cell surface of human umbilical vein endothelial cells (HUVEC) within confluent monolayers. *Biophysical Journal*, **87** (2), 1338–50.

25 Heinrich, V. and Waugh, R.E. (1996) A piconewton force transducer and its application to measurement of the bending stiffness of phospholipid membranes. *Annals of Biomedical Engineering*, **24** (5), 595–605.

26 Huang, H., Kamm, R.D., So, P.T. and Lee, R.T. (2001) Receptor-based differences in human aortic smooth muscle cell membrane stiffness. *Hypertension*, **38** (5), 1158–61.

27 Ohayon, J. and Tracqui, P. (2005) Computation of adherent cell elasticity

for critical cell-bead geometry in magnetic twisting experiments. *Annals of Biomedical Engineering*, **33** (2), 131–41.
28 Sokolov, I., Kievsky, Y. and Kaszpurenko, J.M. (2007) Self-assembly of ultra-bright fluorescent silica particles. *Small*, **3** (3), 419–23.
29 Sokolov, I., Iyer, S., Subba-Rao, V., Gaikwad, R.M. and Woodworth, C.D. (2007) Detection of surface brush on biological cells in vitro with atomic force microscopy. *Applied Physics Letters*, **91**, 023902-1-3.
30 Binnig, G., Quate, C.F. and Gerber, C. (1986) Atomic force microscope. *Physical Review Letters*, **56** (9), 930–3.
31 Sokolov, I. (2007) Atomic force microscopy in cancer cell research, in *Cancer Nanotechnology – Nanomaterials for Cancer Diagnosis and Therapy* (ed. H.S. Nalwa and T. Webster), American Scientific Publishers, Los Angeles, CA, pp. 43–59.
32 Alonso, J.L. and Goldmann, W.H. (2003) Feeling the forces: atomic force microscopy in cell biology. *Life Sciences*, **72** (23), 2553–60.
33 Arzate, H., Alvarez-Perez, M.A., Aguilar-Mendoza, M.E. and Alvarez-Fregoso, O. (1998) Human cementum tumor cells have different features from human osteoblastic cells in vitro. *Journal of Periodontal Research*, **33** (5), 249–58.
34 Barakat, S., Gayet, L., Dayan, G., Labialle, S., Lazar, A., Oleinikov, V., Coleman, A.W. and Baggetto, L.G. (2005) Multidrug-resistant cancer cells contain two populations of P-glycoprotein with differently stimulated P-gp ATPase activities: evidence from atomic force microscopy and biochemical analysis. *Biochemical Journal*, **388** (Pt 2), 563–71.
35 Barrera, N.P., Herbert, P., Henderson, R.M., Martin, I.L. and Edwardson, J.M. (2005) Atomic force microscopy reveals the stoichiometry and subunit arrangement of 5-HT$_3$ receptors. *Proceedings of the National Academy of Sciences of the United States of America*, **102** (35), 12595–600.
36 Bischoff, G., Bernstein, A., Wohlrab, D. and Hein, H.J. (2004) Imaging living chondrocyte surface structures with AFM contact mode. *Methods in Molecular Biology*, **242**, 105–24.
37 Bischoff, R., Bischoff, G. and Hoffmann, S. (2001) Scanning force microscopy observation of tumor cells treated with hematoporphyrin IX derivatives. *Annals of Biomedical Engineering*, **29** (12), 1092–9.
38 Braet, F., Vermijlen, D., Bossuyt, V., De Zanger, R. and Wisse, E. (2001) Early detection of cytotoxic events between hepatic natural killer cells and colon carcinoma cells as probed with the atomic force microscope. *Ultramicroscopy*, **89** (4), 265–73.
39 Brus, C., Kleemann, E., Aigner, A., Czubayko, F. and Kissel, T. (2004) Stabilization of oligonucleotide-polyethylenimine complexes by freeze-drying: physicochemical and biological characterization. *Journal of Controlled Release*, **95** (1), 119–31.
40 Chasiotis, I., Fillmore, H.L. and Gillies, G.T. (2003) Atomic force microscopy measurement of cytostructural elements involved in the nanodynamics of tumour cell invasion. *Nanotechnology*, **14** (5), 557–61.
41 Chen, B., Wang, Q. and Han, L. (2004) Using the atomic force microscope to observe and study the ultrastructure of the living BIU-87 cells of the human bladder cancer. *Scanning*, **26** (4), 162–6.
42 Domke, J., Dannohl, S., Parak, W.J., Muller, O., Aicher, W.K. and Radmacher, M. (2000) Substrate dependent differences in morphology and elasticity of living osteoblasts investigated by atomic force microscopy. *Colloids Surf B Biointerfaces*, **19** (4), 367–79.
43 Drochon, A., Barthes-Biesel, D., Lacombe, C. and Lelievre, J.C. (1990) Determination of the red blood cell apparent membrane elastic modulus from viscometric measurements. *Journal of Biomechanical Engineering*, **112** (3), 241–9.
44 Eibl, R.H. and Moy, V.T. (2005) Atomic force microscopy measurements of protein–ligand interactions on living cells. *Methods in Molecular Biology*, **305**, 439–50.

45 Feng, S. and Huang, G. (2001) Effects of emulsifiers on the controlled release of paclitaxel (Taxol) from nanospheres of biodegradable polymers. *Journal of Controlled Release*, **71** (1), 53–69.

46 Feng, S.S., Mu, L., Win, K.Y. and Huang, G. (2004) Nanoparticles of biodegradable polymers for clinical administration of paclitaxel. *Current Medicinal Chemistry*, **11** (4), 413–24.

47 Fillmore, H.L., Chasiotis, I., Chow, S.W. and Gillies, G.T. (2003) Atomic force microscopy observations of tumour cell invadopodia: novel cellular nano-morphologies on collagen substrates. *Nanotechnology*, **14** (1), 73–6.

48 Fonseca, C., Moreira, J.N., Ciudad, C.J., Pedroso de Lima, M.C. and Simoes, S. (2005) Targeting of sterically stabilised pH-sensitive liposomes to human T-leukaemia cells. *European Journal of Pharmaceutics and Biopharmaceutics*, **59** (2), 359–66.

49 Gliss, C., Randel, O., Casalta, H., Sackmann, E., Zorn, R. and Bayerl, T. (1999) Anisotropic motion of cholesterol in oriented DPPC bilayers studied by quasielastic neutron scattering: the liquid-ordered phase. *Biophysical Journal*, **77** (1), 331–40.

50 Goldmann, W.H. and Ezzell, R.M. (1996) Viscoelasticity in wild-type and vinculin-deficient (5.51) mouse F9 embryonic carcinoma cells examined by atomic force microscopy and rheology. *Experimental Cell Research*, **226** (1), 234–7.

51 Goldmann, W.H., Galneder, R., Ludwig, M., Xu, W.M., Adamson, E.D., Wang, N. and Ezzell, R.M. (1998) Differences in elasticity of vinculin-deficient F9 cells measured by magnetometry and atomic force microscopy. *Experimental Cell Research*, **239** (2), 235–42.

52 Hategan, A., Law, R., Kahn, S. and Discher, D.E. (2003) Adhesively-tensed cell membranes: lysis kinetics and atomic force microscopy probing. *Biophysical Journal*, **85** (4), 2746–59.

53 Horton, M., Charras, G. and Lehenkari, P. (2002) Analysis of ligand–receptor interactions in cells by atomic force microscopy. *Journal of Receptor and Signal Transduction Research*, **22** (1–4), 169–90.

54 Huang, S. and Ingber, D.E. (2005) Cell tension, matrix mechanics, and cancer development. *Cancer Cell*, **8** (3), 175–6.

55 Kusick, S., Bertram, H., Oberleithner, H. and Ludwig, T. (2005) Nanoscale imaging and quantification of local proteolytic activity. *Journal of Cellular Physiology*, **204** (3), 767–74.

56 Lehenkari, P.P., Charras, G.T., Nesbitt, S.A. and Horton, M.A. (2000) New technologies in scanning probe microscopy for studying molecular interactions in cells. *Expert Reviews in Molecular Medicine*, **2000**, 1–19.

57 Lekka, M., Laidler, P., Dulinska, J., Labedz, M. and Pyka, G. (2004) Probing molecular interaction between concanavalin A and mannose ligands by means of SFM. *European Biophysics Journal*, **33** (7), 644–50.

58 Lekka, M., Laidler, P., Gil, D., Lekki, J., Stachura, Z. and Hrynkiewicz, A.Z. (1999) Elasticity of normal and cancerous human bladder cells studied by scanning force microscopy. *European Biophysics Journal with Biophysics Letters*, **28** (4), 312–16.

59 Lekka, M., Laidler, P., Ignacak, J., Labedz, M., Lekki, J., Struszczyk, H., Stachura, Z. and Hrynkiewicz, A.Z. (2001) The effect of chitosan on stiffness and glycolytic activity of human bladder cells. *Biochimica et Biophysica Acta*, **1540** (2), 127–36.

60 Lekka, M., Lekki, J., Marszałek, M., Golonka, P., Stachura, Z., Cleff, B. and Hrynkiewicz, A.Z. (1999) Local elastic properties of cells studied by SFM. *Applied Surface Science*, **141**, 345–9.

61 Liang, H.F., Yang, T.F., Huang, C.T., Chen, M.C. and Sung, H.W. (2005) Preparation of nanoparticles composed of poly(gamma-glutamic acid)–poly(lactide) block copolymers and evaluation of their uptake by HepG2 cells. *Journal of Controlled Release*, **105** (3), 213–25.

62 Mahaffy, R.E., Park, S., Gerde, E., Kas, J. and Shih, C.K. (2004) Quantitative analysis of the viscoelastic properties of thin regions of fibroblasts using atomic

force microscopy. *Biophysical Journal*, **86** (3), 1777–93.

63 Mahaffy, R.E., Shih, C.K., MacKintosh, F.C. and Kas, J. (2000) Scanning probe-based frequency-dependent microrheology of polymer gels and biological cells. *Physical Review Letters*, **85** (4), 880–3.

64 Mu, L. and Feng, S.S. (2003) A novel controlled release formulation for the anticancer drug paclitaxel (Taxol): PLGA nanoparticles containing vitamin E TPGS. *Journal of Controlled Release*, **86** (1), 33–48.

65 Muramatsu, H., Chiba, N., Nakajima, K., Ataka, T., Fujihira, M., Hitomi, J. and Ushiki, T. (1996) Fluorescence imaging and spectroscopy of biomaterials in air and liquid by scanning near-field optical/atomic force microscopy. *Scanning Microscopy*, **10** (4), 975–82.

66 Noll, D.M., Webba da Silva, M., Noronha, A.M., Wilds, C.J., Colvin, O.M., Gamcsik, M.P. and Miller, P.S. (2005) Structure, flexibility, and repair of two different orientations of the same alkyl interstrand DNA cross-link. *Biochemistry*, **44** (18), 6764–75.

67 Paszek, M.J., Zahir, N., Johnson, K.R., Lakins, J.N., Rozenberg, G.I., Gefen, A., Reinhart-King, C.A., Margulies, S.S., Dembo, M., Boettiger, D., Hammer, D.A. and Weaver, V.M. (2005) Tensional homeostasis and the malignant phenotype. *Cancer cell*, **8** (3), 241–54.

68 Pelling, A.E., Sehati, S., Gralla, E.B., Valentine, J.S. and Gimzewski, J.K. (2004) Local nanomechanical motion of the cell wall of *Saccharomyces cerevisiae*. *Science*, **305** (5687), 1147–50.

69 Poole, K. and Muller, D. (2005) Flexible, actin-based ridges colocalise with the beta1 integrin on the surface of melanoma cells. *British Journal of Cancer*, **92** (8), 1499–505.

70 Rabinovich, Y., Esayanur, M., Daosukho, S., Byer, K., El-Shall, H. and Khan, S. (2005) Atomic force microscopy measurement of the elastic properties of the kidney epithelial cells. *Journal of Colloid and Interface Science*, **285** (1), 125–35.

71 Radmacher, M. (1997) Measuring the elastic properties of biological samples with the AFM. *IEEE Engineering in Medicine and Biology Magazine*, **16** (2), 47–57.

72 Radmacher, M. (2002) Measuring the elastic properties of living cells by the atomic force microscope. *Methods in Cell Biology*, **68**, 67–90.

73 Sackmann, E. (1994) The Seventh Datta Lecture. Membrane bending energy concept of vesicle- and cell-shapes and shape-transitions. *FEBS Letters*, **346** (1), 3–16.

74 Sagvolden, G., Giaever, I., Pettersen, E.O. and Feder, J. (1999) Cell adhesion force microscopy. *Proceedings of the National Academy of Sciences of the United States of America*, **96** (2), 471–6.

75 Sen, S., Subramanian, S. and Discher, D.E. (2005) Indentation and adhesive probing of a cell membrane with AFM: theoretical model and experiments. *Biophysical Journal*, **89** (5), 3203–13.

76 Simson, R., Wallraff, E., Faix, J., Niewohner, J., Gerisch, G. and Sackmann, E. (1998) Membrane bending modulus and adhesion energy of wild-type and mutant cells of *Dictyostelium* lacking talin or cortexillins. *Biophysical Journal*, **74** (1), 514–22.

77 Strey, H., Peterson, M. and Sackmann, E. (1995) Measurement of erythrocyte membrane elasticity by flicker eigenmode decomposition. *Biophysical Journal*, **69** (2), 478–88.

78 Szabo, B., Selmeczi, D., Kornyei, Z., Madarasz, E. and Rozlosnik, N. (2002) Atomic force microscopy of height fluctuations of fibroblast cells. *Physical Review. E, Statistical, Nonlinear, and Soft Matter Physics*, **65** (4 Pt 1), 041910.

79 Ushiki, T., Hitomi, J., Umemoto, T., Yamamoto, S., Kanazawa, H. and Shigeno, M. (1999) Imaging of living cultured cells of an epithelial nature by atomic force microscopy. *Archives of Histology and Cytology*, **62** (1), 47–55.

80 Woodworth, C.D., Doniger, J. and DiPaolo, J.A. (1989) Immortalization of human foreskin keratinocytes by various human papillomavirus DNAs

corresponds to their association with cervical carcinoma. *Journal of Virology*, **63** (1), 159–64.

81 Iyer, S., Woodworth, C.D., Kievsky, Y.Y. and Sokolov, I. (2008) Nonspecific detection of malignant cervical cells with fluorescent silica beads (in press).

82 Iyer, S., Gaikwad, R.M., Subba-Rao, V., Woodworth, C.D. and Sokolov, I. (2008) Atomic force microscopy detects differences in the surface brush on normal and cancerous cervical cells. *Nature Nanotechnology* (in press).

83 Park, S., Koch, D., Cardenas, R., Kas, J. and Shih, C.K. (2005) Cell motility and local viscoelasticity of fibroblasts. *Biophysical Journal*, **89** (6), 4330–42.

84 Paulitschke, M., Mikita, J., Lerche, D. and Meier, W. (1991) Elastic properties of passive leukemic white blood cells. *International Journal of Microcirculation, Clinical and Experimental*, **10** (1), 67–73.

85 Sokolov, I., Iyer, S. and Woodworth, C.D. (2006) Recover of elasticity of aged human epithelial cells in-vitro. *Nanomedicine*, **2**, 31–6.

86 Berdyyeva, T.K., Woodworth, C.D. and Sokolov, I. (2005) Human epithelial cells increase their rigidity with ageing *in vitro*: direct measurements. *Physics in Medicine and Biology*, **50** (1), 81–92.

87 Drummond, C.J. and Senden, T.J. (1994) Examination of the geometry of long-range tip sample interaction in atomic-force microscopy. *Colloids and Surfaces A–Physicochemical and Engineering Aspects*, **87** (3), 217–34.

88 Li, Q.S., Lee, G. Y., Ong, C.N. and Lim, C.T. (2008) AFM identation study of breast cancer cells. *Biochemical and Biophysical Research Communications*, **374** (4), 609–13.

89 Shoelson, B., Dimitriadis, E.K., Cai, H., Kachar, B. and Chadwick, R.S. (2004) Evidence and implications of inhomogeneity in tectorial membrane elasticity. *Biophysical Journal*, **87** (4), 2768–77.

90 Rotsch, C., Braet, F., Wisse, E. and Radmacher, M. (1997) AFM imaging and elasticity measurements on living rat liver macrophages. *Cell Biology International*, **21**, 685–96.

91 Wu, H.W., Kuhn, T. and Moy, V.T. (1998) Mechanical properties of L929 cells measured by atomic force microscopy: effects of anticytoskeletal drugs and membrane crosslinking. *Scanning*, **20**, 389–97.

92 Radmacher, M. (1997) Review of AFM mechanical properties of the cells Measuring the elastic properties of biological samples with the AFM. *IEEE Engineering in Medicine and Biology Magazine*, **16**, 47–57.

93 Matzke, R., Jacobson, K. and Radmacher, M. (2001) Direct, high-resolution measurement of furrow stiffening during division of adherent cells. *Nature Cell Biology*, **3**, 607–10.

94 Vinckier, A. and Semenza, G. (1998) Measuring elasticity of biological materials by atomic force microscopy. *FEBS Letters*, **430** (1–2), 12–16.

95 Sneddon, I.N. (1965) The relation between load and penetration in the axisymmetric Boussinesq problem for a punch of arbitrary profile. *International Journal of Engineering Science*, **3**, 47–57.

96 Israelachivili, J. (1992) *Intermolecular and Surface Forces*, 2nd edn, Academic Press, San Diego, CA.

97 Butt, H.J., Kappl, M., Mueller, H., Raiteri, R., Meyer, W. and Rühe, J. (1999) Steric forces measured with the atomic force microscope at various temperatures. *Langmuir*, **15**, 2559–65.

98 Emerson, R.J.t. and Camesano, T.A. (2004) Nanoscale investigation of pathogenic microbial adhesion to a biomaterial. *Applied and Environmental Microbiology*, **70** (10), 6012–22.

99 Frantz, R., Carbonneau, C., Granier, M., Durand, J.O., Lanneau, G.F. and Corriu, R.J.P. (2002) Studies of organic-inorganic solids possessing sensitive oligoarylene-vinylene chromophore-terminated phosphonates. *Tetrahedron Letters*, **43** (37), 6569–72.

100 Leventis, N., Elder, I.A., Rolison, D.R., Anderson, M.L. and Merzbacher, C.I. (1999) Durable modification of silica aerogel monoliths with fluorescent 2,7-diazapyrenium moieties. Sensing

oxygen near the speed of open-air diffusion. *Chemistry of Materials*, **11** (10), 2837–45.
101 Baker, G.A., Pandey, S., Maziarz, E.P. and Bright, F.V. (1999) Toward tailored xerogel composites: local dipolarity and nanosecond dynamics within binary composites derived from tetraethylorthosilane and ORMOSILs, oligomers or surfactants. *Journal of Sol-Gel Science and Technology*, **15** (1), 37–48.
102 Suratwala, T., Gardlund, Z., Davidson, K., Uhlmann, D.R., Watson, J. and Peyghambarian, N. (1998) Silylated coumarin dyes in sol-gel hosts. 1. Structure and environmental factors on fluorescent properties. *Chemistry of Materials*, **10** (1), 190–8.
103 Lin, Y.S., Tsai, C.P., Huang, H.Y., Kuo, C.T., Hung, Y., Huang, D.M., Chen, Y.C. and Mou, C.Y. (2005) Well-ordered mesoporous silica nanoparticles as cell markers. *Chemistry of Materials*, **17** (18), 4570–3.
104 Yang, P., Wirnsberger, G., Huang, H.C., Cordero, S.R., McGehee, M.D., Scott, B., Deng, T., Whitesides, G.M., Chmelka, B.F., Buratto, S.K. and Stucky, G.D. (2000) Mirrorless lasing from mesostructured waveguides patterned by soft lithography. *Science*, **287** (5452), 465–8.
105 Marlow, F., McGehee, M.D., Zhao, D., Chmelka, B.F. and Stucky, G.D. (1999) Doped mesoporous silica fibers: a new laser material. *Adanced Materials*, **11** (8), 632–6.
106 Naik, S.P. and Sokolov, I. (2008) Ultra-bright fluorescent silica particles: physical entrapment of fluorescent dye rhodamine 640 in nanochannels, in *Nanoparticles: Synthesis, Stabilization, Passivation and Functionalization* (ed. R. Nagarajan), Chap. 16, Oxford University Press, Oxford.
107 Ow, H., Larson, D.R., Srivastava, M., Baird, B.A., Webb, W.W. and Wiesner, U. (2005) Bright and stable core–shell fluorescent silica nanoparticles. *Nano Letters*, **5** (1), 113–17.
108 Han, M.Y., Gao, X.H., Su, J.Z. and Nie, S. (2001) Quantum-dot-tagged microbeads for multiplexed optical coding of biomolecules. *Nature Biotechnology*, **19** (7), 631–5.

4
Chiral Molecular Imprinting as a Tool for Drug Sensing
Sharon Marx

4.1
Introduction

Chiral discrimination between enantiomers is becoming an important research field especially in the pharmaceutical industry, as well as in clinical analysis, food inspection and forensic science [1]. The molecular recognition of the target enantiomer is most often performed using natural receptors such as cyclic sugars, antibodies and macrocyclic antibiotics. Artificial receptors have become increasingly important as an alternative to the natural binding systems. The creation of recognition sites is a fundamental issue in the development of a molecular sensor. In many cases, when designing a sensor for a biologically relevant analyte (e.g., a drug, hormone, metabolite, etc.), the natural receptor for the analyte is used: an antibody, a receptor or an enzyme. For many years, there have been thousands of studies on the development of biosensors based on these biorecognition elements. However, there are several limitations to using these biologically originating molecules: the variety is limited, the thermal and physical stability are restricted, and the sensors are functional only in aqueous medium. Molecular imprinting (MI) technology presents an elegant and simple solution to these shortcomings. MI emerged originally in the 1950s as a trial to create recognition sites in silica for chromatographic separations [2, 3]. The basic idea is to create a three-dimensional (3D) cavity within a rigid matrix of a polymer during the polymerization process by using a template molecule. After the polymerization is complete, most often with high degree of cross-linking, the template is extracted from the polymer, leaving a "negative" 3D cavity complementary in shape and size to the original template. The monomers used to create the polymeric cavity usually carry one or more functional groups that can interact noncovalently with the template, to promote further binding interactions that are beneficial in the rebinding process. The noncovalent interactions that can be exploited are, for example, hydrogen bonds, electrostatic attraction, acid–base interactions, van der Vaals and metal–ligand bonds. MI is of course not limited to chiral templates, but chirality adds yet another degree of complexity to the system: the 3D cavity formed in the polymer

4 Chiral Molecular Imprinting as a Tool for Drug Sensing

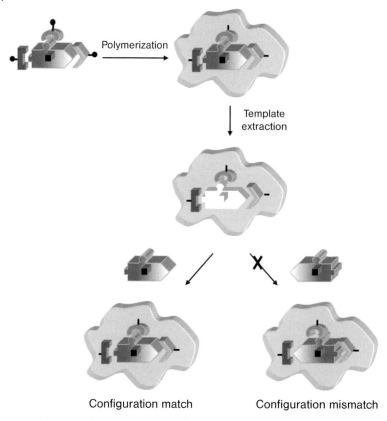

Figure 4.1 Scheme for imprinting a chiral template in a polymer.

is now required to recognize not only the right size and general shape, but also the specific handedness – the formed cavity should be able to distinguish between (R) and (S) enantiomers of the template (Figure 4.1) We would like to point out an interesting feature of the chiral imprinting: while the template used is chiral, the monomers are achiral; yet the polymerization yields a chiral polymer, at least in the vicinity of the template.

The theory and practice behind MI has been extensively reviewed in the literature [4–6], with emphasis on chromatographic applications [7]. Several papers deal with the application of MI technology as recognition elements coupled to sensor surfaces [8–11]. This application is particularly appealing, since the variety of templates that can be used is unlimited, the physical and chemical stability of the imprinted polymer is far greater than immobilized biomolecules, and even the binding affinities between the polymer and template are comparable to antigen–antibody affinities, in most cases.

The coupling between the recognition layer and the sensor surface is a key issue. In order to avoid a long diffusion time, it is desirable that a thin film of the molecu-

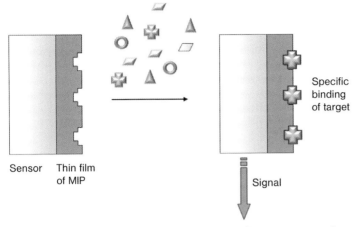

Figure 4.2 Scheme of coupling an imprinted polymer film to a sensor surface.

larly imprinted polymer (MIP) will be coupled to the transducer surface (Figure 4.2). However, in some cases that will be reviewed here, particles of the MIP were coupled to the sensor surface since technically it was impossible to fabricate a film. Coupling was achieved by combining the MIP particles in an electropolymerized film or by adding the particles to a polymer gel membrane. Preparation of thin films is very attractive for chemical and biological sensing applications, since it reduces the response time by significantly shortening the diffusional path length for the reagent [12]. The advantages of using a film are uniform diffusion profile, reduced diffusion of the analyte into and out of the film, and better coupling with the transducer in cases of optical or mass sensors, where such coupling is a necessity. The combination of thin films of imprinted polymers as the sensing element and electrodes as the transduction element was investigated by several groups [13–16]. A sensor based on a MIP coupled to a transducer requires that the analyte contains a chromophore or electrochemical activity, or to be able to generate other readable signals. It is desirable to develop label-free sensors that give a readable signal upon binding of the analyte to the MIP, thus enabling direct quantifying. Examples of such label-free sensors can be found in mass sensors, surface plasmon resonance or photonic crystals (examples of which will be demonstrated).

In this chapter, we review a specific application of the MI technique, namely the use of MI to create recognition elements for chiral drugs. The application for drug monitoring spans from clinical application, where it is desired to accurately titrate the level of a drug administered to a patient, to environmental applications, where it is desired to monitor sewage systems for drug-related pollution, or even forensic detection scenarios are envisaged. We chose to focus on the utilization of the MI techniques on chiral drugs, since the chirality adds yet another dimension to the complexity of the recognition event, and since most of the drugs are optically active and administered as a single enantiomer. The chemical structures of the drugs that were used for the studies reviewed here are portrayed in Figure 4.3.

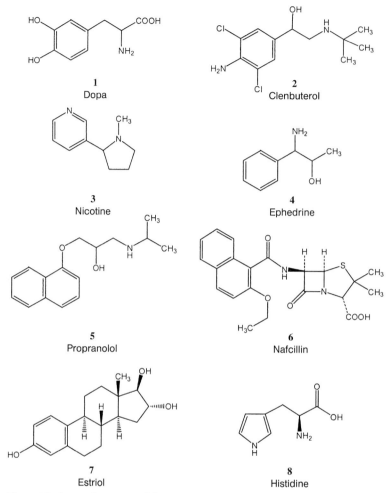

Figure 4.3 Chemical structure of the chiral drugs reviewed.

When reviewing the relevant literature we noticed an interesting common feature of many of the studies: numerous studies investigated the imprinting of a chiral drug molecule in a polymer particle or film, and characterization of specificity and selectivity was performed. However, the authors neglected to look into the obvious aspect of chiral specificity of the MIP. Most often the reason was the absence of a commercially available pure enantiomer of the drug.

The transduction systems used by us and by other groups are very diverse and range from optical (ultraviolet/visible (UV-Vis) absorption, fluorescence or phosphorescence) to electrochemical and mass sensors. In this chapter, various applications of MIP coupled to sensor devices will be discussed, divided by the different mechanisms of transduction.

It is important to note the different polymeric systems used for the preparation of the MIP recognition layers. The first papers on MIPs coupled to sensor surfaces were performed using acrylic polymers [17, 18]. These polymers portrayed excellent baseline separation characteristics when used as a chromatographic separation media for a wide variety of compounds, mainly chiral drugs. However, these polymers suffer from a major drawback regarding sensor development – it is very difficult to cast films from these polymers and most of the papers used crushed particles attached to the sensor surface. This geometry is unsatisfactory since the resulting layer is relatively thick (tens of microns), thus resulting in long diffusion time [18, 19]. Recently, several groups have developed methods to construct films of acrylic-based polymers on the sensor surface [20]. Sol-gel polymers provide a solution to this problem [21], since it is easy to cast films from these polymers and couple the films to sensor surfaces. There are several papers that used electropolymerized imprinted films as well as polyurethane [22] membranes.

4.2
Electrochemical Drug Sensors

The fabrication of an electrochemical sensor for a drug requires that the drug exhibits some degree of redox characteristics. This feature is not very common within the variety of drugs and, especially, the chiral drugs used. However, some examples exist in the literature, portraying the advantages of this method for sensor development.

Ultra-thin films, of around 70 nm, of sol-gel polymer were imprinted with D-dopa, **1**, a neurotransmitter given to Alzheimer's and Parkinson's disease patients [23]. Indium tin oxide electrodes were coated with the imprinted film and used as the working electrode in a standard three-electrode cell. Square wave voltammetry (SWV) was chosen as the electrochemical method used to probe the biding properties of the MIP film. A level of 1 nmol/l of D-dopa was measured after a 10-min incubation period in buffer. Nonspecific binding was very low (5%) and originated from the interaction of dopa with the functional surface groups included in the sol-gel mixture. Nonspecific binding was measured using nonimprinted films, which were prepared from the same sol mixture, but did not contain the template molecule. Enatioselectivity was demonstrated using both D- and L-dopa imprinted films, and measuring the cross-binding of both enantiomers (Figure 4.4). The discrimination factor was 7. The films also exhibited very good selectivity toward closely related compounds such as dopamine, dihydroxy phenyl acetic acid and catechol.

Differential pulse voltammetry was used to evaluate the binding of clenbuterol, **2**, a bronchodilator, in bovine liver samples [24] in a composite MIP–graphite electrode. Clenbuterol-imprinted polymer particles were mixed with graphite powder and used to construct a bulk electrode. The electrode was incubated in clenbuterol for 20 min to achieve a signal. The sensor could be easily regenerated by simple mechanical polishing of the surface. The limit of detection (LOD) was

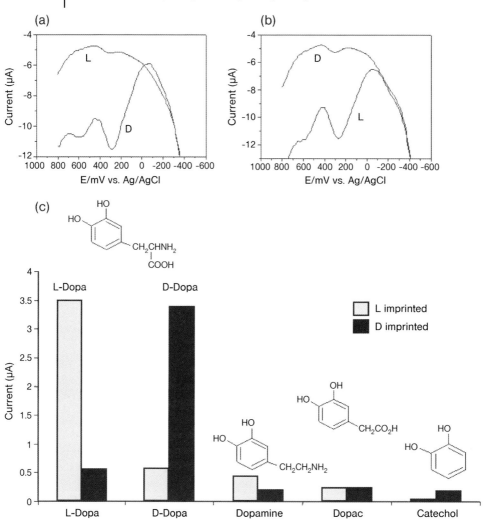

Figure 4.4 SWV of dopa-imprinted sol–gel films (thickness 70 nm): (a) imprinted with D-dopa and (b) imprinted with L-dopa. (a) SWV after rebinding of L-dopa, (b) after rebinding of L-dopa, (c) after rebinding of either enantiomer to a nonimprinted film. Template concentration 1 nmol/l after a 10-min incubation. (c) Selective uptake of D- and L-dopa, dopamine, dopac and catechol from imprinted sol–gel film ($n = 3$).

0.2 nmol/l. In this case, chiral discrimination was not evaluated, since the enantiomer is not available commercially. Selectivity towards closely related compounds, such as salbutamol or terbutaline, was demonstrated. Similarly, MIP particles imprinted for L-nicotine, **3**, were immobilized on electrode surface using poly(2-methoxy-5-(3′,7′-dimethyloctyloxy))-1,4-phenylene vinylene, a soluble polymer. The polymer was used to coat platinum electrodes and the MIP was embedded using soft lithography methods [25]. The binding of nicotine was evaluated using impedance spectroscopy, with a LOD of 10 nmol/l. High selectivity towards cotinine (a metabolite of nicotine, different in one oxygen atom) was observed, as well as low nonspecific binding. No enantioselectivity was measured.

Another example for an electrochemical sensor was recently demonstrated using electropolymerized MIP films and cyclic voltammetry for the detection of ephedrine, **4** [26]. Ephedrine is a stimulating alkaloid and its use by athletes is clinically monitored. Ephedrine imprinted MIP particles were incorporated into a poly(pyrrole) film that was grown electrochemically on a glassy carbon electrode. The film was grown to thickness of 5 μm. Ephedrine binding was detected after an incubation period of 3 h and the LOD was 0.5 mmol/l. The sensor was stable for several days of repetitive use and also after being regenerated by polishing surface of the polymer electrode. Chiral discrimination was not evaluated for this material, but selectivity was found towards common interfering compounds such as urea, ascorbic acid, glucose and sorbitol. It was found that nonspecific binding is a problem in this sensor, since ephedrine binds to the poly(pyrrole) polymer though hydrogen and acid–base bonds between the ephedrine amino group and carbonyl/carboxyl groups situated on the polymer backbone.

4.3
Optical Drug Sensors

The development of an optical sensor for a drug analyte requires the analyte to posses a definite absorption or fluorescence spectrum with relatively high absorption coefficient, or an indirect method for measurement of a signal from the sensor upon binding of the analyte. Usually, fluorescence is favored since it is more sensitive that UV-Vis absorption. In cases where the analyte molecule posses a fluorophore, it is possible to follow the fluorescence from the analyte bound to the polymer surface. An example for a popular fluorogenic analyte is the β-blocker drug. Propranolol, **5**, which is characterized with relatively high emission λ_{em} = 355 nm in aqueous solution. For example, sol-gel films were imprinted with propranolol and the binding was evaluated by the fluorescence arising from the bound analyte. Films with thickness of 700 nm bound around 1.6 nmol/l (R)- or (S)-propranolol. The chiral discrimination ration was 1.6 and nonspecific binding was about 10% [14]. In comparison, acrylic films imprinted for propranolol portrayed similar binding affinities, but were accompanied with higher nonspecific binding [21]. In the case of sol-gel propranolol imprinted films, a film that was imprinted for the (R) enantiomer bound this enantiomer preferentially to the (S) enantiomer, and the mirror image binding profile was achieved by imprinting the (S) enantiomer (Figure 4.5).

Fluorescence anisotropy is yet another fluorescent detecting scheme used to evaluate binding of propranolol to imprinted particles in solution [27]. In this study, the anisotropy in the fluorescence of the MIP-bound propranolol was measured in organic solvents. Some degree of chiral discrimination was found and it was reported that this discrimination was completely lost in aqueous solution.

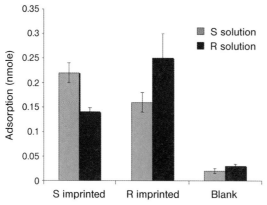

Figure 4.5 Fluorescence assay for binding of propranolol to a sol-gel imprinted film. Film thickness = 700 nm. λ_{ex} = 288 nm, λ_{em} = 335 nm. Black bars: binding of (S)-propranolol; Grey bars: binding of (R)-propranolol. n = 4.

The chiral β-lactam antibiotic drug nafcillin, **6**, is another example for optical transduction of binding to a MIP sensor. Sol-gel films were imprinted with nafcillin and the binding was evaluated using room temperature phosphorescence [28, 29]. It is claimed that room temperature phosphorescence is more selective than fluorescence in a rigid environment, similar to that present in the imprinted polymeric cage, since the nonradiational triplet state decay pathways are minimized. In order to enhance the emission of nafcillin, a heavy atom (iodide) was added to the polymeric mixture. Nonspecific binding was reported to be around 1%, but chiral discrimination was not measured, probably due to the absence of a commercial second enantiomer of the drug. Selectivity of the imprinted polymer was measured not in the film, but in a bulk polymer that was crushed and packed in a flow-through optical cell using a competitive assay. The bulk material used in the flow-injection room temperature phosphoresce assay was used to determine nafcillin in milk samples with excellent recovery. Reaction time of the film was 32 min, including incubation, drying and measurement. The films were stable for 1 month without loss of activity.

Another example for an optical sensor is the elegant use of molecularly imprinted photonic crystals for the detection of L-dopa [30]. A 3D array was constructed by polymerizing methyl methacrylate with a cross-linker and the template (L-dopa) on a surface that contained monodispersed silica nanoparticles. The MIP filled the spaces between the particles and a 2-μm film of a photonic MIP was created after removal of the particles by hydrolysis with hydrofluoric acid. The analyte, L-dopa, interacts with the acrylic acid residues of the polymer backbone and these interactions are reflected in the fluorescence spectrum of dopa. A red shift in the spectrum occurs after binding of the template to the film. No shift was observed when the film was challenged with D-dopa, thus exhibiting chiral discrimination. The LOD was 10 nmol/l. The response time of this sensor was 20 s and the change in the visible spectrum was evident to the naked eye.

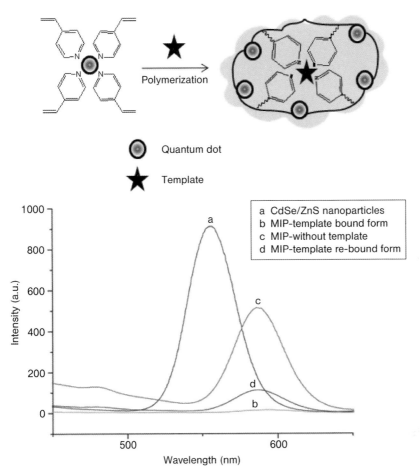

Figure 4.6 Top: Scheme of using QDs as reporting elements in a fluorescent assay of chiral templates. Bottom: Comparison of photoluminescence emission spectrum of CdSe/ZnS (a) with caffeine-imprinted MIP in the template-bound (b), free and (c) rebound forms (d). (With permission from [31]).

Another example of a sophisticated way to exploit nanoparticles in detection was demonstrated recently by the incorporation of quantum dots (QDs) in MIPs [31]. The MIPs contained CdSe core/ZnS shell particles that were functionalized with 4-vinyl pyridine. The nanoparticles were thus incorporated in the polymerization process in the vicinity of the template, Figure 4.6. Template/analyte rebinding was evaluated by the photoluminescence quenching. Several solvents were evaluated and water was found to yield the best effect, which is an advantage for biosensing. The mechanism of the sensing is quenching of the photoluminescence of the QDs by fluorescence resonance energy transfer between the analyte and the immobilized QDs. This scheme was used to detect 200 ppm of estriol, **7**, in water after 20 min, with no detectable nonspecific binding.

4.4
Mass Drug Sensors

The application of an imprinted film or particles on mass sensors is very appealing. The binding is measured as a result from the accumulation of mass on the sensor without labeling, and with relatively simple and inexpensive instrumentation. Mass sensors, such as quartz crystal microbalance (QCM) sensors, are based on gold-coated piezoelectric quartz crystal that vibrates in a specific frequency upon application of voltage. When mass is loaded on the sensor surface the frequency changes and this change can be translated to the amount of bound material using the Sauerbrey equation [32]. The first report of a chirally discriminating mass sensor was for (S)-propranolol imprinted in a film of an acrylic-based polymer film. The LOD was 50 mmol/l, for film thickness of 2 mm. The (S)-propranolol imprinted film was challenged with (R)-propranolol and only 10% of cross-enantiomer binding was observed (Figure 4.7). Nonspecific binding was around 50%. The authors suggest that the performance of the sensor can be improved by using a higher-frequency sensor and by minimizing the nonspecific binding [33].

Propranolol was also used in another study of MIP-QCM chiral sensor [20]. A very thin film, 24–30 nm, was grafted on the gold layer of the QCM sensor using an elegant technique to bind the radical initiator to the sensor surface to control film formation. In this study, propranolol was detected at submillimole per liter concentrations after less than 1 min and chiral discrimination was observed in drug concentration greater than 0.4 mmol/l, with a chiral discrimination factor of around 1.2.

Another example of chiral discrimination on a mass sensor is for histidine, **8**, which although is not a drug, has biological importance. Sol-gel films were imprinted with L-histidine and the films were applied as coating to a 9-MHz QCM

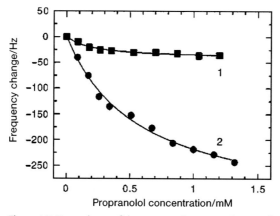

Figure 4.7 Dependence of the resonant frequency change of the QCM sensor coated with (S)-propranolol-imprinted polymer film on the concentration of propranolol: 1-(R)-propranolol, 2-(S)-propranolol. (With permission from [33]).

sensor [34]. The film thickness was estimated to be around 216 nm. A LOD of 2.5×10^{-8} mol/l was observed, with high chiral discrimination (25% binding of D-histidine to L-histidine-imprinted films). In this study, the binding was evaluated also by using the electrochemical impedance technique. This sensor exhibited very good recovery – the sensor was used for several repeated cycles of binding/extraction and performance was consistent.

4.5
Conclusions and Summary

The combination of MI and sensing has been successfully combined for sensitive and selective detection of chiral drugs. Several examples of the exploitation of the high selectivity, and especially the enatioselectivity, of the imprinted polymers were demonstrated using one enantiomer of a commercial drug as an example. The polymers were applied on sensor surfaces as films or as particles coupled to the surface using various techniques. Three general methods for signal transduction were shown: optical, electrochemical and mass sensing. Other signal transduction methods are of course applicable (e.g., surface plasmon resonance); however, no examples of utilization of these methods for chiral drug sensing were found in the literature (to the best of our knowledge) in conjunction to chiral drug sensing. In cases where both enantiomers were challenged with the sensor, excellent enantioselectivity was detected, regardless of the analytical method. LODs of the drugs were most often in the nanomole per liter range. This LOD is suitable for most clinical applications. From the reviewed examples, it is clear that the MIP–sensor combination will be able to replace biochemical assays based on antibodies for the evaluation of drugs and other biomolecularly relevant species in clinical samples with high selectivity, sensitivity and specificity.

References

1. Izake, E.I. (2007) Chiral discrimination and enantioselective analysis of drugs: an overview. *Journal of Pharmacological Science*, **96**, (7), 1659–76.
2. Dickey, F.H. (1949) The preparation of specific adsorbents. *Proceedings of the National Academy of Sciences of the United States of America*, **35**, (5), 227–9.
3. Curti, R. and Colombo, U. (1952) Chromatography of stereoisomers with "tailor made" compounds. *Journal of the American Chemical Society*, **74**, 3961.
4. Wei, S., Jakusch, M. and Mizaikoff, B. (2006) Capturing molecules with templated materials – analysis and rational design of molecularly imprinted polymers. *Anaytica Chimica Acta*, **578**, 50–8.
5. Alexander, C., Davidson, L. and Hayes, W. (2003) Imprinted polymers: artificial molecular recognition materials with applications in synthesis and catalysis. *Tetrahedron*, **59**, 2025–57.
6. Whitcombe, M.J. and Vilfson, E.N. (2001) Imprinted polymers. *Advanced Materials*, **13**, (7), 467–78.
7. Ansell, R.J. (2005) Molecularly imprinted polymers for the enantioseparation of chiral drugs.

Advances in Drug Delivery Reviews, **57**, 1809–35.
8 Haupt, K. and Mosbach, K. (1999) Molecularly imprinted polymers in chemical and biological sensing. *Biochemical Society Transaction*, **27**, 344–50.
9 Dickert, F.L. and Hayden, O. (1999) Imprinting with sensor development – on the way to synthetic antibodies. *Fresenius Journal of Analytical Chemistry*, **364**, 506–11.
10 Haupt, K. and Mosbach, K. (2000) Molecularly imprinted polymers and their use in biomimetic sensors. *Chemical Reviews*, **100**, 2495–504.
11 Arnold, B.R., Euler, A.C., Jenkins, A.L., Uy, M. and Murrey, G.M. (1999) Progress in the development of molecularly imprinted polymer sensors. *Johns Hopkins APL Technical Digest*, **20**, (2), 190–8.
12 He, J., Ichinose, I. and Kunitake, T. (2001) Imprinting of coordination geometry in ultrathin films via the surface sol-gel process. *Chemical Letters*, **30**, 850–1.
13 Gutierrez-Fernandez, S., Lobo-Castanon, M.J., Miranda-Ordieres, A.J., Tunon-Blano, P., Carriedo, G.A., Garcia-Alonso, G.J. and Fidalgo, J.I. (2001) Molecularly imprinted polyphosphazene films as recognition element in voltammetric rifamycyn SC sensor. *Electroanalysis*, **13**, (17), 1399–404.
14 Fireman-Shoresh, S., Avnir, D. and Marx, S. (2003) General method for chiral imprinting of sol gel thin films exhibiting enantioselectivity. *Chemistry of Materials*, **15**, (19), 3607–13.
15 Shustak, G., Marx, S., Turyan, I. and Mandler, D. (2003) Application of sol gel technology for electroanalytical sensing. *Electroanalysis*, **15**, (5–6), 398–408.
16 Marx, S., Zaltsman, A., Turyan, I. and Mandler, D. (2004) Parathion sensor based on molecularly imprinted sol gel films. *Analytical Chemistry*, **76**, (1), 120–6.
17 Hedborg, E., Winquist, F., Lundstrom, I., Andersson, L.I. and Mosbach, K. (1993) Some studies of molecularly imprinted polymer membranes on combination with field effect devices. *Sensors and Actuators A*, **37–38**, 796–9.
18 Kriz, D., Ramstrom, O., Svensson, A. and Mosbach, K. (1995) Introducing biomimetic sensors based on molecularly imprinted polymers as recognition elements. *Analytical Chemistry*, **67**, 2142–4.
19 Kriz, D., Kempe, M. and Mosbach, K. (1996) Introduction of molecularly imprinted polymers as recognition elements in conductometric sensors. *Sensors and Actuators B*, **33**, 178–81.
20 Piacham, T., Josell, A., Arwin, H. and Prachayasittikul, V. (2005) Molecularly imprinted polymer thin film on quartz crystal microbalance using a surface bound photo-radical initiator. *Analytica Chimica Acta*, **536**, 191–6.
21 Marx, S. and Liron, Z. (2001) Molecular imprinting in thin films of organic–inorganic hybrid sol gel and acrylic polymers. *Chemistry of Materials*, **13**, 3624–30.
22 Dickert, F.L., Besenbock, H. and Tortschanoff, M. (1998) Molecular imprinting through van der vaals interactions: fluorescence detection of PAHs in water. *Advanced Materials*, **10**, (2), 149–51.
23 Fireman-Shoresh, S., Turyan, I., Mandler, D., Avnir, D. and Marx, S. (2005) Chiral electrochemical recognition by very thin molecularly imprinted sol-gel films. *Langmuir*, **21**, 7842–7.
24 Pizzariello, A., Stred'ansky, M., Stred'ansky, S. and Miertus, S. (2001) A solid binding matrix/molecularly imprinted polymer based sensor system for the determination of clenbuterol in bovine liver using differential-pulse voltammetry. *Sensors and Actuators B*, **76**, 286–94.
25 Thoelen, R., Vansweevelt, R., Duchateau, J., Horemans, F., D'Haen, J., Lusten, L., Vanderzande, D., Ameloot, M., vandeVen, M., Cleij, T.J. and Wagner, P. (2008) A MIP-based impedimetric sensor for the detection of low-MW molecules. *Biosensors and Bioelectronics*, **23**, (23), 913–18.

26 Mazzota, E., Picca, R.A., Malitesta, C., Piletsky, S.A. and Piletska, E.V. (2008) Development of a sensor prepared by entrapment of MIP particles in electrosynthesized polymer films for electrochemical detection of ephedrine. *Biosensors and Bioelectronics*, **23**, 1152–6.

27 Hunt, C.E. and Ansell, R.J. (2006) Use of fluorescence shift and fluorescence anisotropy to evaluate the rebinding of template to (S)-propranolol imprinted polymer. *Analyst*, **131**, 678–83.

28 Fernandez-Gonzalez, A., Badia Laino, R., Diaz-Garcia, M.E. and Viale, A. (2004) Assessment of molecularly imprinted sol gel materials for selective room temperature phosphorescence recognition of nafcillin. *Journal of Chromatography B*, **804**, 247–54.

29 Guardia, L., Badia, R. and Diaz-Garcia, M.E. (2006) Molecular imprinted ormosils for nafcillin recognition by room temperature phosphorescence optosensing. *Biosensors and Bioelectronics*, **21**, 1822–9.

30 Hu, X., An, Q., Li, G., Tao, S. and Liu, J. (2006) Imprinted photonic polymers for chiral recognition. *Angewandte Chemie (International Edition in English)*, **45**, 8145–8.

31 Lin, C.I., Joseph, A.K., Chang, C.K. and Lee, Y.D. (2004) Molecularly imprinted polymeric film on semiconductor nanoparticles analyte detection by quantum dot photoluminescence. *Journal of Chromatography A*, **1027**, 259–62.

32 Sauerbrey, G. (1959) The use of quartz crystal oscillators for weighing thin layers and for micro-weighing. *Zeitschrift für Physik*, **155**, 206–22.

33 Haupt, K., Nowortya, K. and Kunter, W. (1999) Imprinted polymer-based enantioselective acoustic sensor using a quartz crystal microbalance. *Analytical Communications*, **36**, 391–3.

34 Zhang, Z., Liao, H., Li, H., Nie, L. and Yao, S. (2005) Stereoselective histidine senor based on molecularly imprinted sol gel film. *Analytical Biochemistry*, **336**, 108–16.

5
Catalytic Antibodies for Selective Cancer Chemotherapy
Roy Weinstain and Doron Shabat

5.1
Introduction

Over two decades ago, in 1986, the first literature reports of catalytic antibodies sparked excitement and imagination among the world scientific community [1, 2]. It was realized that by an appropriate design, monoclonal antibodies could be generated against specific small molecules with a stable structure of the transition state analog of a chemical reaction. These antibodies had the ability to catalyze that chemical reaction similarly to enzymes. The technological and intellectual advance of this field has come into realization with the development of the first commercial catalytic antibody (antibody 38C2). This aldolase antibody is being sold by Aldrich as a standard chemical reagent and was the first commercial protein offered for sale by Aldrich.

The intriguing concept of using catalytic antibodies as therapeutic agents became even more appealing when it was shown that most of the amino acids in a mouse antibody molecule could be replaced with human sequences and thereby make it compatible for *in vivo* treatment in humans [3]. Furthermore, it is a molecule synthesized by a highly evolved biological system that was naturally designed for *in vivo* activity applications. Since catalytic antibodies are kinds of artificial enzymes, they can be designed to catalyze specific chemical reactions that are not catalyzed by natural enzymes. Therefore, they can be used for selective activities, avoiding specific catalytic competition by natural proteins. The major part of the therapeutic potential of catalytic antibodies lies within the concept of selective prodrug activation for antitumor therapy [4–8].

5.2
Catalytic Antibodies Designed for Prodrug Activation

Prolonged administration of effective concentrations of chemotherapeutic agents is usually not possible because dose-limiting systemic toxicities and strong side-effects involving nonmalignant tissues are observed. Thus, new strategies to

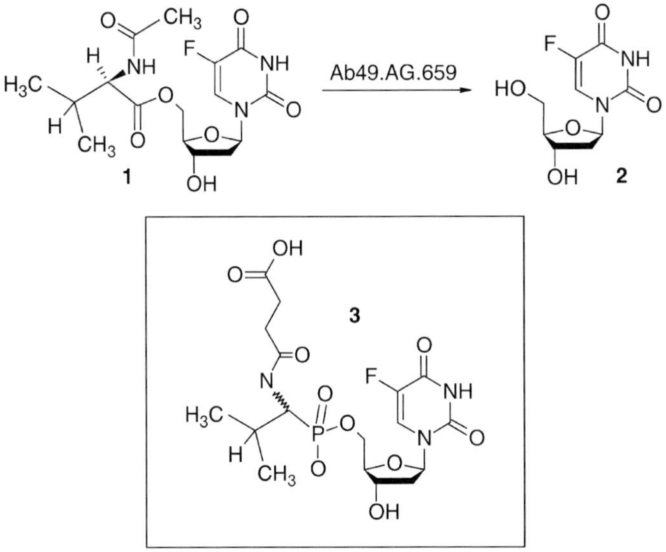

Figure 5.1 Monoclonal antibody 49.AG.659 elicited against phosphonate hapten **3** catalyzes the hydrolysis of ester **1** prodrug to 5-fluorodeoxyuridine **2**.

target cytotoxic agents specifically to sites of metastatic or solid cancer are required. One such targeting approach involves selective enzymatic activation of a nontoxic prodrug to a toxic drug. The enzyme is either directed to the tumor by a targeting device (i.e., specific monoclonal antibody) or selectively secreted by the tumor cells. This strategy has been exploited within the application of prodrug activation by a catalytic antibody.

In 1994, Schultz et al. [9] utilized phosphonate **3** to elicit antibody 49.AG.659.12 that catalyzes the hydrolysis of ester prodrug **1** to release 5-fluorodeoxyuridine **2**, a known anticancer drug that inhibits thymidylate synthetase (Figure 5.1). The rate enhancement of the catalyzed reaction was 968-fold over the uncatalyzed reaction. The free drug completely inhibits the growth of *Escherichia coli* HB101 at concentration of 20 µM, whereas prodrug **1** does not affect the growth of bacteria at a concentration of 400 µM. Incubation of prodrug **1** (400 µM) with antibody 49.AG.659.12 (20 µM) led to complete inhibition of *E. coli* HB101 growth, showing the effective activation of the prodrug in this system.

The next progress in prodrug activation by antibody catalysis was reported by Blackburn et al. in 1996 [10]. They used the nitrogen mustard anticancer drug **5** to generate a carbamate prodrug **4**, which could be hydrolyzed to form the parent drug and glutamic acid. Phosphoamidate **6** was used as a stable transition state analog for the purpose of immunization (Figure 5.2). One monoclonal antibody, EA11-D7, was found to efficiently catalyze the hydrolysis of prodrug **4** and was selected for growth inhibition studies. Incubation of the antibody (1 µM) and prodrug **4** (100 µM) led to significant *in vitro* cell kill of human colorectal

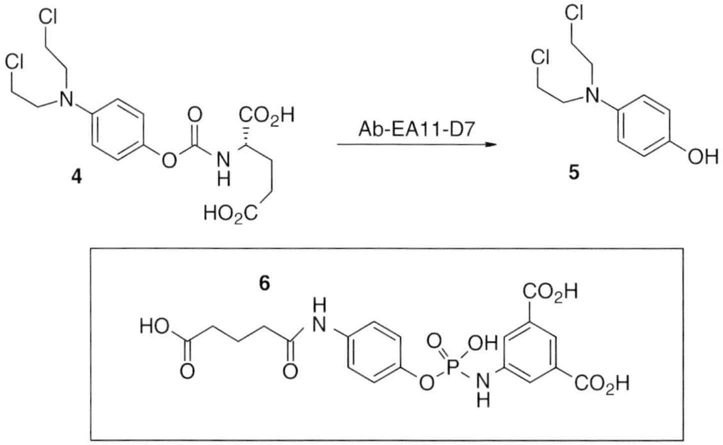

Figure 5.2 Monoclonal antibody EA11-D7 elicited against phosphonamidate hapten **6** catalyzes the hydrolysis of carbamate **4** prodrug to release drug **5**.

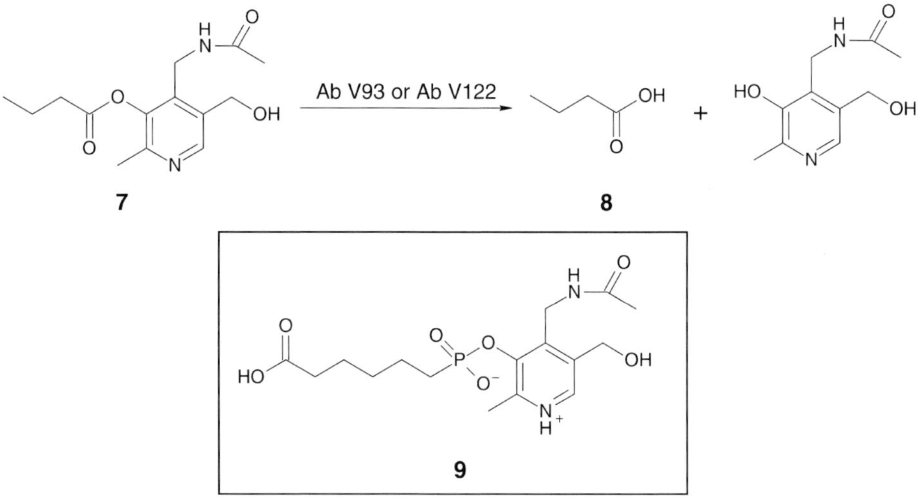

Figure 5.3 Monoclonal antibodies V93 and V122 elicited against hapten **9** catalyze the hydrolysis of prodrug **7** to release drug **8**.

tumor cell line (LoVo), whereas incubation of prodrug **4** alone resulted in no toxicity.

In 2002, Fujii et al. elicited antibodies V93 and V122, which were raised against haptene **9** [7]. The antibodies catalyzed the hydrolysis of ester prodrug **7** to release butyric acid **8** (Figure 5.3). In this study, the authors showed the activation of several prodrugs based on B_6 ester derivatives. These esters were shown to be very stable against hydrolytic enzymes in serum. Upon incubation with either antibody V93 or V122, the hydrolysis of the B$_6$ derivative ester was observed, leading to the release of the appropriate drug with a carboxylic acid moiety. Growth inhibition assay using HeLa cell line and prodrug **7** has shown that in the presence of antibody, the activity was more then 5 times the activity without the antibody.

5.3
Catalytic Antibody 38C2 and Cancer Therapy

5.3.1
General Approach for Prodrug Activation with Antibody 38C2

Barbas and Lerner achieved a major breakthrough with the development of a new immunization concept. Instead of immunizing against transition state analogs, they immunize with a compound that is highly reactive in order to create a chemical reaction during the binding of the antigen to the antibody. The same reaction becomes part of the mechanism of the catalytic event. In other words, the antibodies are elicited against a chemical reaction instead of a transition state analog. This strategy was termed reactive immunization [11]. The 1,3-diketone **10** was used as a trap for an amino lysine residue in the antibody active site (Figure 5.4). Two antibodies, 38C2 and 33F12, that contained the desired lysine were found to mimic type I aldolases very efficiently.

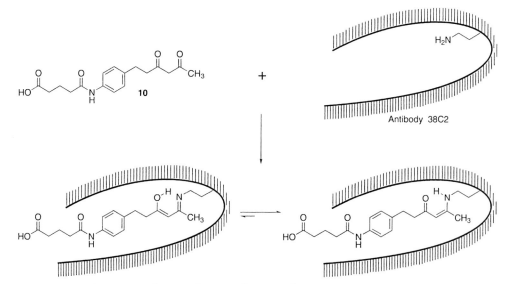

Figure 5.4 Mechanism of trapping the essential ε-amino group of a lysine residue in the antibody's binding pocket by using the 1,3-diketone hapten **10**.

Figure 5.5 Prodrug activation via a tandem *retro*-aldol-*retro*-Michael reaction. X = heteroatoms N, O or S.

Antibody 38C2 has a major advantage regarding the approach of prodrug activation. Since it has the capability to accept a broad variety of substrates, the antibody may potentially activate any prodrug with the specific trigger on it. A general prodrug masking chemistry was developed, and was designed to take advantage of the broad scope and mechanism of catalytic antibody 38C2. The drug masking/activation concept was based on a sequential *retro*-aldol-*retro*-Michael reaction catalyzed by antibody 38C2 (Figure 5.5) [12].

This reaction sequence is not known to be catalyzed by any other enzyme and has a very low background (i.e., the reaction is very slow in the absence of the catalyst). This chemistry was first demonstrated in 1999 with the anticancer drugs doxorubicin and camptothecin [12]. The doxorubicin prodrug **11** (Figure 5.6) was constructed by masking the amine functionality with the *retro*-aldol-*retro*-Michael

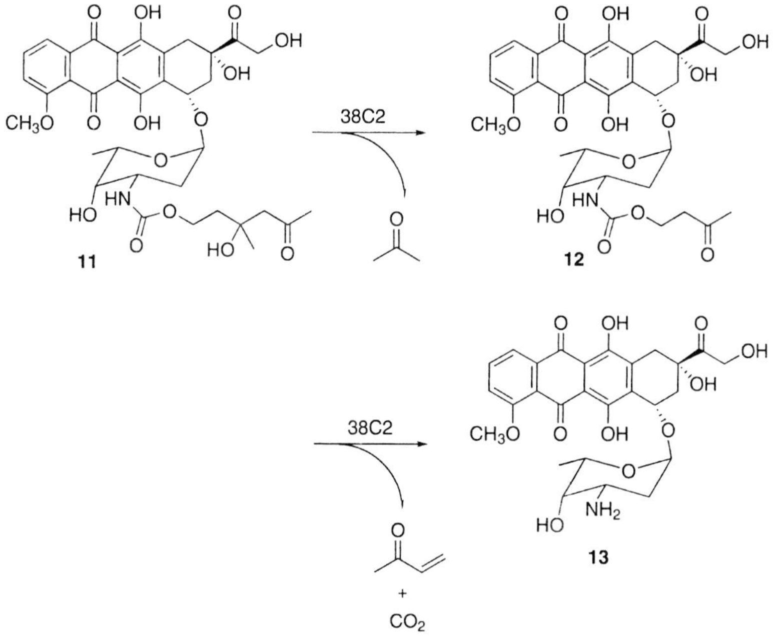

Figure 5.6 Antibody 38C2 catalyzes the activation of doxorubicin prodrug **11** via the tandem *retro*-aldol-*retro*-Michael reaction to release doxorubicin **13** (through intermediate **12**).

linker. The rate enhancement of the doxorubicin prodrug **11** activation reaction was more than 10^5-fold over the uncatalyzed reaction.

It was shown that weakly or nontoxic concentrations of the corresponding prodrug **11** can be activated by therapeutically relevant concentrations of antibody 38C2 to kill colon and prostate cancer cell lines. To further test the therapeutic relevance of this model system, it was shown that antibody 38C2 remained catalytically active over weeks after intravenous injection into mice.

5.3.2
Bifunctional Antibodies for Targeted Chemotherapy

The development of strategies that provide selective chemotherapy presents significant multidisciplinary challenges. Selective chemotherapy might, in the case of cancer, be based on the enzymatic activation of a prodrug at the tumor site. The enzymatic activity must be directed to the site with a targeting molecule, usually an antibody, that recognizes a cell surface molecule selectively expressed at the tumor site. Since a single molecule of enzyme catalyzes the activation of many molecules of prodrug, a localized and high drug concentration may be maintained at the tumor site. This concept of antibody-directed enzyme prodrug therapy (ADEPT) holds promise as a general and selective chemotherapeutic strategy if several specific criteria can be met [13, 14].

A number of antigens that are expressed on the surface of tumor cells or in their supporting vasculature have been shown to be effective targets for antibody-mediated cancer therapy. Thus, for the most part, the targeting antibody component of this strategy is not limiting. By contrast, the requirements of the enzyme component and complementary prodrug chemistries for ADEPT are difficult to achieve. First of all, selective prodrug activation requires the catalysis of a reaction that must not be accomplished by endogenous enzymes in the blood or normal tissues of the patient. Enzymes of nonhuman origin that meet these needs are, however, likely to be highly immunogenic – a fact that makes repeated administration impossible. Finally, the chemistry used to convert a drug into a prodrug should be versatile enough to allow for the modification of many drug classes while not interfering with the operation of the enzyme so that a single enzyme could be used for the activation of a multiplicity of prodrugs.

The limitations of the ADEPT complex encourage scientists to suggest that a catalytic antibody might replace the enzyme component for ADEPT. The potential of catalytic antibodies for ADEPT is indeed compelling. The catalysis of reactions that are not catalyzed by human enzymes and minimal immunogenicity through antibody humanization is feasible. Combining these features, the ADEPT conjugate translates into a bispecific antibody consisting of a targeting arm and catalytic arm.

The general concept of bispecific antibodies is illustrated in Figure 5.7. Two parent antibodies are combined in a manner such that each antibody contributes one light chain and one heavy chain. The bispecific construct contains two different binding regions. One originates from the targeting antibody and can bind

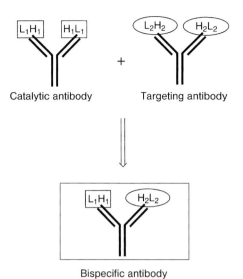

Figure 5.7 Generation of bispecific antibody hybrid originating from a corresponding catalytic and a targeting antibody.

specifically to antigens that are expressed specifically on tumor cells, and the other originates from the catalytic antibody and is used to activate a prodrug.

Figure 5.8 illustrates how such bispecific antibody could be used for selective chemotherapy. This possible future treatment can consist of two steps. First, a dose of the bispecific antibody can be administered and the targeting arm can locate and attach to specific antigens on tumor cell surface. Excess of the antibody can be cleared after a limited time from the patient's blood, therefore preventing nonspecific prodrug activation. In the second step, several doses of prodrug can

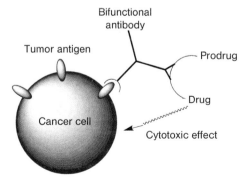

Figure 5.8 Schematic illustration of targeting chemotherapy selectively to a tumor cell. The prodrug is transformed to the active drug by the catalytic arm of the bifunctional antibody. The other arm binds to an antigen on a tumor cell.

be administered on suitable time gaps. The prodrug can reach the tumor site through the blood circulation and can be activated by the catalytic arm at a very close proximity to the tumor cell. The damage to noncancer cells should be therefore minimized and the free drug can specifically target cancer cells.

5.3.3
In Vitro and *In Vivo* Evaluations of Antibody 38C2-Catalyzed Prodrug Activation

In order to test this approach, Barbas *et al.* performed and reported in 2001 the first *in vivo* study of prodrug activation by antibody catalysis [15]. A new etoposide (VP16) prodrug **14** was designed and synthesized for the purpose of this study. The phenol functionality was masked by the *retro*-aldol-*retro*-Michael trigger, which was attached to N,N-dimethyl-ethylenediamine – a self-immolative extension. The prodrug activation was achieved after cleavage of the trigger followed by spontaneous cyclization to release the free etoposide drug **15**, as shown in Figure 5.9.

To evaluate the efficacy of antibody 38C2-mediated etoposide prodrug activation, the activity of the prodrug was tested *in vitro* in the presence or absence of 1 µM antibody 38C2 with cultured murine NXS2 neuroblastoma cells. Interestingly, the prodrug revealed a clearly reduced potential for growth inhibition; greater than 100-fold (Figure 5.10a). Activation of the prodrug with antibody 38C2 resulted in growth profiles similar to the unmodified etoposide control.

The efficacy of localized catalytic antibody-mediated etoposide prodrug activation against syngeneic murine NXS2 neuroblastoma was evaluated on established primary tumors (Figure 5.10b). Primary tumors were established in the lateral flank of animals and studies were initiated when an average tumor size of 100 mm [3] was obtained. Each experimental group consisted of eight animals. Catalytic antibody 38C2 was delivered locally by intratumoral injection. Prodrug or etoposide was delivered systemically by intraperitoneal injection at three time

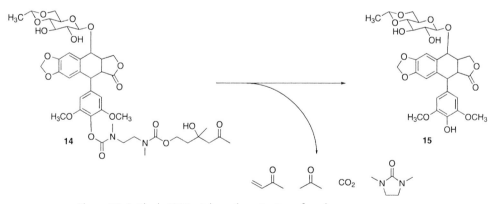

Figure 5.9 Antibody 38C2 catalyzes the activation of prodrug **14** via the tandem *retro*-aldol-*retro*-Michael reaction and a spontaneous cyclization to release free etoposide **15**.

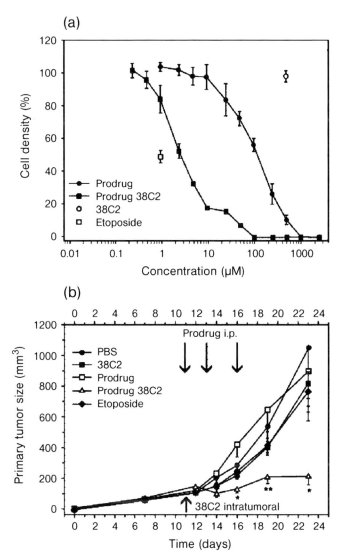

Figure 5.10 (a) Growth inhibition activity of etoposide prodrug **14** in the presence and absence of catalytic antibody 38C2. The growth response of NXS2 neuroblastoma cells to a 72-h incubation with increasing concentrations of etoposide prodrug **14** in the presence and absence of 1 mM 38C2 catalytic antibody was analyzed by using a standard lactate dehydrogenase release assay. A serial dilution of etoposide and wells containing 1 mM 38C2 only are shown as controls.
(b) Effect of catalytic antibody 38C2-mediated etoposide prodrug **14** activation on the growth of primary neuroblastoma tumors. Treatment of mice bearing established subcutaneous primary neuroblastoma tumors induced by subcutaneous injection with 2×10^6 NXS2 neuroblastoma cells was initiated 11 days after tumor cell inoculation. The treatment consisted of intratumoral injection of 38C2 (0.5 mg) followed by three intraperitoneal etoposide prodrug **14** (total dose 1250 mg/kg) injections. Arrows indicate days of treatment with each reagent, respectively. One control group received unmodified etoposide (total dose 40 mg/kg).

points. Two major findings were obtained in this experiment. First, an increased antitumor efficacy of locally activated etoposide prodrug over systemically applied unmodified etoposide was observed at the maximum tolerated dose of etoposide. In fact, a dramatic 75% reduction in subcutaneous tumor growth was observed only in the group of mice receiving both intratumoral injections of catalytic antibody 38C2 and systemic treatments with etoposide prodrug. This is in contrast to control groups receiving each agent as monotherapy or injections with phosphate-buffered saline. As all animals that received both catalytic antibody 38C2 and prodrug survived the 24-day experiment, they were treated at day 25 with a second cycle of catalytic antibody 38C2 and prodrug therapy. After this, three of eight mice revealed the complete absence of a primary tumor, underlining the efficacy of the principle of catalytic antibody-mediated prodrug activation *in vivo*. Furthermore, etoposide prodrug demonstrated dramatically reduced toxicity *in vivo* as compared with unmodified etoposide itself. In fact, no toxicity was observed at the dosage administered to the animals (1250 mg/kg), as defined by the absence of any decrease in body weight in contrast to a 20% weight loss observed in mice treated with the maximal tolerated dose of etoposide (40 mg/kg).

5.3.4
Polymer Directed Enzyme Prodrug Therapy: An Approach to Target Antibody 38C2 to a Tumor Site

The study above has shown that prodrug activation by antibody catalysis is feasible *in vivo*. Next, it was necessary to attach targeting devices to the catalytic antibody in order to direct it selectively to the tumor site. Our group reported the first targeting device conjugated to a catalytic antibody. We have used a water-soluble synthetic polymer, N-(2-hydroxypropyl)methacrylamide (HPMA), which is biocompatible, nonimmunogenic and nontoxic. Moreover, its *in vivo* body distribution is well characterized and it is known to accumulate selectively at tumor sites due to the enhanced permeability and retention (EPR) effect [16]. This effect occurs due to the difference between the vasculature physiology of solid tumors and normal tissues. Compared with the regular ordered vasculature of normal tissues, blood vessels in tumors are often highly abnormal. The growth of the tumor creates a constant need for the continuous supply of new blood vessels. This process, termed angiogenesis, often results in the construction of vessels with leaky walls, which allows enhanced permeability of macromolecules within the tumor. In addition, poor lymphatic drainage at the tumor site promotes accumulation of large molecules.

Similarly to the previous described ADEPT system, polymer molecules can be used as passive targeting devices instead of specific monoclonal antibodies. polymer-directed enzyme prodrug therapy (PDEPT) is a two-step antitumor approach in which both the prodrug and the enzyme are targeted to the tumor site with a polymer molecule (Figure 5.11) [17, 18]. In the first step, a polymer–pro-

Step I. Administration of polymer-drug

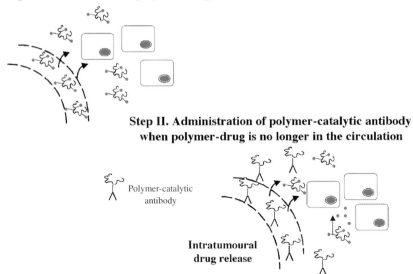

Step II. Administration of polymer-catalytic antibody when polymer-drug is no longer in the circulation

Figure 5.11 Schematic representation of PDEPT.

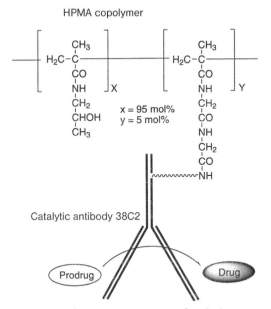

Figure 5.12 Schematic representation of antibody 38C2–HPMA copolymer conjugate, designed for selective prodrug activation by the PDEPT approach.

drug conjugate is administered and trapped in tumor tissues through the EPR effect. The excess of the conjugate is cleared out from the blood in a relatively short time. In the second step, a polymer–enzyme conjugate is injected. The polymer molecule carries the enzyme to the tumor site, where it releases the drug from the polymer.

We have prepared a conjugate of catalytic antibody with the targeting moiety HPMA copolymer, based on amide bond formation between an external lysine of the antibody and an active ester of the HPMA copolymer (Figure 5.12) [19]. The conjugation yield was very high and the antibody retained most of its catalytic activity in the conjugate. Furthermore, we have shown that the antibody–polymer conjugate can activate an etoposide prodrug *in vitro* and consequently inhibit proliferation of two different cancer cell lines (MOLT-3 T cell leukemia and NXS2 neuroblastoma).

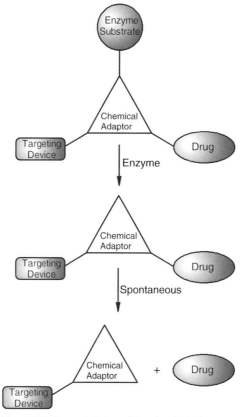

Figure 5.13 General design of the chemical adaptor system. Cleavage of the enzyme substrate generates an intermediate that spontaneously rearranges to release the drug from the targeting device.

5.3.5
Chemical Adaptor Concept

In order to improve the drug delivery selectivity, our group has developed a new concept that combines a tumor-targeting device, a prodrug and a prodrug activation trigger in a single entity (Figure 5.13) [20–22].

We designed a generic module chemical adaptor that is based on three chemical functionalities as shown in Figure 5.14. The first functionality is attached to an active drug and, thereby, masks it to yield a prodrug, the second is linked to a targeting moiety, which is responsible for guiding the prodrug to the tumor site, and the third is attached to an enzyme substrate. When the corresponding enzyme cleaves the substrate, it triggers a spontaneous reaction that releases the active drug from the targeting moiety. As a result, prodrug activation will preferentially occur at the tumor site.

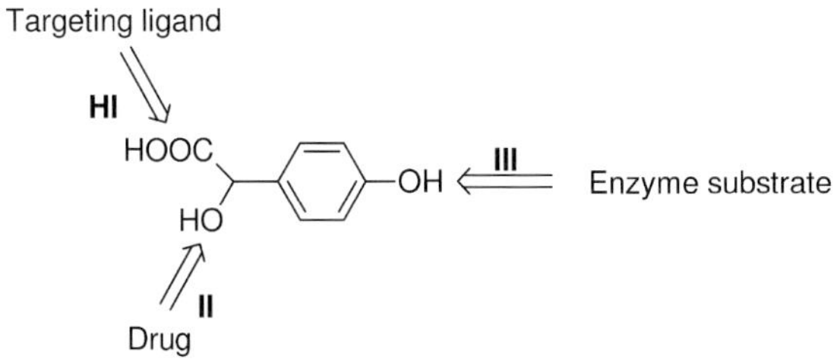

Figure 5.14 4-Hydroxy-mandelic acid – the central core of the chemical adaptor system.

The design of the generic module allows us to potentially link any targeting device to a variety of drugs and to release them with any enzyme by using the corresponding substrate as a trigger. As proof of concept, a pilot system was designed for which catalytic antibody 38C2 was chosen as the cleaving agent, a polymer molecule as the targeting device and etoposide as the drug. Next, it was tested whether the etoposide drug can be released from complex **16** by the catalytic activity of antibody 38C2. According to the design, the drug should be spontaneously released after the generation of phenol **17** as illustrated in Figure 5.15. Complex **16** was incubated with catalytic antibody 38C2 in phosphate-buffered saline (pH 7.4) at 37 °C and the appearance of etoposide was monitored using a high-performance liquid chromatography assay. As a positive control, a previously described etoposide prodrug **14** that is activated by antibody 38C2 was used. As Figure 5.15 shows, etoposide was released by the catalytic activity of antibody 38C2 to form compound **18** and the free drug. The rate of drug release was similar to the activation rate of the known etoposide prodrug. No spontaneous etoposide release was observed in the absence of the antibody.

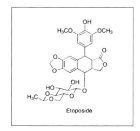

Figure 5.15 Mechanism of etoposide drug release from the HPMA copolymer, using catalytic antibody 38C2 as the triggering enzyme.

5.3.6
Self-Immolative Dendrimers Concept

Since the amount of the targeted catalytic antibody at the tumor site is limited, we recently developed unique drug delivery platform that can amplify cleavage signals into a multiple drug release [23, 24]. The platform is based on a self-immolative dendritic prodrug system that could be activated by 38C2 under physiological conditions. It was designed to conduct a cleavage signal through its structure, allowing the release of the reporter units upon a single cleavage of the trigger. By applying anticancer drug molecules as the reporter units and an enzyme substrate as the trigger, a dendritic multi-prodrug was obtained. We synthesized dendritic prodrugs based on AB_2 and AB_3 units that can generate double- or triple-drug doses upon a single cleavage event (Figure 5.16) [25, 26].

A trimeric prodrug system was synthesized by linking three molecules of the anticancer drug camptothecin (CPT) (pro-tCPT) to a *retro*-aldol *retro*-Michael trigger as a substrate for catalytic antibody 38C2. In addition, we prepared a monomeric CPT prodrug (pro-mCPT) with an identical linker (Figure 5.17a). Both prodrugs were activated upon incubation with antibody 38C2 and the CPT release was confirmed by high-performance liquid chromatography analysis.

Next, we examined whether the trimeric prodrug system had an advantage over the monomeric one in a cell growth inhibition assay. We evaluated the ability of the prodrugs to inhibit cell proliferation in the presence of catalytic antibody 38C2 using three different cell lines: the human T-lineage acute lymphoblastic leukemia cell line MOLT-3, the human erythroleukemia cell line HEL and the human acute myeloid leukemia cell line HL-60. The data from the MOLT-3 cell line are presented in Figure 5.17b. The trimeric prodrug is more potent than the monomeric one when incubated with the antibody, as expected, since the total amount of CPT release is tripled in comparison with the release from an equivalent concentration of the monomeric prodrug.

In the trimeric system, one cleavage by the antibody releases 3 times the amount of CPT than a cleavage in the monomeric prodrug system. We selected one cell line (MOLT-3 leukemia) for further studies with fixed concentrations of the prodrug and varying concentrations of antibody 38C2. In order to have equal amounts of CPT, the monomeric prodrug concentration was 3 times that of the

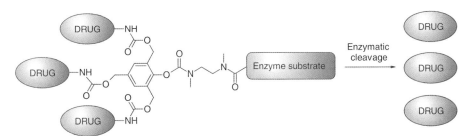

Figure 5.16 General design of a single triggered trimeric prodrug (AB_3).

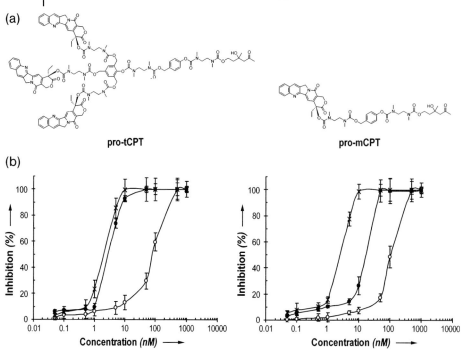

Figure 5.17 (a) Molecular structures of a single triggered CPT trimeric prodrug versus CPT classic monomeric prodrug with identical trigger. (b) Growth inhibition assay of human MOLT-3 leukemia cell line, in the presence and absence of catalytic antibody 38C2. Left: ○ pro-tCPT; ● pro-tCPT + 38C2; ×, CPT. Right: ○ pro-mCPT; ● pro-mCPT + 38C2; ×, CPT.

trimeric one. The results are shown in Figure 5.18. The trimeric prodrug inhibited cell growth up to 3 times more effectively than the monomeric one in the range of 15–150 nM antibody. In other words, the antibody concentration needed to achieve 50% cell growth inhibition with the pro-tCPT is about 3 times less than that used in the pro-mCPT system. It should also be noted that the cytotoxicity of the platform degradation products was previously evaluated in cell growth inhibition. It was found to have negligible or no toxicity at all within the drug concentration range of the cell assay.

5.3.7
Prodrugs of Dynemicin and Doxorubicin Analogs

In 2004, Sinha *et al.* reported the activation of dynemicin prodrug analogs for selective chemotherapy by 38C2 [27]. The proposed mechanism for activation begins with a *retro*-aldol *retro*-Michael reaction sequence of prodrug **19** to produce amine

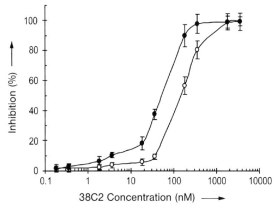

Figure 5.18 Growth inhibition assay of the human MOLT-3 leukemia cell line with a fixed concentration of prodrugs and varying concentration of catalytic antibody 38C2. Cells were incubated for 72 h: ○ 36 nM pro-mCPT; ● 12 nM pro-tCPT.

21. The lone pair on the nitrogen atom triggers the opening of the epoxide function in intermediate **22**. The product **23** then undergoes Bergman cyclization to generate the phenyl diradical **24**. The diradical is thought to react with double-strand DNA and cleave it (Figure 5.19). To study the efficiency of the prodrugs, an *in vitro* cell growth inhibition assay using the human colon carcinoma cell line LIM1215 was conducted. Two prodrugs of dynemicin [one that can be activated by 38C2 (**21a**) and one that cannot (**26**)] were tested in the presence and absence of 38C2. While prodrug **26** showed no activity difference in cell proliferation inhibition in the presence or absence of the antibody, a significant proliferation inhibition was observed with rising prodrug (**21a**) concentrations in presence of 38C2.

Barbas *et al.* showed in 2004 the activation of unique doxorubicin analogs prodrugs [28]. Aldolase Ab93F3 catalyzes a *retro*-aldol reaction of the β-hydroxyethyl ketone function in **27a** and **27b**. The aldehyde intermediate **28** formed by the *retro*-aldol reaction undergoes an intramolecular cyclization to afford the corresponding cabinolamine derivative and then dehydrates to produce the imminium intermediate **29**. These imminium intermediates are highly reactive towards nucleophilic attacks. Such an attack by the 2-amino group of a guanine residue of a DNA strand leads to the formation of a mono-DNA adduct. A further hydrolysis of the glycoside in the DNA adduct followed by subsequent reaction with another DNA molecule forms DNA adducts **30** (Figure 5.20).

A cell-growth inhibition study of prodrugs **27a** and **27b** has shown that the prodrugs were 20–30 times more toxic in the presence of the antibody then in its absence. These doxorubicin analogs (**27a** and **27b**) were shown to successfully inhibit the proliferation of human breast cancer cells (MDA-MB-435) and Kaposi's sarcoma cells (SLK) in the presence of the antibody.

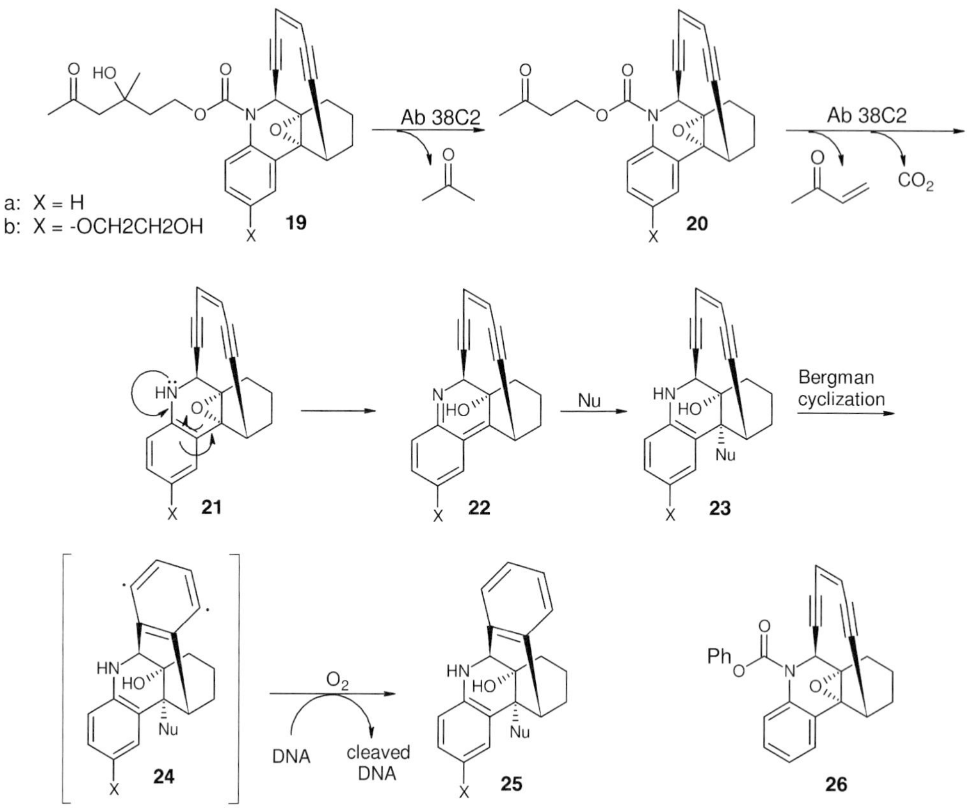

a: X = H
b: X = -OCH2CH2OH

Figure 5.19 Suggested activation pathway of dynemicin analogs **a** and **b** by 38C2.

5.4
Chemically Programmed Antibodies

A novel targeting device, that combines the merits of small synthetic molecules with that of antibodies was reported in 2003 by Barbas et al. [29]. The device is based on the formation of a reversible covalent bond between a diketone derivative of a targeting small synthetic molecule and the reactive lysine of 38C2 (Figure 5.21). Formation of such a linkage provides a complex with joint features: it joins together the unlimited structural diversity of small synthetic molecules and their targeting ability with the antibody's long serum half-life and effector function of the Fc region. In addition, the linkage of a targeting synthetic molecule to the antibody's binding site, essentially allows chemically programming of the binding site. Such an antibody was termed a chemically programmed antibody (cpAb). 38C2 was linked to a 1,3-diketone derivative of integrin $\alpha_v\beta_3$ and $\alpha_v\beta_5$ targeting the Arg–Gly–Asp peptidomimetic. The complex (termed SCS-873) was shown to: (i) form spontaneously *in vitro* and *in vivo*, (ii) selectively target 38C2 to the surface of cells expressing $\alpha_v\beta_3$ and $\alpha_v\beta_5$ integrins, (iii) increase the circulatory half-life of

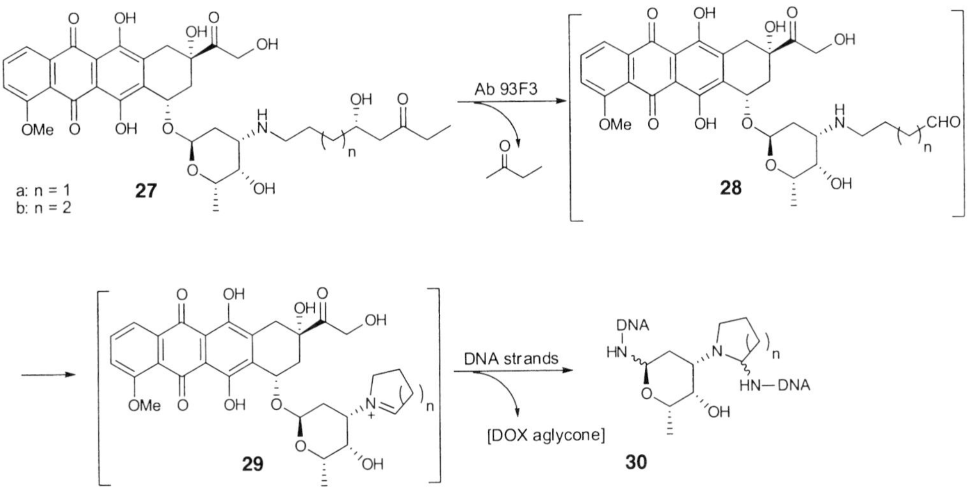

Figure 5.20 Proposed mechanism for doxorubicin analogs **27a** and **27b** activation and cytotoxicity. Antibody 93F3 catalyzes a *retro*-aldol reaction on doxorubicin analogs **27** forming intermediate aldehyde **28**, which undergoes an intramolecular cyclization to form imminium **29**. A nucleophilic attack of DNA strands lead to DNA adduct formation **30**.

Arg–Gly–Asp peptidomimetic, and (iv) effectively reduce tumor growth in animal models of human Kaposi's sarcoma and colon cancer.

A tumor growth inhibition study was conducted in nude mice using xenografts from human Kaposi's sarcoma cell line SLK, expressing $\alpha_v\beta_3$ and $\alpha_v\beta_5$ integrins, and human colon cancer cell line SW1222, that is not expressing the integrins. In both cases, treatment with SCS-873 complex (**32**) revealed a significant decrease in tumor growth compared with small molecule alone or 38C2 alone.

In 2006, Sinha et al. elaborated their targeting device cpAb into a novel construct [30]. In the new device, the diketone function on the targeting molecule was replaced with a proadapter function. In this way, the antibody catalyzes its own adapter formation from the less reactive proadapter. The proadapter inertness allows the programming agent to present chemical functions that are reactive to diketone. In the first step, 38C2 catalyzes a *retro*-aldol reaction of the proadapter **33**, to revel a reactive vinyl ketone **34** that is subsequently covalently attached to the antibody through a Michael addition reaction **35** (Figure 5.22).

To test their new approach, they have synthesized an integrin-targeted cpAb 38C2 **38**. 38C2 was added to an integrin binding agent, carrying the proadapter unit **36**. The antibody catalyzes a *retro*-aldol reaction to release the reactive vinyl ketone **37**, which subsequently is attached with the antibody through a Michael addition reaction to form integrin-targeted cpAb 38C2 **38** (Figure 5.23).

In vivo breast cancer metastasis prevention was studied in a mouse model. Three groups of six immunodeficient SCID mice were intravenously injected with

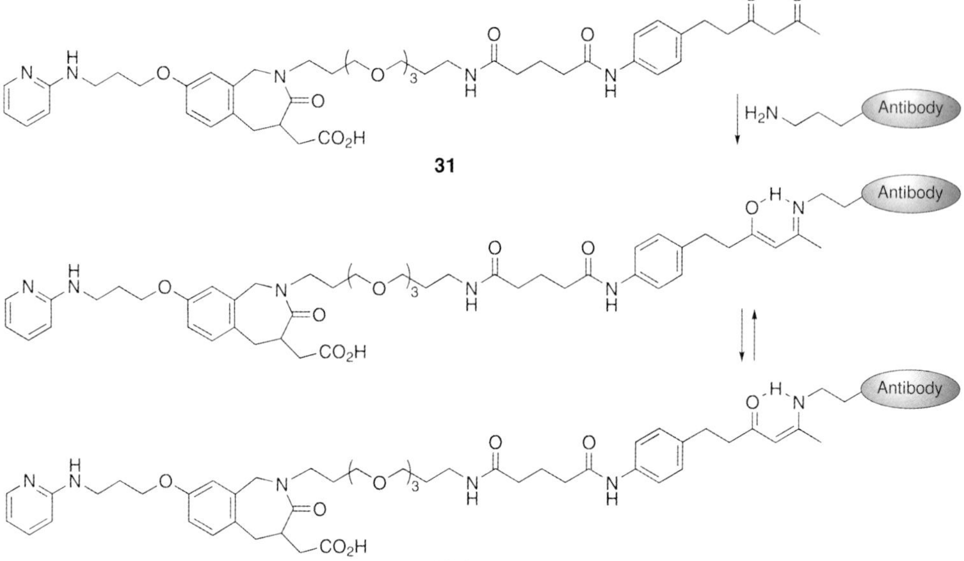

Figure 5.21 A reversible covalent bond formation between a 1,3-diketone derivative of an integrin $\alpha_v\beta_3$ and $\alpha_v\beta_5$ targeting Arg–Gly–Asp peptidomimetic (**31**) and the reactive lysine of 38C2, forming an antibody–small molecule complex (**32** SCS-873).

1×10^6 MDA-MB-231 cells pretreated with integrin-targeted cpAb 38C2, integrin-targeting small molecule and 38C2, and were continuously treated with the same compounds. Results showed that animals treated with integrin-targeted cpAb 38C2 had significantly less metastatic foci then the ones treated with either integrin-targeting small molecule or 38C2 alone [31].

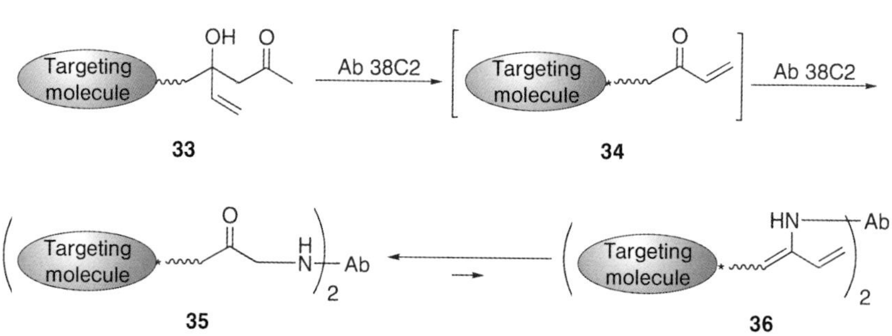

Figure 5.22 Schematic diagram showing cell-targeting antibody complex formation. First, 38C2 catalyzes a *retro*-aldol reaction on the proadapter **33**, unveiling the adapter as a reactive vinyl ketone **34**, which then reacts with the antibody's lysine to form the complex **35**.

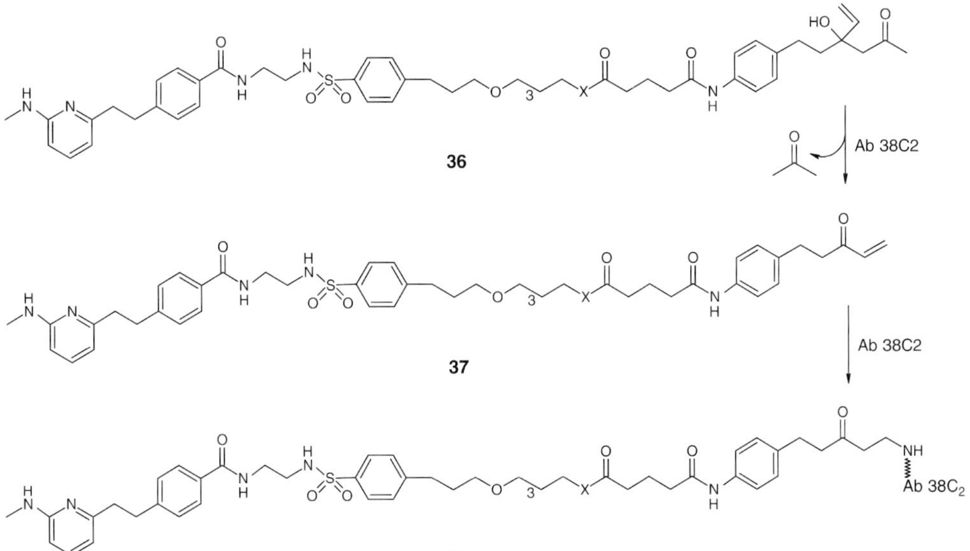

Figure 5.23 Formation of a cpmAb. α$_v$β$_3$ integrin binding agent, carrying a spacer and a proadapter unit **36**, undergoes a *retro*-aldol catalysis by 38C2, unveiling a reactive vinyl ketone **37** that is subsequently covalently attached to the antibody through a Michael addition reaction to form integrin-targeted cpAb 38C2 **38**.

Later that year, Doppalapudi *et al.* expanded their technology of chemically programmed antibodies by synthesizing a conjugate of endothelin-A (ET$_A$) antagonist and 38C2 [32]. The conjugate was termed CovX-Body. The ET$_A$ antagonist applied as a targeting prototype was the sulfonamide Edonentan (**39**), developed by Bristol-Myers Squibb, and recently was used in phase III clinical studies for hypertension. A β-diketone, incorporated into the extended ET$_A$ antagonist **40**, was reacted with the lysine in the binding site of 38C2 forming a reversible, covalent enaminone bond in a well-controlled fashion (Figure 5.24).

The ET$_A$ receptor binding affinity of CovX-Body was measured *in vitro* using ^{125}I-labeled ET-1 and CHO cells that express human ET$_A$ receptors. Compound **40** showed an affinity for ET$_A$ similar to that of the parent compound **39** (IC$_{50}$ 1.6 versus 1.0 nM). CovX-Body, with an IC$_{50}$ of 8.6 nM, showed a marginal loss in the ET$_A$ binding affinity. In order to evaluate *in vivo* antitumor activity of CovX-Body, animal studies were conducted in a nude mouse/human xenograft model. In a prostate cancer xenograft (PC-3) study, a 10 mg/kg once weekly dose of the CovX-Body inhibited the tumor volume growth by greater than 45% by day 31, compared to the vehicle control. In comparison, a 2 mg/kg daily dose of compound **39** showed 50% tumor growth inhibition by day 31. The individual components of CovX-Body, 38C2 alone (20 mg/kg/once a week) and compound **40** alone (0.25 mg/kg/once a week) did not inhibit the tumor growth.

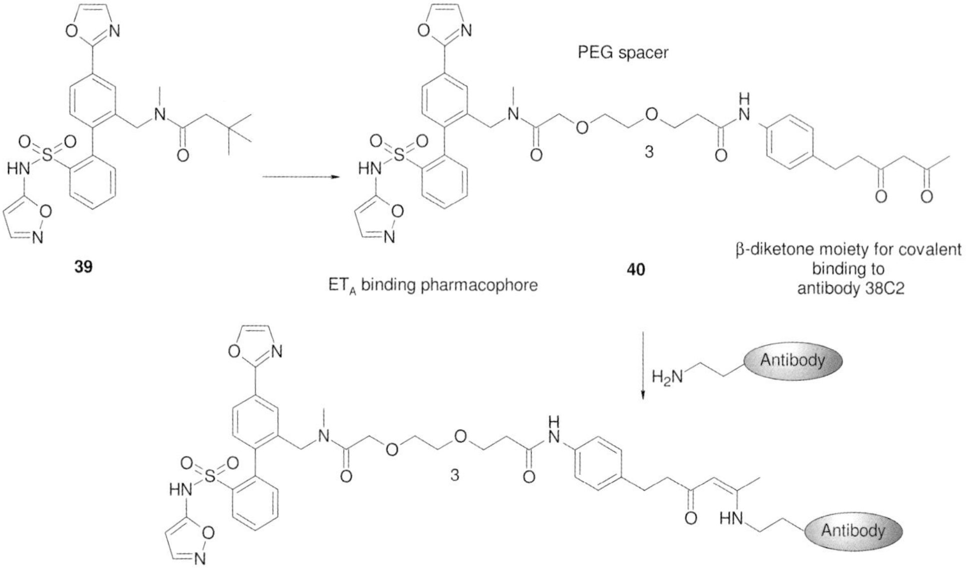

Figure 5.24 Design of β-diketone containing, ET$_A$-binding pharmacophore.

In 2007, Sinha et al. reported the generation of a catalytic antibody that targets tumor cells with undiminished prodrug activation capability [33]. In a previous work, conjugation of the targeting small molecule with 38C2 was accomplished by reaction of a β-diketone moiety, incorporated into the targeting molecule, and the active site lysine in the antibody. This has allowed the chemical programming of the antibodies, but diminished their catalytic capabilities. In the current work, conjugation of integrin $\alpha_v\beta_3$-targeting antagonists **41** or **42** to 38C2 was made through a cysteine thiol or a lysine amine residue located outside the binding site, thereby leaving the active site available for prodrug activation (Figure 5.25).

Flow cytometry studies were conducted in order to evaluate the binding efficiencies of conjugates 38C2-41 and 38C2-42 to integrin $\alpha_v\beta_3$-expressing human breast cancer cell line MDA-MB-231. The results showed that conjugates 38C2-41 and 38C2-42 bind to the integrin $\alpha_v\beta_3$-expressing cells at the same efficiency as the previously described SCS-873 complex (**32**), while 38C2 alone did not show any binding to these cells. Catalytic activity tests comparing the conjugates to free 38C2 showed that the conjugates retained around 50% catalytic activity with respect to the free 38C2. To complete their work, Sinha et al. have also developed a new set of doxorubicin prodrugs, and evaluated their toxicity and activation by 38C2. Two prodrugs were selected for further evaluation (prodrugs **43** and **44**, Figure 5.26).

To determine the efficiency for cell killing of the antibody conjugates that contained the targeting moiety, a cell growth inhibition study, comparing 38C2 with the conjugates 38C2-41 and 38C2-42, was conducted. Results showed 85–90% growth inhibition of MDA-MB-231 human breast cancer cells, with both 38C2

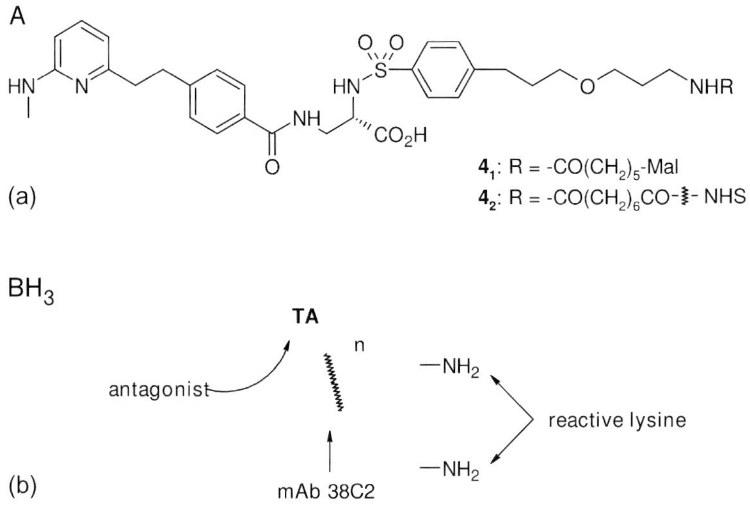

Figure 5.25 Catalytic antibody conjugates. (a) Structure of small-molecule Arg–Gly–Asp peptidomimetic antagonists equipped with a linker. (b) Schematic drawing of a catalytic cell-targeting antagonist–38C2 conjugate.

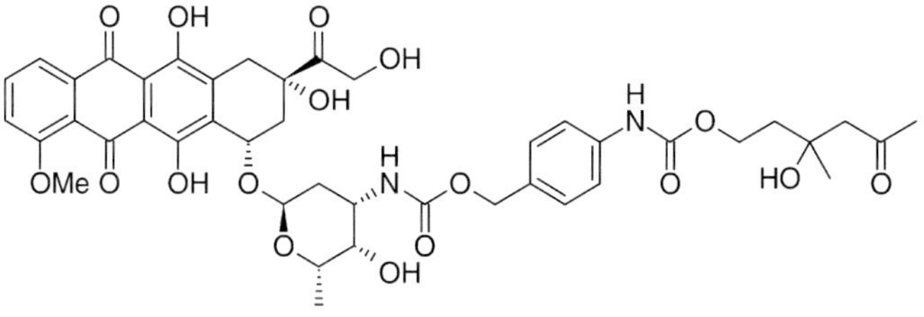

4_1: R = -CO(CH$_2$)$_5$-Mal
4_2: R = -CO(CH$_2$)$_6$CO-⅔- NHS

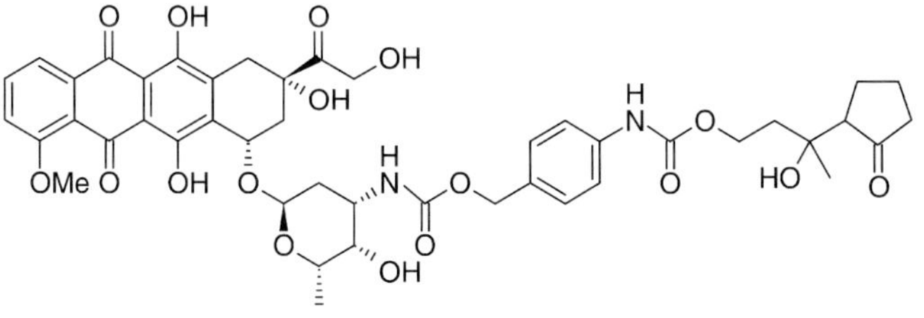

Figure 5.26 Chemical structures of doxorubicin prodrugs.

or its conjugates, when prodox 44 was used. The use of 38C2 conjugates together with the doxorubicin prodrugs revealed that cell-targeting and prodrug activation capabilities could be efficiently combined.

5.5
Outlook

Catalytic antibodies were generated by the immune system and adjusted for *in vivo* activity by nature. Most of the applied research studies performed so far with catalytic antibodies have dealt with the function of prodrug activation. The concept of a nontoxic prodrug, which is transformed to a toxic drug selectively at the tumor site, is a promising approach for chemotherapy treatment. Catalytic antibodies have proven themselves to be efficient reagents for the function of prodrug activation. However, in order to make this approach practical, the catalytic antibody needs to be targeted to the tumor site. The first step towards targeting was already reported when an HPMA copolymer was conjugated with catalytic antibody 38C2. The goal of targeting an antibody molecule to a tumor with polymers may be achieved, similarly as enzymes are targeted in the PDEPT approach. The recent approach of Barbas and Lerner to chemically program the antibody's binding site may open new methods for tumor treatment with catalytic antibodies. Finally, it should be noted that in a relative young research field like this, a new breakthrough can always appear and strike new exciting progress towards a practical medicinal application.

References

1 Pollack, S.J., Jacobs, J.W. and Schultz, P.G. (1986) Selective chemical catalysis by an antibody. *Science*, **234**, 1570–3.
2 Tramontano, A., Janda, K.D. and Lerner, R.A. (1986) Catalytic antibodies. *Science*, **234**, 1566–70.
3 Gussow, D. and Seemann, G. (1991) Humanization of monoclonal antibodies. *Methods in Enzymology*, **203**, 99–121.
4 Amir, R. and Shabat, D. (eds) (2005) *Catalytic Antibodies*, Wiley-VCH, Weinheim.
5 Blackburn, G.M., Datta, A. and Patridge, L.J. (1996) The medical potential of catalytic antibodies. *Pure and Applied Chemistry*, **68**, 2009–16.
6 Jones, L.H. and Wentworth, P., Jr. (2001) The therapeutic potential for catalytic antibodies: from a concept to a promise. *Mini-Reviews in Medicinal Chemistry*, **1**, 125–32.
7 Kakinuma, H., Fujii, I. and Nishi, Y. (2002) Selective chemotherapeutic strategies using catalytic antibodies: a common promoiety for antibody-directed abzyme prodrug therapy. *Journal of Immunological Methods*, **269** (1–2), 269–81.
8 Nishi, Y. (2003) Enzyme/Abzyme prodrug activation systems: potential use in clinical oncology. *Current Pharmaceutical Design*, **9**, 2113–30.
9 Campbell, D.A., Gong, B., Kochersperger, L.M., Yonkovich, S., Gallop, M.A. and Schultz, P.G. (1994) Antibody-catalyzed prodrug activation. *Journal of the American Chemical Society*, **116**, 2165–6.
10 Wentworth, P., Datta, A., Blakey, D., Boyle, T., Partridge, L.J. and Blackburn, G.M. (1996) Toward antibody-directed "abzyme" prodrug therapy, ADAPT: carbamate prodrug activation by a catalytic antibody and its *in vitro* application to human tumor cell killing. *Proceedings of the National Academy of Sciences of the United States of America*, **93**, (2), 799–803.
11 Wagner, J., Lerner, R.A. and Barbas, C.F., III. (1995) Efficient aldolase

catalytic antibodies that use the enamine mechanism of natural enzymes. *Science*, **270**, 1797–800.

12 Shabat, D., Rader, C., List, B., Lerner, R.A. and Barbas, C.F., III. (1999) Multiple event activation of a generic prodrug trigger by antibody catalysis. *Proceedings of the National Academy of Sciences of the United States of America*, **96**, (12), 6925–30.

13 Bagshawe, K.D. (1990) Antibody-directed enzyme/prodrug therapy. *Biochemical Society Transactions*, **18**, (5), 750–2.

14 Bagshawe, K.D., Sharma, S.K., Springer, C.J. and Rogers, G.T. (1994) Antibody-directed enzyme/prodrug therapy(ADEPT). A review of some theoretical, experimental and clinical aspects. *Annals of Oncology*, **5**, (10), 879–91.

15 Shabat, D., Lode, H.N., Pertl, U., Reisfeld, R.A., Rader, C., Lerner, R.A. and Barbas, C.F., III. (2001) In vivo activity in a catalytic antibody–prodrug system: antibody catalyzed etoposide prodrug activation for selective chemotherapy. *Proceedings of the National Academy of Sciences of the United States of America*, **98**, (13), 7528–33.

16 Maeda, H., Wu, J., Sawa, T., Matsumura, Y. and Hori, K. (2000) Tumor vascular permeability and the EPR effect in macromolecular therapeutics: a review. *Journal of Controlled Release*, **65**, 271–84.

17 Duncan, R., Gac-Breton, S., Keane, R., Musila, R., Sat, Y.N., Satchi, R. and Searle, F. (2001) Polymer–drug conjugates, PDEPT and PELT: basic principles for design and transfer from the laboratory to clinic. *Journal of Controlled Release*, **74**, 135–46.

18 Satchi, R., Connors, T.A. and Duncan, R. (2001) PDEPT: polymer-directed enzyme prodrug therapy. *British Journal of Cancer*, **85**, 1070–6.

19 Satchi-Fainaro, R., Wrasidlo, W., Lode, H.N. and Shabat, D. (2002) Synthesis and characterization of a catalytic antibody–HPMA copolymer-conjugate as a tool for tumor selective prodrug activation. *Bioorganic and Medicinal Chemistry*, **10**, 3023–9.

20 Gopin, A., Pessah, N., Shamis, M., Rader, C. and Shabat, D. (2003) A chemical adaptor system designed to link a tumor-targeting device with a prodrug and an enzymatic trigger, *Angewandte Chemie (International Edition in English)*, **42**, 327–32.

21 Gopin, A., Rader, C. and Shabat, D. (2004) New chemical adaptor unit designed to release a drug from a tumor targeting device by enzymatic triggering. *Bioorganic and Medicinal Chemistry*, **12**, (8), 1853–8.

22 Shabat, D., Amir, R.J., Gopin, A., Pessah, N. and Shamis, M. (2004) Chemical adaptor systems. *Chemistry – A European Journal*, **10**, 2626–34.

23 Amir, R.J., Pessah, N., Shamis, M. and Shabat, D. (2003) Self-immolative dendrimers, *Angewandte Chemie (International Edition in English)*, **42**, 4494–9.

24 Shabat, D. (2006) Self-immolative dendrimers as novel drug delivery platforms. *Journal of Polymer Science, Part A: Polymer Chemistry*, **44**, 1569–78.

25 Haba, K., Popkov, M., Shamis, M., Lerner, R.A., Barbas, C.F., III and Shabat, D. (2005) Single-triggered trimeric prodrugs, *Angewandte Chemie (International Edition in English)*, **44**, 716–20.

26 Shamis, M., Lode, H.N. and Shabat, D. (2004) Bioactivation of self-immolative dendritic prodrugs by catalytic antibody 38C2. *Journal of the American Chemical Society*, **126**, 1726–31.

27 Sinha, S.C., Li, L.S., Miller, G.P., Dutta, S., Rader, C. and Lerner, R.A. (2004) Prodrugs of dynemicin analogs for selective chemotherapy mediated by an aldolase catalytic Ab. *Proceedings of the National Academy of Sciences of the United States of America*, **101**, 3095–9.

28 Sinha, S.C., Li, L.S., Watanabe, S., Kaltgrad, E., Tanaka, F., Rader, C., Lerner, R.A. and Barbas, C.F., III. (2004) Aldolase antibody activation of prodrugs of potent aldehyde-containing cytotoxics for selective chemotherapy. *Chemistry – A European Journal*, **10**, 5467–72.

29 Rader, C., Sinha, S.C., Popkov, M., Lerner, R.A. and Barbas, C.F., III (2003)

Chemically programmed monoclonal antibodies for cancer therapy: adaptor immunotherapy based on a covalent antibody catalyst. *Proceedings of the National Academy of Sciences of the United States of America*, **100**, 5396–400.

30 Guo, F., Das, S., Mueller, B.M., Barbas, C.F., III, Lerner, R.A. and Sinha, S.C. (2006) Breaking the one antibody–one target axiom. *Proceedings of the National Academy of Sciences of the United States of America*, **103**, 11009–14.

31 Popkov, M., Rader, C., Gonzalez, B., Sinha, S.C. and Barbas, C.F., III (2006) Small molecule drug activity in melanoma models may be dramatically enhanced with an antibody effector. *International Journal of Cancer*, **119**, 1194–207.

32 Doppalapudi, V.R., Tryder, N., Li, L., Aja, T., Griffith, D., Liao, F.F., Roxas, G., Ramprasad, M.P., Bradshaw, C. and Barbas, C.F., III (2007) Chemically programmed antibodies: endothelin receptor targeting CovX-Bodies. *Bioorganic and Medicinal Chemistry Letters*, **17**, (2), 501–6.

33 Abraham, S., Guo, F., Li, L.S., Rader, C., Liu, C., Barbas, C.F., III, Lerner, R.A. and Sinha, S.C. (2007) Synthesis of the next-generation therapeutic antibodies that combine cell targeting and antibody-catalyzed prodrug activation. *Proceedings of the National Academy of Sciences of the United States of America*, **104**, (13), 5584–9.

6
Natural and Synthetic Stimulators of the Immune Response

Marine C. Raman and Dominic J. Campopiano

6.1
Introduction

Mammals have evolved sophisticated early-warning, self-defense systems to be ready to defend themselves against incoming pathogenic organisms. Once a pathogen has been detected, a vast battery of weaponry can be sent out to tackle the invader. This natural innate immunity is required to sense a range of pathogen-specific signals and set off an immunological response cascade that will lead to controlled killing and mediated cell death. Controlled inflammation is required and delicate balances must be maintained, since unnecessary stimulation and inflammation is highly dangerous and detrimental to the mammalian host. A range of stimuli are recognized by the innate immune system and in this chapter we concentrate on two that provide fascinating fields of study for microbiologists, immunologists, clinicians and chemists alike. Detailed molecular knowledge of "natural" stimulators and their protein receptors has begun to impact on the synthesis of chemical tools with highly desirable immunomodulatory properties which could have great clinical use.

6.2
Lipopolysaccharide Endotoxin – A Potent Immunostimulatory Molecule

Bacterial membranes are composed of lipids and proteins. The selective recognition of bacterial versus mammalian membrane components (nonself and self) is an important function of the immune system. Specific bacterial components can induce acute inflammatory responses via a number of sensitive cascades. One important molecule that has fascinated scientists for many years is lipopolysaccharide (LPS), also known as "endotoxin". An excellent review of the story of the discovery, characterization and biological analysis of this molecule can be found in [1]. Endotoxin is a highly abundant proportion of the outer membrane of the Gram-negative bacterial cell wall and can also be shed by growing as well as dying

Cellular and Biomolecular Recognition: Synthetic and Non-Biological Molecules. Edited by Raz Jelinek
Copyright © 2009 WILEY-VCH Verlag GmbH & Co. KGaA, Weinheim
ISBN: 978-3-527-32265-7

cells. The discovery and characterization of endotoxin dates back over 100 years. At the end of the 1800s, Richard Pfeiffer showed that cholera bacteria that had been killed by heat retained their toxic potential. This experiment proved that the poison was not a classical protein toxin and led him to conceive the idea that *Vibrio cholerae* harbored a heat-stable toxic substance that was associated with the insoluble part of the bacterial cell. The active substance was named endotoxin (taken from the Greek *endo* meaning "within"). Pfeiffer postulated that these endotoxins were fundamental constituents of bacterial species. Chemical characterization of the endotoxins involved contributions from many researchers who devised methods to extract, purify and analyze this complex natural product (Osborn, Nikaido, Lüderitz, Westphal, Boivin and Rietschel).

Detailed chemical analysis ultimately revealed LPS to be organized into three parts – lipid A, core and O-antigen. There are around 10^6 lipid A residues and 10^7 glycerophospholipids per *Escherichia coli* cell [2]. The LPS from *E. coli* has been studied extensively to provide a model endotoxin architecture for comparative studies with those from other species. The lipid A (Kdo_2-lipid A) from a deep rough *E. coli* strain is shown in Figure 6.1 (compound **1**).

The lipid A is an usual glucosamine-based β1,6-linked disaccharide with long-chain fatty acids attached to the 2 and 2′ positions via amides. The 3 and 3′ positions are also derivatized as fatty acid esters. Monophosphate esters are found at the 1 and 4′ positions, and can be the site for further functionalization (see below). The 6′ position links the glucosamine disaccharide to an unusual 3-deoxy-D-manno-octulosonic acid (Kdo) disaccharide. To this Kdo is the site of attachment of heptose sugars that form part of the core region. Finally, the characteristic, bacterial-specific oligosaccharide O-specific chain ("O-antigen") is attached to this core.

A seminal breakthrough in the field of endotoxin research came in the 1980s when Kusumoto and Shiba synthesized *E. coli* lipid A, thereby confirming the structure of the isolated natural product. More importantly, this synthetic lipid A had identical endotoxic activity against mammalian cells as its bacterial counterpart [3]. Thereafter began a series of studies to probe the structure–functional relationships that explored the nature of the inflammatory potential of this important bacterial toxin [4]. A large number of synthetic lipid A derivatives were prepared, but none were as active as the parent molecule.

6.3
LPS Recognition

Various proteins have been sequentially discovered that generate a circuit that connects LPS recognition through cell surface receptor binding and intracellular signaling [5]. LPS-binding protein (LBP) is an approximately 60-kDa serum binding protein that belongs to the lipid transfer family. It is the LBP that binds the LPS in the blood and delivers it to the CD14 receptor that is present on many cell types, including macrophages and dendritic cells [6]. CD14 is a glycosyl phosphatidylinositol-anchored, high-affinity membrane protein that can also circulate in a soluble form. Another protein involved in passing this signal onto cell-surface receptors is myeloid differentiation-2 (MD-2) (see Sections 6.9.2 and 6.9.3). It is thought that CD14 concentrates LPS for binding to a Toll-like receptor (TLR) TLR4–MD-2 complex. How this complex recognizes lipid A and signals across the plasma membrane is still not completely understood [7]. Toll is a *Drosophila* fruit-fly receptor essential for the production of antimicrobial peptides in response to fungal pathogens that is also essential for fly development. This link in flies led to the isolation of a human homolog then the discovery of a family of mammalian transmembrane, Toll-like pattern recognition receptors. The Toll receptor has the characteristic modular structure of a type I transmembrane receptor: an ectodomain containing blocks of leucine-rich repeats (LRRs) and a conserved intracellular region, the Toll/interleukin (IL)-1 receptor domain, that is required for downstream signal transduction. In 1998, the direct link between LPS and the TLR4 receptor was made by Beutler *et al.* [who had previously isolated tumor necrosis factor (TNF)-α] [8] and the expanding TLR family has been intensively studied over the past decade [7–12]. Specific TLRs have been shown to interact with RNA, DNA and many other ligands. The TLR4 pathway activates the key transcriptional regulator NF-κB, resulting in the production of proinflammatory cytokines and the progression to adaptive immunity. A complicated series of intracellular proteins are then expressed including MyD88, although a MyD88-independent pathway has also been discovered. The complex gene expression networks are highly regulated, and are now the target of various anti-inflammatory and autoimmune disease drugs. The details of LPS binding and signaling, and discussions of the TLR family have been the subject of a number of recent excellent reviews [5, 9, 12–15].

6.4
Septic Shock

Sepsis is a systemic inflammatory response syndrome triggered by an infection and is the major cause of death in patients in intensive care units [16]. In severe cases it can lead to organ failure or dysfunction, and severe sepsis with hypotension unresponsive to fluid resuscitation defines septic shock. Sepsis syndrome results from the body's systemic inflammatory response to any of several infectious stimuli. Endotoxins, such as LPS from Gram-negative bacteria and other antigens from infectious agents (e.g., DNA, peptidoglycan and flagellan, lipotechoic acid, mannan) stimulate macrophages and monocytes to release TNF-α, resulting in a cascade of cytokine release [17]. Following the release of TNF-α, other proinflammatory cytokines, including IL-1 and IL-6, are released into the circulation that, in turn, trigger numerous additional proinflammatory events within endothelial cells and leukocytes. Although these proinflammatory responses are vital to the host defense against infection, TNF-α acts in conjunction with IL-1 to produce the clinical signs of the systemic inflammatory response syndrome, and their synergistic effects are probably responsible for the hypotension and resultant organ dysfunction seen early in the course of severe sepsis. Sepsis syndrome afflicts almost 750 000 patients in the United States each year, costing almost $17 billion and causing 210000 deaths annually. Despite increasing research and accrued knowledge about the pathophysiological pathways and processes involved in sepsis, morbidity and mortality remain unacceptably high. A large number of natural and synthetic immunomodulatory agents have been studied in experimental and clinical settings in an attempt to find an efficacious anti-inflammatory drug that reduces mortality [18].

6.5
LPS Biosynthesis

The efforts of Raetz et al. have delineated the biosynthetic steps involved in the synthesis of the lipid A core using E. coli as a model (see excellent reviews [19, 20]). There are nine enzymes required to convert the UDP-Glc-NAc sugar nucleotide precursor into Kdo_2-lipid A. Each enzyme has been named, the gene cloned and characterized to some extent at a biochemical or genetic level: LpxA, LpxC, LpxD, LpxH, LpxB, LpxK, WaaA, LpxL and LpxM. In the first step, the UDP-Glc-NAc building block is acylated with an acyl-carrier protein substrate in a thermodynamically reversible step. Thereafter, an N-acyl deactylase irreversibly removes the acetate group which commits the UDP-sugar to the pathway. The N-acetyl glucosamine disaccharide is generated by cleavage of the UDP-pyrophosphate of 1 mol of bis-acylated UDP-2,3-diacylglucosamine to give 2,3-diacylglucosamine-1-phosphate (lipid X). Then another mole of bis-acylated UDP-2,3-diacylglucosamine is condensed with lipid X to give the disaccharide. Phosphorylation of the 4′ position is then followed by the addition of the Kdo residue(s). This step is bacteria-specific and Kdo-transferase isozymes can add one, two or three Kdo residues. The product in E. coli, Kdo_2-IVa, with four fatty acids is then converted to the hexa-acylated Kdo_2-lipid A by two acyl-transferases. Since the early genes in the pathway are essential they have been identified as good targets for the development of new lipid A-specific antibiotics. One such antibiotic, CHIR-90, has been shown to be active against *Pseudomonas aeruginosa* and *E. coli*. This is specific for the second enzyme, the deacetylase LpxC, and a recent structure of the LpxC/CHIR-90 complex has revealed the molecular details of the mechanism of inhibition thus paving the way for development of more potent agents [21].

6.6
Minimal, Modified Lipid A

Until recently, it was thought that the minimal LPS required for the growth of *E. coli* consisted of the lipid A core and disaccharide made up of two Kdo sugars (e.g., Figure 6.1). However, recent work by Woodard et al. has challenged this dogma. Meredith et al. revealed that an *E. coli* K-12 nonconditional suppressor strain, KPM22, disrupted in D-arabinose-5-phosphate isomerase activity (through

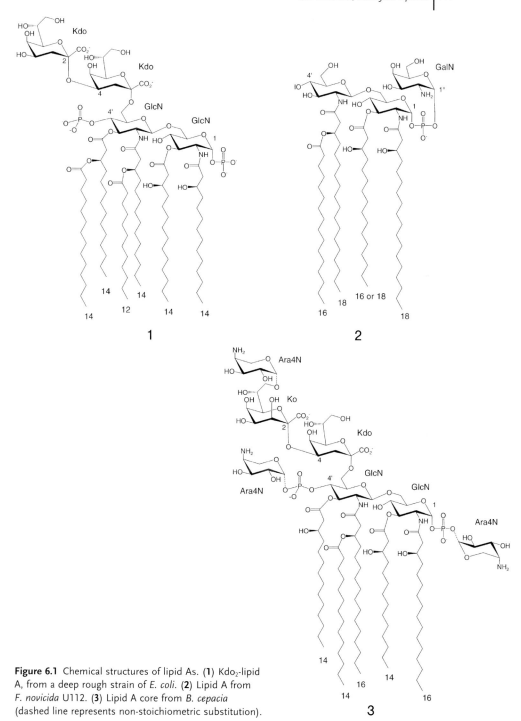

Figure 6.1 Chemical structures of lipid As. (**1**) Kdo$_2$-lipid A, from a deep rough strain of *E. coli*. (**2**) Lipid A from *F. novicida* U112. (**3**) Lipid A core from *B. cepacia* (dashed line represents non-stoichiometric substitution).

manipulation of *kdsD* and *gutQ*) could not produce a LPS containing Kdo and thus generated the endotoxically inactive LPS precursor lipid IVa [22]. These cells grew and were viable, albeit that they were hypersensitive to large, hydrophobic antibiotics (e.g., rifampin, fusidic acid, novobiocin and erythromycin) and sodium dodecyl sulfate. The cell wall of the mutant strain also showed unusual altered morphology in the inner and outer membrane cell walls. Interestingly, inclusion of the inner membrane LPS transporter MsbA on a multicopy plasmid partially suppressed the lethal Kdo knockout phenotype directly in the auxotrophic parent strain, suggesting increased rates of nonglycosylated lipid A transport can, in part, compensate for Kdo depletion.

It appears that bacterial species use a range of biosynthetic strategies to build lipid A derivatives that allows them to escape detection by the mammalian innate immunity surveillance system. This is exemplified by a recent study that revealed *Francisella tularensis*, the causative agent of tularemia—a highly contagious disease of animals and humans, and the related mouse pathogen *Francisella novicida* synthesize unusual lipid A molecules lacking the 4′-monophosphate group typically found in the lipid A of Gram-negative bacteria (Figure 6.1, compound 2). During lipid A biosynthesis LpxF, a selective phosphatase located on the periplasmic surface of the inner membrane, removes the 4′-phosphate moiety in the late stages of *F. novicida* lipid A assembly. To evaluate the relevance of the LpxF 4′-phosphatase to pathogenesis, Raetz et al. constructed a *lpxF* deletion mutant and compared its virulence with the wild-type organism [23]. Intradermal injection of 1×10^6 wild-type *F. novicida* cells is lethal to mice, but 100-fold (1×10^8) mutant is not. The rapid clearance of the *lpxF* mutant was associated with a stronger local cytokine response and a greater influx of neutrophils compared with wild-type. Furthermore, the *lpxF* mutant was highly susceptible to the cationic antimicrobial peptide polymyxin. The authors conclude that LpxF represents a kind of virulence factor that confers a distinct lipid A phenotype, preventing *Francisella* from activating the host innate immune response and preventing the bactericidal actions of host cationic peptides. Here, the loss of a simple phosphate moiety allows the bacterial pathogen to survive. This emphasizes how chemically sensitive the mammalian LPS detection is, but also suggests it can be "blindfolded" by a small modification [24].

It is now apparent that there is an increasingly diverse range of covalent modifications that various bacteria use to decorate their lipid A core. This allows them to escape the mammalian immune system but also protects them from killing by lethal agents such as cationic antimicrobial peptide antibiotics (cAMPs) including mammalian defensins and bacterial polymyxins [polymixin B (PMB)] [25]. Moreover, bacteria can rapidly react to changes in their external environment (e.g., pH, NaCl concentration, as well as Mg^{2+}, Ca^{2+}, Fe^{3+}) and alter their metabolism to express lipid A-modifying enzymes. These enzymes have been shown to catalyze the transfer of, among other things, phosphoethanolamine (e.g., EptA and EptB) and the unusual sugar 4-amino-4-deoxy-L-arabinose (L-Ara4N; catalyzed by ArnT) onto the lipid A phosphate residues. Alteration of the fatty acid content of lipid A can also be achieved using acyl-transferases (e.g., PagP uses palmitate).

An excellent, in-depth review of this field has recently been published [26]. The sensing of environmental change that controls expression of the modification genes is complex, involving the interesting a two-component PhoP/PhoQ system. Detailed molecular, genetic and immunological analyses of this system and its role in *Salmonella* pathogenesis have been carried by out Miller *et al.* over a number of years [27]. Unusual anomalies in lipid A biosynthesis continue to spring up – we, in collaboration with others, have studied the pathogen *Burkholderia cepacia* and related species, and their unusually high resistance to cAMPs and other clinically useful antibiotics. These organisms cause a range of clinical problems, especially to Cystic Fibrosis sufferers, and infection with certain strains leads to chronic lung inflammation and increased mortality. *Burkholderia* LPS is also highly inflammatory and has unusual fatty acid and sugar content including L-Ara4N modification on the Ko octulosonic acid sugar (Figure 6.1, compound **3**) [28, 29]. *Burkholderia* are able to modify both their lipid A core sugar phosphates and their inner core sugars by Ara4N presumably with two different transferases. It is interesting to also note that *Burkholderia* LPS cannot be neutralized by polymyxin and indeed the LPS/PMB complex has unusual properties [30]. Covalent L-Ara4N modification of LPS has been linked to increased cAMP resistance in *E. coli* and *Pseudomonas aeruginosa*, but the L-Ara4N biosynthetic pathway in these bacteria is not essential since deletion causes an increased cAMP sensitivity. In contrast, we found that the operon encoding the Ara4N biosynthetic genes in *B. cepacia* is constitutively expressed and that deletion of the L-Ara4N biosynthetic genes in strain K56-2 was lethal, suggesting that this pathogen has evolved to have an intrinsic requirement for this sugar [31]. We could not isolate sufficient quantities of LPS from the mutant strains to characterize the L-Ara4N content, but there appears to be no logical biochemical explanation as to why *Burkholderia* should need its lipid A to be covalently modified. It could be that L-Ara4N plays roles in other, as yet undiscovered, metabolic pathways in this highly resistant pathogen.

6.7
Isolation of "Natural" Kdo$_2$-Lipid A

Such is the importance of LPS in human biology that the National Institutes of Health and National Institute of General Medical Sciences awarded a 5-year, $35 million grant in 2003 to fund a Lipid Metabolites And Pathways Strategy (LIPID MAPS) consortium led by Edward A. Dennis which included 18 universities, medical research institutes and companies across the United States [32]. They have worked together in a detailed analysis of the structure and function of lipids, and their work is published in major journals and online (see www.lipidmaps.org and www.sphingomap.org). Their goal is to provide comprehensive procedures to the community for identifying all lipids produced by mammalian cells. They have begun with the macrophage, following activation by endotoxin, aiming to quantify temporal and spatial changes in lipids that occur with cellular metabolism, as well as to develop bioinformatic approaches that establish dynamic lipid networks. To achieve these aims, a natural endotoxin of the highest possible analytical specification was crucial. The LPS used by researchers in the immunology field are typically chemically heterogeneous mixtures obtained from bacterial strains such as *E. coli* or *Salmonella typhimurium*. These preparations have high inflammatory properties, but display batch-to-batch variability. An extremely useful tool has recently become available from the LIPID MAPS consortium. They recently reported the isolation of Kdo_2-lipid A, a nearly homogeneous rough (Re) LPS substructure with endotoxin activity equal to LPS [33]. The LPS was extracted from 2 kg cell paste of a heptose-deficient *E. coli* mutant WBB06. This material is composed of six lipid A molecules with Kdo_2-lipid A present in 91% abundance (compound **1**, Figure 6.1). The structure and purity were evaluated by electrospray ionization mass spectroscopy, liquid chromatography mass spectroscopy and ^{1}H-nuclear magnetic resonance (NMR), and its activity was compared with LPS in RAW 264.7 cells and bone marrow macrophages from wild-type and TLR4-deficient mice. Cytokine and eicosanoid production (e.g., TNF-α), in conjunction with gene expression profiling were employed as readouts, and it was found that this material was as potent as LPS. The possibility of generating pure ^{13}C-labeled Kdo_2-Lipid A should greatly facilitate NMR solution studies of complexes with MD-2 or CD14. This material is commercially available through Avanti Polar Lipids.

6.8
Synthetic LPS, Lipid A and Their Uses

As knowledge of the molecular details of the host/pathogen interaction is increased, it is hoped that this could be transferred into useful clinical applications. One such expanding area is the adjuvant field. An effective adjuvant should help modulate or enhance the effects of an antigen (vaccine) and promote an immune response, but have no toxic side-effects itself. A recent reviews of this area was recently published [34]. These adjuvants could act upon a number of cells and receptors to illicit a range of responses, and have taken the form of TLR agonists. Others such as liposomes and mineral salts (e.g., aluminum hydroxide, alum) have been used extensively where their exact mode of action is still unclear. Again, an understanding of what makes a pathogen able to escape detection as well as those with potent inflammatory properties could aid in adjuvant design.

Since natural preparations are inherently heterogeneous (even the *E. coli* Kdo_2-lipid A has five other lipid A derivatives), synthetic preparations of single compounds have proven useful tools to tease out the details of LPS recognition. A number of groups have developed synthetic routes to various lipid A derivatives and tested their activity against a range of cell types. It is around 20 years since Shiba *et al.* made synthetic lipid A derivatives and tested their activity [3]. A non-toxic lipid A derivative from *Salmonella Minnesota* R595 lacking the 1-phosphate and fatty acid esters on the first glucosamine was chemically synthesized and this monophosphoryl lipid A (MPL) was shown to have low cytokine stimulation activity (Figure 6.2, compound **4**). This further emphasized that hexa-acylation is a

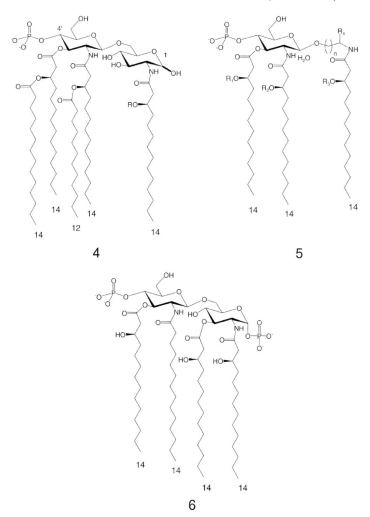

Figure 6.2 Various synthetic lipid A analogs. (**4**) MPL (R = H or C16). (**5**) Generic AGPs (R_1, R_2, R_3, R_4 range in change length, $n = 1–7$). In adjuvant RC529 $R_1 = R_2 = R_3 = n\text{-}C_{13}H_{27}CO$, $R_4 = H$, $n = 1$. (**6**) Synthetic lipid IVa.

prerequisite for endotoxin activity and that underacylated lipid As tend to be less immunologically active. The authors also probed fatty acid chain length, and used their strategy to prepare the MPL containing the all C14 hexa-acyl lipid A which had "intermediate" activity and toxicity. Stöver et al. used transcriptional profiling of human monocytes and responses of TLR4 transfectants to demonstrate a clear dependence of the length of the secondary acyl chain on receptor activation [35]. Also of interest is a MPL derivative with seven fatty acids that is being used as an adjuvant in therapeutic cancer vaccine trials [36].

A range of monosaccharide lipid A derivatives have also been prepared and tested for biological activity (reviewed in [18].). In some cases the glucosamine saccharide core has been retained, whereas in others it has been replaced altogether by an acylic derivative. Although penta-acylated monosaccharides tend to be inactive, some containing three acyl residues on a nonreducing monosaccharide are endotoxic, but still 100 times less active than E. coli lipid A in TNF-α production, possibly due to its biophysical properties. To overcome these problems, a series of synthetic aminoalkyl glucosaminide 4-phosphonates (AGPs) has been investigated (Figure 6.2, compound 5). These glycolipid/LPS "hybrids" are based on a serine scaffold and it appears that the carboxyl group is a good isostere for the 1-phosphate of lipid A. A number of academic and industrial groups have explored the structure–activity relationships of these AGPs by changing esters for ethers, moderating chain length and optimization of the distances between functional groups. This has led to the clinical use of RC-529 as a safe and effective adjuvant in a hepatitis B vaccine.

Boons et al. have also developed elegant, highly convergent synthetic strategies to generate a range of useful lipid A analogs. They generate a "prototypical" lipid A from E. coli that is hexa-substituted in an asymmetrical fashion, as well as a derivative with several of its acyl groups having been shortened. They also synthesized hepta-acylated lipid As derived from S. typhimurium LPS that differ in lipid length and phosphorylation pattern. Their strategy employed an advanced disaccharide intermediate building block that could then be selectively modified with any lipid at the desired position [37]. The authors then exposed mouse macrophages to their synthetic lipid As and E. coli 055:B5 LPS, and the resulting supernatants were examined for mouse TNF-α, interferon (IFN)-β, IL-6, IP-10, RANTES (Regulated upon Activation, Normal T cell Expressed and Secreted) and IL-1β1. They found that particular modifications had different effects on the potencies and efficacies of induction of the various cytokines. However, no bias toward a MyD88- or TRIF (Toll/IL-1 receptor domain-containing adaptor)-dependent response was observed, which emphasizes the fact that lipid A derivatives modulate innate immune responses in a highly complex manner.

As well as tackling lipid A targets, Boons et al. have also examined the role of the Kdo moieties in enhancing the biological activities of LPS. They have also optimized the synthesis of a lipid A derivative containing Kdo and the strategy was employed for the synthesis of Neisseria. meningitidis lipid A containing Kdo on the nonreducing sugar [38]. Mouse macrophages were exposed to the synthetic compound and E. coli lipid A and a hybrid derivative that has the asymmetrical acylation pattern of E. coli lipid A, but the shorter lipids of meningococcal lipid A. The resulting supernatants were examined for TNF-α and IFN-β production. The lipid A derivative containing Kdo was much more active than lipid A alone and just slightly less active than its parent LPS, indicating that one Kdo moiety is sufficient for full activity of TNF-α and IFN-β induction. The lipid A of N. meningitidis was a significantly more potent inducer of TNF-α and IFN-β than E. coli lipid A, which is due to a number of shorter fatty acids. The compounds did not demonstrate a bias towards a MyD88- or TRIF-dependent response.

A number of Japanese groups have also contributed greatly to the development of synthetic tools for studying lipid A activity [4, 39]. Various lipid A analogs, as well as their radiolabeled derivatives and more complex partial structures of LPS have also synthesized. For example, a Kdo_2-lipid A tetrasaccharide (Re LPS) has been synthesized and its IL-6-inducing activity tested. The synthetic Kdo_2-lipid A clearly showed that the addition of the Kdo residues enhances the potency of lipid A to activate the innate immune system. Their strategy also allowed the synthesis of ^{3}H-labeled LPS to study the stoichiometry of binding to various receptors (e.g., MD-2, TLR4). These synthetic LPS derivatives including molecules such as lipid IVa have also been used in high resolution structural studies (Figure 6.2, compound **6**, and Section 6.9.2) [40].

The neutralization of LPS to prevent toxic shock has also been a goal of synthetic chemists. There is growing literature in this field but, a recent example took a lead from the fact that the bacterial peptide natural product PMB binds and neutralizes LPS toxicity and is commonly used by immunologists. PMB is too toxic to be used clinically; however, based on the NMR-derived model of a PMB–LPS complex the authors designed a small molecule (a linked mono-alkyl spermine derivative) that binds LPS and neutralizes its toxicity with a potency indistinguishable from that of PMB in a wide range of *in vitro* assays. It also affords complete protection in a murine model of LPS-induced lethality and is apparently nontoxic in vertebrate animal models [41, 42].

Finally, LPS has been an attractive target for synthetic chemists to target sensitive sensors to detect the presence of contaminating endotoxin in water and other fluids. For example, one group have prepared fluorescently labeled analogs of two peptide variants derived from the putative ligand-binding domain of the LBP CD14 that detect and discriminate LPS and lipids down to the submicromolar concentration range [43]. In a different sensor approach fluorescent quantum dots coated with a zinc complex can selectively stain a rough LPS *E. coli* mutant and permit optical detection in a living mouse leg infection model [44].

6.9
Structural Studies

6.9.1
LBP, CD14 and FhuA

It is hoped that knowledge of the high-resolution three-dimensional structures of various LBPs (e.g., MD-2, CD14 and TLR4) in complex with synthetic LPS/lipid A ligands and their natural product ancestors will further guide the design of more specific, clinically useful molecules [45]. Furthermore, the structures of protein complexes will shed light on how these proteins signal to each other and generate the interlinked networks. A number of structural studies have been carried; these are too extensive to cover in this chapter, but we aim to pick out personal highlights. Efforts to reveal the molecular structures of LPSs from various organisms and their interaction with bacterial and mammalian protein targets are time consuming due to the inherent heterogeneity of natural LPS and the fact that many of the protein receptors are membrane bound.

The importance of LBP as the initial LPS binder has meant it has undergone a range of structural studies, but a high-resolution LBP/LPS structure has not been realized to date. However, the structure of a related human bactericidal/permeability-increasing protein has been determined and has been used as a model for LBP [46]. A group of cationic residues at the N-terminus of LBP have been shown to be essential for LPS binding. Synthetic fragments of LBP (14 amino acids) have been used in NMR studies with LPS to reveal the LPS-binding epitope and aid design inhibitors [47].

The first structure of mouse CD14 (residues 5–313) was also recently published, albeit without a bound ligand (PDB code: 1WWL) [48]. The monomeric subunit of CD14 contains 13 β-strands and 11 of them overlap with conserved LRRs. The concave surface of the horseshoe-shaped structure consists of a large β-sheet of 11 parallel and two antiparallel β-strands. The binding sites for LPS in CD14 have been intensively studied by mutagenesis and by epitope mapping of antibodies that block LPS binding, and four regions have been identified within the N-terminal 65 residues of CD14. The most striking feature of the structure of CD14 is the N-terminal pocket that is located on the side of the horseshoe near the N-terminus and it is entirely hydrophobic except for the rim. The main pocket is both wide and deep with dimensions 8 Å wide × 13 Å long × 10 Å deep; thus, overall, the pocket has a total volume of around 820 Å^3 and is thus large enough to accommodate at least part of the lipid chains of LPS. The authors propose that LPS would induce only small changes to the CD14 structure and we await the structure of such a complex.

It was thought that the inherent conformational flexibility (and heterogeneity) would prevent crystallization of LPS, but over 10 years ago a breakthrough was made due to the serendipitous cocrystallization of LPS with a protein, FhuA, which is the receptor for ferrichrome-iron (Figure 6.3) [49]. FhuA is found in the outer membrane of *E. coli*, where, in addition to its crucial role in iron metabolism, it also functions as the primary receptor for the structurally related antibiotic albomycin and for several bacteriophages. The authors were primarily interested in the structure and mechanism of action of FhuA, and to facilitate this they overexpressed the FhuA gene in *E. coli* with a surfaced-exposed six His-tag and extracted the protein from the outer membrane C12-DAO as a solubilizing agent. It turned out that crystallization of FhuA is dependent on the presence of stoichiometric amounts of LPS. In fact, if LPS is completely removed from FhuA protein preparations or if an excess of LPS is present in such preparations, the growth of FhuA crystals is inhibited. They propose that LPS remained bound to FhuA throughout the process of purification and crystallization, and that it did not adsorb to FhuA during isolation. Since it is known that LPS is localized to the outer leaflet of the outer membrane, the location of bound LPS marks its position relative to the upper aromatic girdle of FhuA and to the outer membrane. The single LPS molecule that is noncovalently associated with the membrane-embedded outer surface of

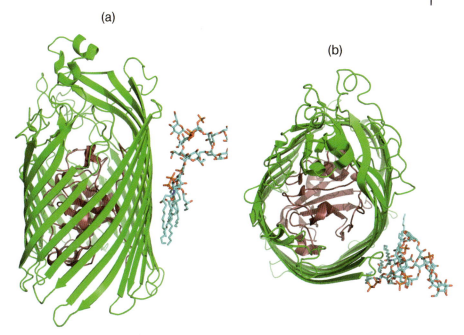

Figure 6.3 (a) Side view of FhuA with bound LPS. (b) Top view of same complex.

FhuA exhibits the expected chemical structure for *E. coli* K-12 LPS (PDB code: 1FCP, Figure 6.3).

6.9.2
MD-2/Lipid IVa Complex

Recent structural analyses by Ohto *et al.* have begun to reveal the key features of LPS recognition by essential components of the mammalian innate system. They presented crystal structures of human MD-2 alone and its complex with *E. coli* lipid IVa [40]. Lipid IVa inhibits the effects of bacterial LPS since it is similar in structure to lipid A (it is a lipid A precursor), but lacks two of the six fatty acids present on lipid A that is found in the LPS of *E. coli* or *S. typhimurium*. LPS is recognized by the receptor complex of MD-2 and TLR4. MD-2 is a 160-amino-acid glycoprotein and member of the MD-2-related lipid recognition family that forms a complex with TLR4 on the cell surface [50]. It has been shown that MD-2 and the MD-2/TLR4 complex can bind LPS with nanomolar affinity although LPS can bind to TLR4 alone. The recombinant human MD-2 used in the study was isolated from *Pichia pastoris* and treated with a glycosidase to produce a homogenous, monomeric protein. The structure of MD-2 alone and in complex with lipid IVa revealed that MD-2 displays a deep hydrophobic cavity sandwiched by two β-sheets, in which four acyl chains of the lipid IVa ligand are fully confined (Figure 6.4a and b).

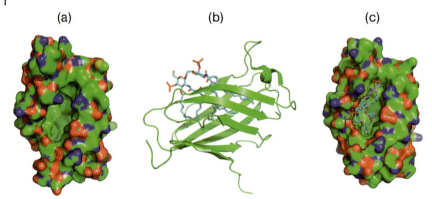

Figure 6.4 (a) Structure of ligand-free MD-2 rendered as a molecular surface showing cavity. (b) Structure of MD-2 with lipid IVa bound with MD-2 displaying β-sheet fold and lipid in chemical structure format. This shows penetration of fatty acids into cavity. (c) Structure of this complex rendered in molecular surface.

The phosphorylated glucosamine disaccharide parts are located at the entrance to the cavity. MD-2 is folded into a single domain consisting of two sets of β-sheets in a common immunoglobulin fold. The structures suggest that MD-2 plays a principal role in LPS recognition. These two sheets (made up of a three and six antiparallel strands) generate a large and deep hydrophobic cavity with a volume big enough to accommodate the LPS molecule (approximate dimensions of $15 \times 8 \times 10$ Å). The ligand-free MD-2 cocrystallized with bound lipid molecules (which were presumed to be $3 \times C14$ myristic acids). In the MD-2/lipid IVa complex electron densities bound in the cavity were assigned to glucosamine disaccharide, 1- and 4'-phosphates, and four fatty acid chains. Comparison of the complex structure with the "unbound" (remembering that it has myristic acid bound) revealed very little conformational change upon LPS binding apart from a lysine residue at the cavity entrance. Overall, the MD-2 is cationic (pI 8.7) and two positively charged residues are at the entrance where they interact with hydrophilic parts of the lipid IVa. This positively charged entrance leads to the hydrophobic cavity (Figure 6.4c) that accommodates the four fatty acid chains. The 1- and 4'-phosphates do not directly interact with MD-2 residues, which is a surprise since these ligands are essential for the inflammatory properties of LPS. MD-2 residues that are essential to the interaction with the TLR4 receptor are located at the cavity entrance. The lipid IVa-binding cavity of MD-2 shows some similarity to the larger hydrophobic cavity of CD14 [48], the protein that transfers LPS to MD-2. Of interest, it appears that CD14 can accommodate lipid A (with six fatty acid chains) due to this expanded cavity, whereas MD-2 cannot. Although lipid IVa is not a potent immunostimulatory molecule since it lacks the six fatty acids of lipid A, the MD-2/lipid IVa complex provides a useful insight into the molecular details of the immune response and begin to answer the question "How does the mammalian immune response recognize such diverse LPSs from the thousands of bacterial species?".

In a complementary paper in the same issue of *Science*, Mata-Haro et al. also investigate the molecular basis of the interactions between LPS and its receptors [51]. The exact molecular details are not yet known as to why certain lipid molecules trigger an inflammatory response as opposed to others that do not. They report that, in mice, MPL interacts with the TLR4–MD-2 complex differently from LPS – they suggest that it is likely caused by the active suppression, rather than passive loss, of proinflammatory activity of this LPS derivative; it is thought that it only activates one part of the inflammatory signal cascade [TRIF and TRAM (TRIF-related adaptor molecule)] and results in T cell activation. They also suggest that MPL could be used as an adjuvant to stimulate the immune response but not cause adverse inflammation.

6.9.3
TLR4–MD-2 Complex

Another recent breakthrough in the understanding of the molecular interactions between the partners involved in LPS detection has recently been published. Eritoran is a synthetic lipid A analog based on the LPS analog from *Rhodobacter sphaerodies*. It is a strong antagonist of the TLR4–MD-2 complex and is currently in phase III clinical trails for severe septic shock. The structure of the full-length ectodomain of the mouse TLR4 alone and in complex with MD-2 was determined using protein expressed in insect cells [52]. The TLR4–MD-2 form a 1:1 complex in solution and in the crystals. The crystal structure shows that TLR4 is an unusual member of the "typical" subfamily of the LRR superfamily with a characteristic horseshoe-like structure whose concave surface is formed by parallel β-strands and whose convex surface is formed by loops and 3_{10} helices. Unlike other typical family members, analysis of the β-sheet conformation of TLR4 demonstrates that it can be divided into N-, central and C-terminal domains and undergoes sharp structural transitions at the domain boundaries. The MD-2 adopts a β-cup fold with two antiparallel β-sheets that contain three and six β-strands, respectively, similar to that found for the human MD-2. In comparison to the human protein the mouse MD-2 pocket is narrow and deep with a total surface area of around 1000 Å^2 that could accommodate LPS. The structure of the complex provided insight into the surface of TLR4 that interacts with MD-2 – it has a long and narrow shape with dimensions 40 × 20 Å. It can be divided into two chemically and evolutionarily distinct areas – the A and B patches. The A patch is negatively charged and evolutionarily conserved across other species, whereas the B patch is positively charged and located in a less-conserved area, although the residues directly interacting with MD-2 are strictly conserved. The A and B patches of TLR4 are composed of the residues in the concave surface derived from the "LxLxxN" part of the LRR modules in the N-terminal domain and of the central domain, respectively The interaction between TLR4 and MD-2 is also mediated by an extensive network of charge-enhanced hydrogen bonds.

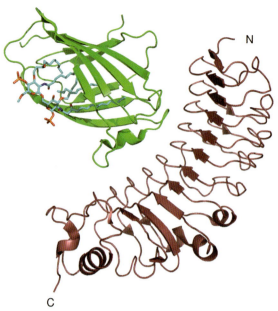

Figure 6.5 The TLR4–MD-2–Eritoran complex. The TLR4 is in red, the MD-2 in green, and Eritoran in ball and stick. The TLR4 contains a TV3 hybrid domain at the C-terminus.

To facilitate soluble expression and crystallization of the TLR4–MD-2 complex with bound ligands, the authors also developed a novel technique that they termed the "hybrid LRR technique". They produced fusion proteins – the variable lymphocyte receptor (VLR) proteins of hagfish were chosen as fusion partners because all VLR proteins have canonical LRR structures with sequence diversity in their variable regions maximized. Among the hybrids, MD-2-bound TV3 was successfully crystallized with Eritoran and their complex structure was solved. They found that Eritoran binds to the hydrophobic pocket in human MD-2, and that there is no direct interaction between Eritoran and TLR4. The structure formed by the four acyl chains of Eritoran complements the shape of the hydrophobic pocket and is a very good fit since it occupies almost 90% of the solvent-accessible volume of the pocket, leaving only a narrow groove near its opening (PDB codes: 2Z65 and 2Z64) (Figure 6.5). They used their fusion structures to also assemble a model of the human TLR4 protein and also used biochemical studies to study cross-linking of the complexes. This allowed them to propose three models for LPS-induced dimerization of the TLR4–MD-2 complex and favor one that fits with the biochemical data. Although Eritoran has a clear structural resemblance to LPS, it is an antagonist, whereas LPS is an agonist. Therefore, there must be significant differences in their modes of interaction. The molecular details of these differences await further structural analysis.

6.10
Bacterial LPS-Binding Proteins

As well as components of mammalian LPS detection and signaling, another area of intensive research is to understand the molecular details of processing of the lipid A and LPS intermediates during the bacterial biosynthetic pathway described by Raetz. Recent breakthroughs have come in the understanding how the lipid A is transported from where it is synthesized in the cytoplasm through the inner to the outer membrane in bacteria. MsbA is a member of the ABC transporter superfamily and is a lipid flippase that transports lipid A and LPS from the cytoplasmic leaflet (inward-facing) to the periplasmic leaflet (outward-facing) of the inner membrane. The structures of the MsbA from three different bacteria (*E. coli, V. cholerae* and *S. typhimurium*) in four various open and closed conformational states has begun to reveal the how bacteria process the LPS intermediates [53]. More recently the structure of a bacterial periplasmic transporter LptA has been determined – this protein is believed to act as a LPS shuttle (with MsbA, LptB and Imp/RlpB) between the inner and outer membranes [54].

6.11
Sphingolipids – Essential Membrane Components of Mammals, Plants, Fungi, Yeast and Bacteria

The sphingolipids were discovered over 100 years ago by Johann Thudichum who described an enigmatic "Sphinx-like" molecule, and since then numerous studies of their structure, function and metabolism have been described in the literature [55]. It was not until the middle of the last century that the chemical structure of the family of sphingolipids was determined by Carter *et al.* [56]. They represent a large family of bioactive molecules, and are ubiquitous constituents of all eukaryotic and some prokaryotic membranes. In association with cholesterol, they form lipid rafts that are implicated in signal transduction and membrane trafficking [57]. Their metabolites (such as ceramides) are also known to be involved in several cellular events such as proliferation, differentiation and apoptosis [58, 59]. Furthermore, sphingolipid metabolites play a crucial action in various pathological processes including tumor cell angiogenesis [60]. For example, sphingosine-1-phosphate (S1P) is a potent signaling molecule in eukaryotic cells and a focus of intense research since it has have been linked to various human diseases [61, 62]. The sphingolipids consists of a long-chain base (LCB), fatty acid and a polar head group. This amino alcohol core is decorated with a large variety of functionality and hundreds of sphingolipids are now known.

Sphingolipids are found in several species and so assume that all of these species possess their own sphingolipid biosynthetic pathway (Figure 6.6). Mammalian, plant, yeast, fruit-fly, parasite and fungal sphingolipid pathways are well described, and many of the intermediates, genes and encoded enzymes have been

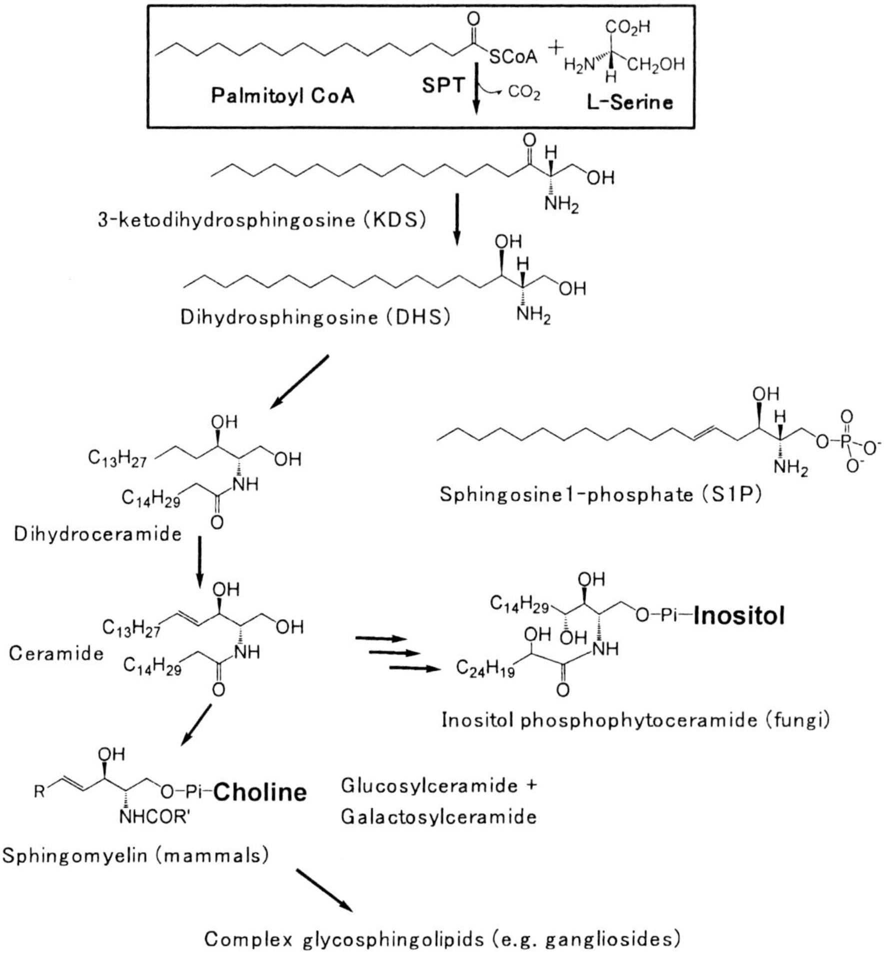

Figure 6.6 Abbreviated sphingolipid biosynthetic pathway showing the early steps and key molecules such as S1P.

identified. Comprehensive reviews of the yeast and plant pathways were recently published [63, 64]. Thematic reviews have recently been published in the *Journal of Lipid Research* (from May 2008 onwards) in honor of Professor Carter and also in *Nature Reviews Molecular Cell Biology* (February 2008).

In summary, the early steps of sphingolipid biosynthesis appear to be highly conserved and the species-specific features occur after attachment of various N-acyl chains, hydroxylation at specific carbons and addition of complex sugars. Of intense recent interest has been the characterization of an unusual, growing family of Gram-negative bacteria that have an unusual membrane composition – it was discovered that instead of the LPS described above for *E. coli*, their membranes consist of a variety of glycosphingolipids (GSLs) [65]. The "Sphingomonads"

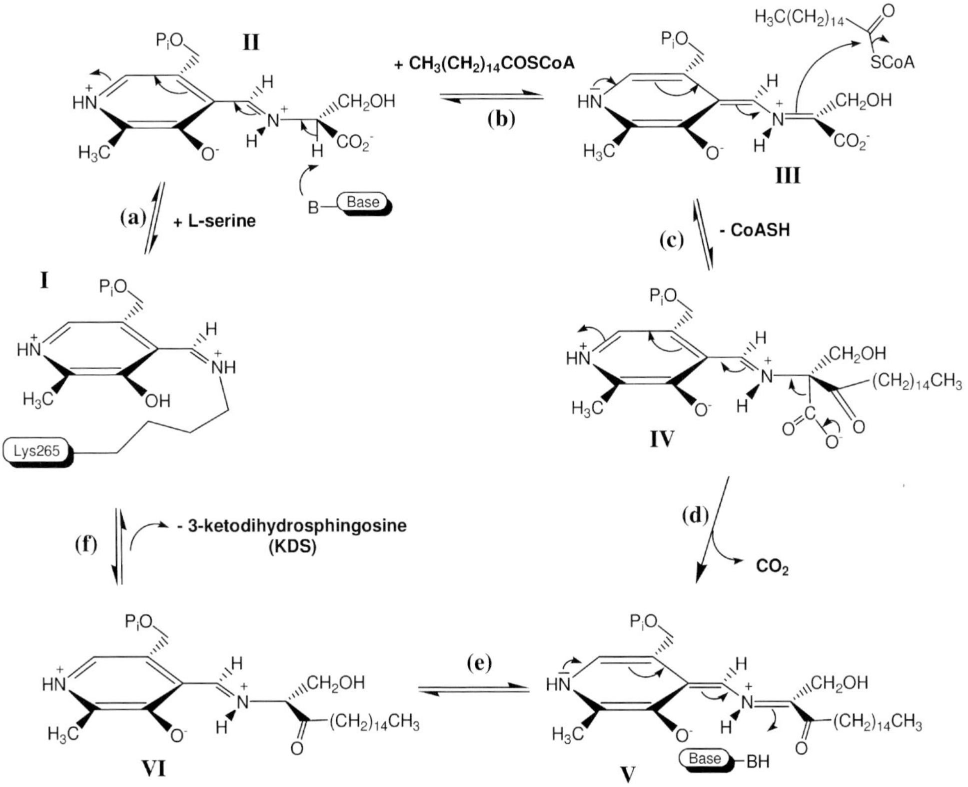

Figure 6.7 Abbreviated reaction mechanism of SPT.

include organisms such as *Sphingomonas paucimobilis*, *Sphingobacterium multivorum* and *Sphingobium yanoikuyae* [66]. Since bacterial GSLs have similar core structures to their eukaryotic homologs it was assumed that the pathway would begin with the production of the common intermediate, 3-ketodihydrosphingosine (KDS) from L-serine and palmitoyl-CoA. This reaction is catalyzed by the enzyme serine palmitoyltransferase (SPT; EC 2.3.1.50). Indeed, a bacterial SPT was isolated from *S. paucimobilis* by Ikushiro et al. in 2001 [67].

SPT catalyses the first and rate-limiting step of the sphingolipid biosynthetic pathway in all organisms studied to date [68]. The reaction is a pyridoxal-5′-phosphate (PLP)-dependent, decarboxylative, Claisen condensation of the amino acid L-serine and the long chain (C16) fatty acid palmitoyl-CoA, which produces KDS. SPT is a member of the α-oxoamine synthase subfamily of PLP-dependent enzymes, which contains three other enzymes: 8-amino-7-oxononanoate synthase, 5-aminolevulinate synthase and 2-amino-3-ketobutyrate-CoA ligase [69–74]. These enzymes generally catalyze the Claisen-like condensation between an amino acid and a CoA-thioester [75]. A common mechanism has been suggested [68, 74, 76–79] comprising the following steps (Figure 6.7): (a) formation of an external aldimine (**II**) via displacement of the lysine-PLP internal aldimine (**I**, holo-SPT) by the incoming amino acid substrate; (b) formation of a quinonoid intermediate (**III**) by abstraction of α-proton from the PLP-amino acid external aldimine; (c) a Claisen condensation with the fatty acid-CoA substrate followed by displacement of the CoA to form a β-ketoacid (**IV**); (d) decarboxylation of this species to form a product quinonoid (**V**); (e) protonation of this quinonoid to form the product external aldimine (**VI**); and, finally (f) release of the α-oxoamine product and regeneration of the enzyme PLP-internal aldimine. The enzyme is assumed to go through a number of conformational changes as it proceeds through the reaction cycle of C–C formation, C–S bond cleavage and decarboxylation. More recent evidence for the reaction sequence shown was obtained using the elegant use of a CoA thioether analog, S-(2-oxoheptadecyl)-CoA. L-Serine deprotonation was accelerated more than 100-fold when this CoA substrate was added, suggesting remarkable substrate synergism [80].

SPT isozymes have been discovered in numerous organisms, such as mammals, fungi, yeast, bacteria and viruses, and sequence analysis of the encoded SPTs from various organisms suggests structural diversity across different isozymes [68]. For example, in humans and the yeast *Saccharomyces cerevisiae* the SPT is formed by two different subunits (SPT1 and SPT2, encoded by the genes *lcb*1 and *lcb*2, respectively), both of which are required for a functionally active enzyme. These two eukaryotic SPTs are predicted to contain large hydrophobic domains at their N- and C-termini, and are membrane-bound proteins, which has made their isolation in high yield and purity particularly difficult. Sequence analysis revealed the SPT1 subunit to be devoid of active site residues, suggesting the SPT1/SPT2 dimer has only one active site, with the SPT1 subunit playing a regulatory role. The recent identification of a LCB3 gene expressed mainly in placenta (with homology to LCB2) suggests greater complexity to this important first enzyme in the pathway [81]. A viral SPT was also recently discovered where it appears that the LCB2 and LCB1 genes are fused head-to-tail to give a large SPT enzyme [82, 83].

Fortunately, a soluble homodimeric enzyme was isolated from *S. paucimobilis* – the structure of this enzyme was determined in our group (the first high-resolution structure of an enzyme from the sphingolipid biosynthetic pathway) and provides a useful tool for structural and mechanistic studies of the SPTs [67, 84]. The bacterial SPT consists of two monomers of 45 kDa and possesses two active sites at the subunit interface (Figure 6.8a). The cofactor PLP is bound into the active site through the formation of a Schiff base with a specific lysine residue (Lys265). The internal aldimine and other intermediates are stabilized by a group of specific catalytic residues. These residues (His159, Asp231 and His234) interact with PLP via π–π stacking, hydrogen bonding or salt bridge (Figure 6.8b). Each monomer is able to bind a molecule of PLP, but the two subunits are required to obtain the active enzyme.

The structure of the bacterial enzyme allowed a model of the human SPT1/SPT2 heterodimer to be built using powerful bioinformatics methods. This model gave molecular insight into the cause of heredity sensory and autonomic neuropathy type I, a rare genetic disease that leads to progressive loss of peripheral nerve

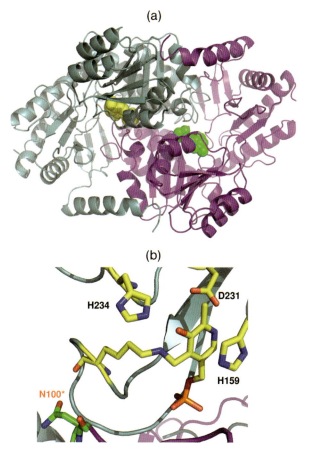

Figure 6.8 Structure of the homodimeric *S. paucimobilis* SPT [84]. (a) Overall dimeric structure with monomer A in blue and monomer B in purple. The PLP internal aldmine is represented in both active sites in solid yellow and green space fill. (b) The SPT PLP-binding site is at the dimer interface. Highlighted is the N100 residue from the opposite monomer.

function. Mutations causing this disease are clustered on the *lcb1* gene and encoded SPT1 subunit, which is surprising since this subunit lacks catalytic residues [85, 86]. Studies of how these mutations impact SPT activity and sphingolipid metabolism are underway in a number of laboratories.

Sphingolipid biosynthesis is a multistep pathway, with many species-specific branch points catalyzed by specific enzymes, so inhibitors of each step have the potential to be powerful drugs. A range of natural product inhibitors are known to inhibit the sphingolipid biosynthetic pathway. Myriocin inhibits directly SPT since it is thought to perfectly mimic the intermediate in the catalytic reaction between L-serine and palmitoyl-CoA. L-Cycloserine has also been shown to inhibit SPTs prokaryotes and eukaryotes. Other enzymes in the pathway from other organisms are also inhibited by interesting natural products – for example fumonisins inhibit ceramide synthase and inositol phosphorylceramide synthase (AUR1) is a target for the peptide aureobasidin in fungal cells. In fact, this pathway is a good antifungal target in general [87]. The regulation of S1P metabolism in mammals is also a target for a number of potential therapies.

6.12
Natural Killer T Cells and GSLs

As well as playing roles as membrane building blocks and signaling molecules, GSLs have recently been shown to be potent activators of the mammalian immune system via specific antigen receptors. The mechanism is somewhat different to the TLR4–LPS interaction. Natural killer T (NKT) cells are a sublineage of natural memory T cells that share properties of both T lymphocytes that were discovered over 10 years ago. Excellent, comprehensive reviews of the NKT field have been published [88, 89]. GSLs are presented by CD1d molecules on antigen-presenting cells (e.g., dendritic cells) to NKT T cell receptors (TCRs), leading to production of several cytokines including IFN-γ [T helper type (T_h1) response] and IL-4 [T helper type 2 (T_h2) response] (Figure 6.9). CD1d is part of the CD1 family, major histocompatibility complex-like molecules that evolved to capture lipid antigens for display at the surface of antigen presenting cells. CD1 genes encode five CD1 molecules in human (CD1a, CD1b, CD1c, CD1d and CD1e) and two homologs in mice (CD1d1 and CD1d2). CD1d, involved in GSL transport, associate with conserved semi-invariant human $V_\alpha 24$–$J_\alpha 18/V_\beta 8$ (V = "variable" and J = "joining") or mouse $V_\alpha 14$–$J_\alpha 18/V_\beta 8$ TCR.

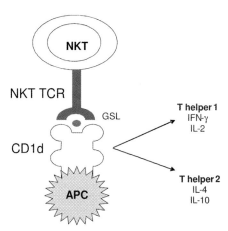

Figure 6.9 Schematic of the interaction between the CD1d receptor found on antigen-presenting cells (APC, e.g., dendritic cells), with presented antigen (GSL) and the NKT TCR, and examples of released inflammatory cytokines.

6.13
GSL Antigens

Screening for products with antitumor activities, Kirin Pharmaceuticals identified the first ligand for the NKT TCR. This was a natural product (Agelasphin 9b, Figure 6.5) extracted from the marine sponge *Agelas mauritianu* and structural studies of the molecule identified it as a GSL [90, 91]. This compound was unusual due to its α-glycosidic linkage instead of the β-linkage present in mammalian glycolipids. In order to increase its potency, structure–activity relationship studies were carried out on a range of agelasphin derivatives. They simplified the sphinganine chain and added two more carbons on the acyl chain. After modification of the saccharidic hydroxyl groups they found that the 4′-OH played a minor role and that the 3′-OH was necessary for optimal activity. Extension of the N-acyl fatty acid chain led to the final compound named KRN7000 or α-galactosylceramide (α-GalCer) (Figure 6.10) [92]. This GSL is still one of the most potent agonists known for NKT TCR. A number of synthetic GSL variants were also made (including both mono- and disaccharides) to begin to explore the nature of the receptor–ligand interaction [93]. Since then extensive synthetic libraries have probed the molecular details to an even greater extent). Very recently, Chung et al. explored the stereoselectivity involved by the synthesis of each of all eight stereoisomers of KRN7000 and presentation of their biological activities [94].

The fact that α-GalCer was extracted from a marine sponge raised doubts about its physiological significance and it seems illogical that the mammalian immune system would evolve a defense mechanism against a product from a marine species. It is interesting to speculate that since the sponge *Algea mauritianus* could have been contaminated by α-proteobacteria, the original glycolipid α-GalCer isolated could be in fact of bacterial origin from a symbiotic organism. Numerous studies searched for other natural product antigens for TCRs, which led to the

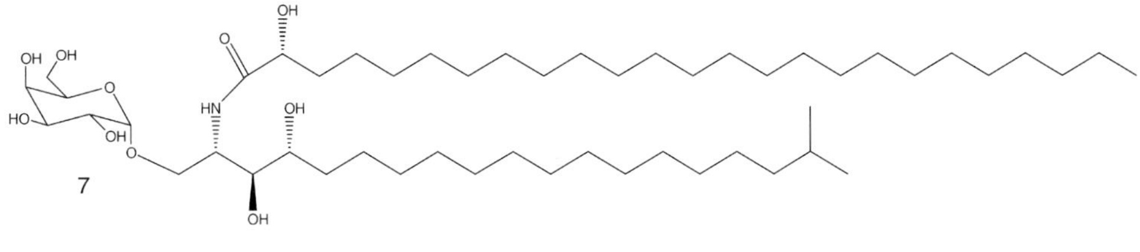

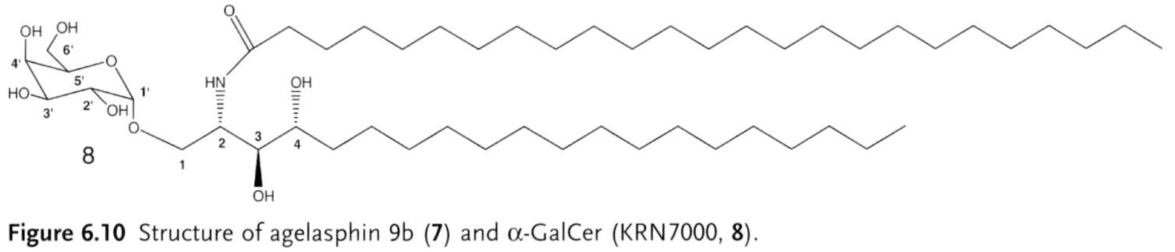

Figure 6.10 Structure of agelasphin 9b (**7**) and α-GalCer (KRN7000, **8**).

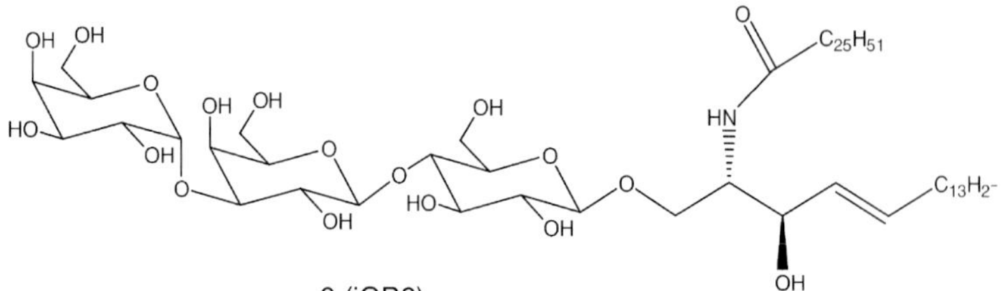

Figure 6.11 Structures of endogenous and exogenous ligands **9** (iGB3) and GSLs of bacterial origin (**10–12**).

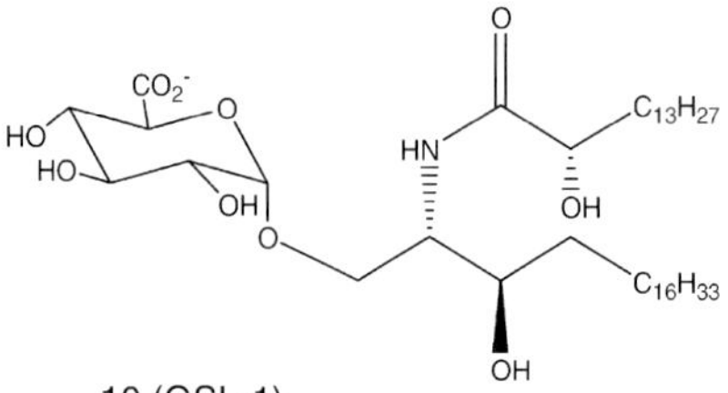

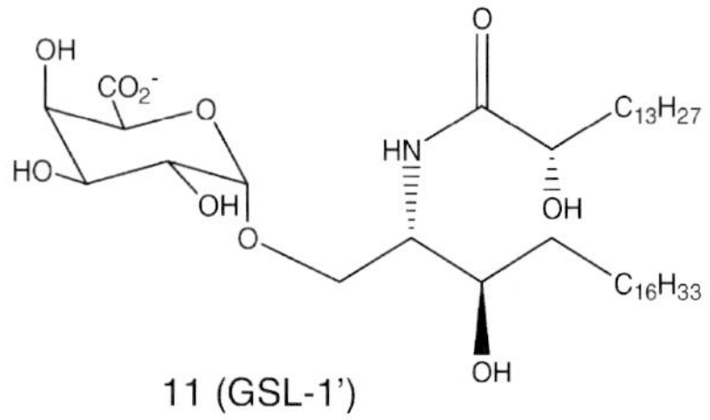

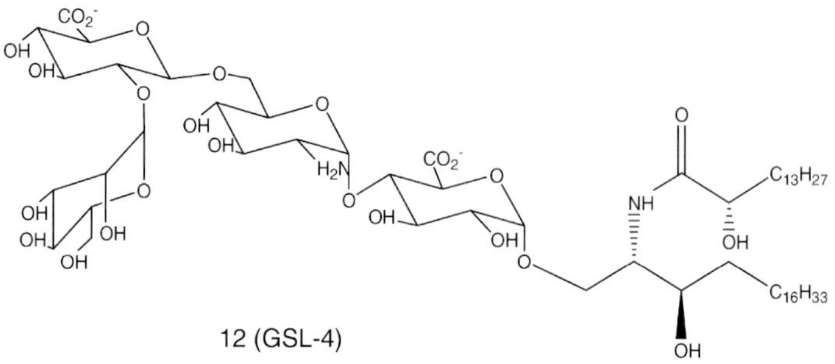

discovery of an endogenous ligand, isoglobotrihexosyl ceramide (iGB3), as well as exogenous ligands from bacteria (Figure 6.11).

iGB3 was identified as the endogenous ligand by various studies on mutant mice. The first study by Stanic et al. concerned mice deficient in β-D-galactosylceramide (β-D-GalCer) or β-D-glucosylceramide (β-D-GlcCer) [95]. Surprisingly, β-D-GalCer-deficient mice (produced by knocking out β-D-GalCer synthase) developed normal NKT cells and were able to activate NKT hybridomas. In contrast, β-D-GlcCer-deficient cells failed to react with the same hybridomas and the addition of β-D-GlcCer synthase restored the recognition. However, direct addition of β-D-GlcCer to NKT TCR did not result in the activation of the receptor. Work by Zhou et al. showed that mice deficient in β-hexosaminidase B were also deficient in NKT cells. β-Hexosaminidases B are responsible for degradation of GB4 and iGB4 to GB3 and iGB3 respectively. NKT cell stimulatory properties were tested with the two last molecules and only iGB3 led to a positive stimulation [96].

These analyses of self/nonself stimulators greatly expanded using the fact that sphingomonad-derived GSLs and mycobacterial lipids had been shown to activate the immune system via NKT cells. Lipids from various species and strains have been tested, for example, the mycobacterial lipid phosphatidylinositol mannoside was reported by Fischer *et al.* to stimulate a modest cytokine release from NKT cells [97]. More efficient NKT activators isolated from *Sphingomonas* possesses GSLs with mono-, di-, tri- and tetrasaccharide units. They present an α-glycosidic linkage similar to the α-GalCer isolated in *A. mauritianus*. GSL-1 can contain galactose or a glucose sugar and unlike α-GalCer can also have a carboxylic acid functional group on the saccharidic C6 instead of a hydroxyl group. GSL-1 and GSL-4 (Figure 6.11, compounds **10** and **12**) have shown high stimulatory activity against NKT TCR.

6.14
Structure–Activity Relationships

Since it is now known that natural endogenous and exogenous GSLs are able to activate the mammalian immune system, it makes these molecules attractive targets as therapy against numerous diseases. However, care must be taken to generate a molecule having only therapeutic benefits and no stimulatory side-effects. NKT cells have been implicated in several immune responses, including pathogens, tumors, tissues grafts, allergens and autoantigens. From these observations, several groups have worked on the modification of the α-GalCer agonist in order to increase the affinity/activity of this compound. Furthermore, activation of NKT cells leads to production of two types of biological response. When α-GalCer binds to TCR, IFN-γ (T_h1) and IL-4 (T_h2) are released in the cell. Several studies have shown that specific modification of the GSL chemical structure can lead to specific activation of one or the other response.

The α-GalCer structure can be separated in two parts: the sugar moiety and the lipid chains. Both of them contain functional groups that can be modified and play a role in the interaction with CD receptors and NKT TCRs. Over the years, a multitude of modifications have been tested and the results obtained permit a better understanding of the structure–activity relationship between the GSL ligands and their receptor. An excellent recent review of this growing literature has recently appeared [98]. The α-GalCer designed by Kirin group possesses a galactose sugar moiety, a sphingosine chain ($C_{18}H_{34}$) and acyl chain ($C_{26}H_{51}$) as the lipid part. The hydroxyl groups as well as the glycosidic linkage are numerous points of modification for the polar head. Concerning, the lipid chains, the length, the degree of unsaturation, as well as the two hydroxyls carried by the sphingosine part allow several possibilities of modification.

6.15
Saccharide Modification

A relatively simple change replaces the galactose by a glucose or a mannose [99] (Figure 6.12, compounds **13** and **14**). The anomeric configuration was tested as well, by synthesis of an β-GalCer (compound **15**). It was observed that the glucose

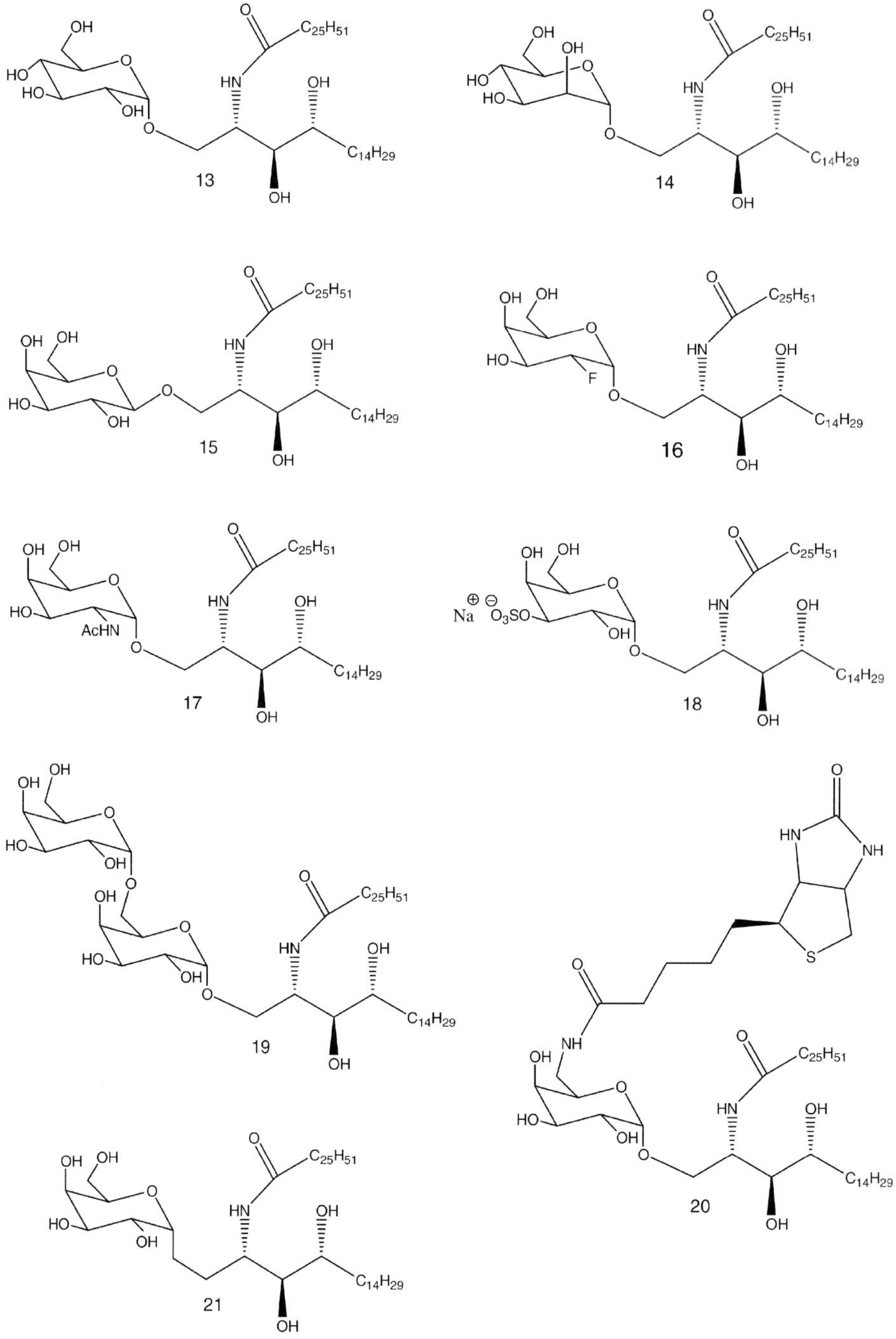

Figure 6.12 GSL variations (**13–21**).

analog has a weak agonist activity, and mannose and β-GalCer analogs have no activity at all. The different OH positions have been investigated and each of them gave rise to different results. The 2′-OH position has been substituted by a diverse range of functional groups (compounds **16** and **17**) and in every case the activity toward TCRs is lost [100, 101].

In contrast, the 3′-OH can accommodate changes such as the sulfate (compound **18**) [101]. The 6′-OH position is quite tolerant of other functional groups (COOH) (Figure 6.11, bacterial GSL-1, **10**), a second sugar (compound **19**) or affinity probes (biotin, compound **20**).The diverse modifications made on the sugar moiety described above affected the recognition with the CD1 receptor part of the complex. The degree of activity was measured against the α-GalCer "standard". No specificity toward T_h1 or T_h2 release was noticed. However, Schmieg et al. and Yang et al. synthesized two C-glycoside analogs and observed that these compounds were able to stimulate specifically a T_h1 response (compound **21**) [102, 103]. Furthermore, this compound showed an activity 1000 times superior compared with α-GalCer and this increase of potency was explained by the longer stability of the compound in the cells. Effectively, α-GalCer and other O-glycosides can be cleaved by glycosidase which is not the case for C-glycosides.

6.16
Ceramide Modification

In their library of synthetic GSLs, Sidobre et al. also probed the 3- and 4-sphingosine positions by substitution with H (Figure 6.13, compounds **22** and **23**) [99]. They reported that they were both involved in CD1d interaction and their substitution leads to destabilization of the complex, with a total loss of activity concerning the substitution of the 3-OH group. Miyamoto et al. first examined the effect of lipids chain truncation and synthesized three new compounds [100]. One of them, OCH (compound **24**), induced a T_h2 bias response of NKT cells. Several other truncated α-GalCer derivatives have been synthesized over the years with the addition of small modifications such as insertion of unsaturation [101, 104–106]. Two particular compounds (**25** and **26**) have shown similar behavior than OCH, leading to a specific T_h2 response. It has been suggested that this specificity could be explained by the fact that IFN-γ (T_h1) needs longer stimulation of NKT cells than IL-4 (T_h2) to be released [107]. Shorter chain analogs of α-GalCer or removal of OH groups leads to destabilization and shorter life time of the glycolipids–CD1d complexes, resulting in shorter stimulation of NKT cells.

6.17
High-Resolution Structural Analysis of Receptor–GSL Complexes

Recent breakthroughs in the determination of the structures of the receptor–GSLs complexes has greatly increased our understanding of the nature of the

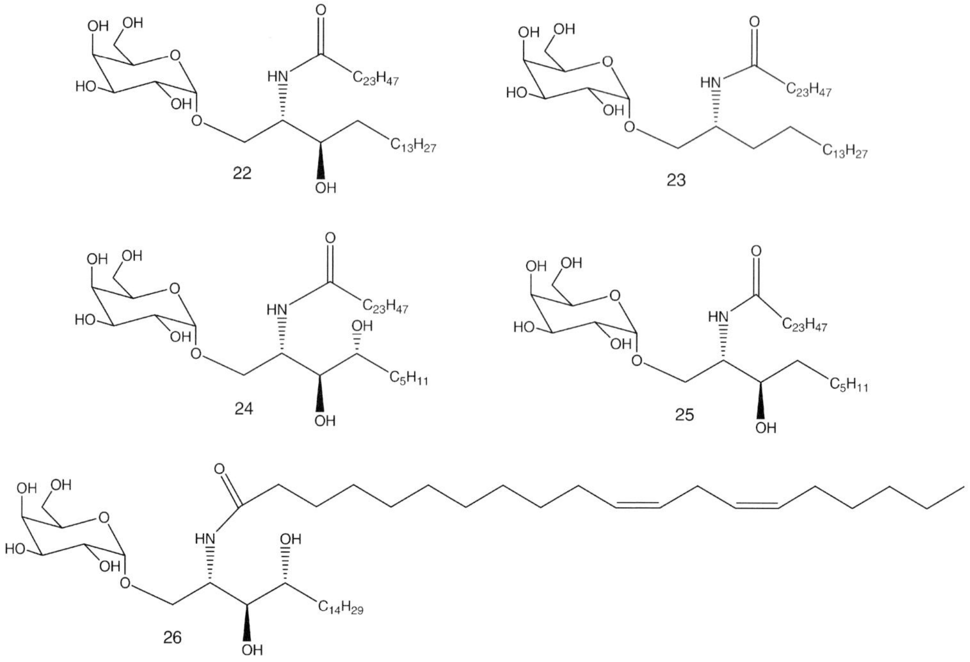

Figure 6.13 Ceramide modifications (**22–26**).

immunostimulatory properties of the various antigens. Numerous structural analyses have been performed on CD1d and NKT TCR. Both have been studied alone, as well as with ligand-bound and in complex with each other. The different interactions and recognition systems between the CD1d–TCR and different glycolipids have been investigated by various groups [108–111]. Presented here are the structures of the human CD1d–α-GalCer complex (exogenous ligand, PDB code: 1ZT4) [109] and CD1d–α-GalCer–TCR [111], the mouse CD1d–iGB3 complex (endogenous ligand, PDB code: 2Q7Y) and a model of its interactions with TCR [112].

The human CD1d protein structure was reported in 2005 by Koch *et al.* They discuss the structural changes between the ligand-free and ligand-bound protein, as well as the similarities existing between the mouse and the human CD1d. The CD1d protein consists of two large domains (α and $β_2M$). Three helices α1, α2 and α3 form the heavy chain (α) of CD1d that is responsible for GSL binding. The antigen-binding site is composed of a large hydrophobic groove allowing the fitting of α-GalCer lipid chains. The acyl chain (C26) binds into pocket A′ in a counterclockwise circular curve, filling the whole of the pocket. A second channel C′ (human) or F′ (mouse) allows the binding of the sphinganine chain (C18) in a more linear conformation (Figure 6.14a).

The galactose ring is located at the surface of the protein and stabilized by specific interactions. Three hydrogen bonds are involved in the stabilization of the

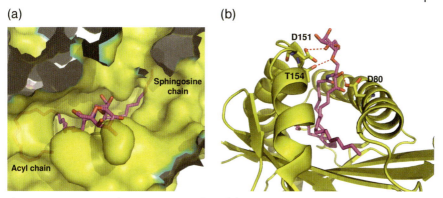

Figure 6.14 Human CD1d structure. (a) Binding of the GSL. (b) Direct interaction of the GSL 2′-OH with Asp151.

antigen. The 3′OH of the sphingosine chain interacts with Asp80 and the glycol linkage 1′-O forms a hydrogen bond with Thr154. Finally, the galactose ring is correctly oriented for TCR recognition by interaction of the 2′OH with Asp151 (Figure 6.14b). The "ligand-free" CD1d is highly similar to the α-GalCer-bound complex. The main difference concerns the binding groove which is noticeably wider (1 Å difference between the Cα chains) than in the lipid-bound form. It may be possible that CD1d adopts a unique "open" conformation to interact and load lipids. The comparison of human and mouse structures showed that they were highly similar, although some subtle differences in the mouse structure were observed, the ability to specifically bind α-GalCer and its recognition by TCR are not compromised. The ligand-binding pockets are similar in size and hydrophobicity, and residues involved in galactose and alkyl chain binding are conserved.

Recently, Rossjohn *et al.* successfully crystallized the human complex CD1d–α-GalCer–TCR (PDB code: 2PO6) giving a first insight into the GSL, TCR and CD1 interactions [111]. The TCR is a semi-invariant protein that comprises an invariant α-chain and a restricted TCR β-chain repertoire. The NKT TCR adopts an unusual parallel docking mode in its interactions with CD1d (Figure 6.15a). Contacts between the two proteins are made via the V_α chain and the V_β chain of TCR and α1-helix and α2-helix of CD1d. More specifically, two TCR loops, CDR3α and CDR2β, are responsible for the CD1d-restricted response. A number of arginine residues from both partners, Arg79 (CD1d), Arg103 and Arg95 (CDR3α), are involved in a complex hydrogen bond network permitting the specific recognition/activation of NKT TCR by CD1d–α-GalCer. The strong positive charged created by the group of arginines is dissipated by negative residues including Asp94 (CDR3α) and Asp80 (CD1d). Previous mutational studies of these residues on mouse CD1d led to the inactivation of NKT TCR [113–115].

α-GalCer protrudes outside of CD1d, and interacts only with the CDR1α and CDR3α loops of TCR. The galactose ring is sandwiched between Trp153 of CD1d and the aliphatic moiety of Arg95 (CDR3α). Moreover, the guanidinium group of

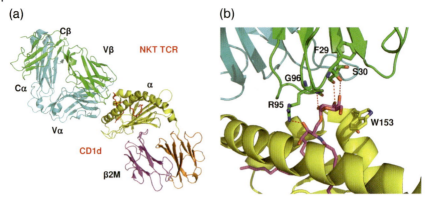

Figure 6.15 (a) Structure of the human NKT TCR–α-GalCer–CD1d complex (NKT TCR: cyan and green; CD1d: yellow, purple and orange; ligand: ball and stick. β2M = $β_2$-microglobulin. (b) Close-up view of the interface between the proteins and ligand [colors same as in (a)].

Arg95 is also involved in stabilization of the lipid part by forming a hydrogen bond with the 3′-OH group of the sphingosine chain. The three sugar hydroxyl groups are involved in specific hydrogen bonds with the α invariant α-chain of TCR. The 2′-OH interacts with Gly96α (CDR3α), 3′-OH and 4′-OH have contact with Ser30α and Phe29α (CDR1α). This specific hydrogen bond system is in correlation with α-GalCer high affinity for TCR (Figure 6.15b). The α-glycosidic linkage allows a strategic orientation of the sugar head and the presence of a β-linkage would lead to a more perpendicular orientation creating perturbations in CDR1α interactions.

Curiously, a cross-reactivity between human TCR and mouse CD1d (and the converse) have been observed. Indeed, it was proved that mouse CD1d–α-GalCer were able to bind and activate human NKT TCR and human CD1d–α-GalCer bind to mouse TCR. This cross-reactivity is easily understandable by comparison of both human and mouse α-GalCer–TCR interactions. All residues and interactions involved in antigen recognition are conserved.

Comparison of the TCR structure with and without CD1d–α-GalCer reveals that no large conformational changes takes place. The TCR protein appears to present a rigid "lock and key" conformation, attributable to strong intermolecular interactions between the CDR loops in the ligand-free TCR.

Most recently, Zajonc et al. have solved the structures of the mouse CD1d–iGB3 complex and the mouse TCR (PDB codes: 2Q7Y and 2Q86 for CD1d–iGb3 and TCR, respectively). By analogy with the human CD1d–α-GalCer–TCR, they proposed two models to explain the interactions between the endogenous ligand and TCR [112].

The stabilization of the lipid chains in CD1d–iGB3 complex is very similar to that observed in CD1d–α-GalCer. The most differences are, as expected, located at the interface between CD1 and the saccharidic moiety. The polar head group is still solvent exposed and only the proximal glucose is well ordered. The second

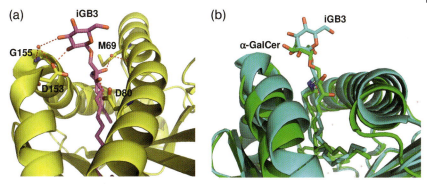

Figure 6.16 (a) CD1d and iGB3 specific interactions. (b) Superposition of human CD1d–α-GalCer (green) and mouse CD1d–iGB3 (cyan) structures.

glucose and the galactose are highly flexible, and do not interact with CD1d. The group of residues involved in the first glucose stabilization is totally different to those used for the galactose binding in CD1d–α-GalCer complex. The two hydroxyl groups, 2′-OH and 3′-OH, form H-bonds with Asp153 and Gly155 (water mediated), whereas the lipid chains interact with Asp80 and Met69 (Figure 6.16a).

The TCR crystal obtained by Zajonc et al. possesses two TCR molecules in the asymmetric unit. Superposition of the molecules reveals a change in the conformations of CDR loops, especially CDR3α and CDR3β. Furthermore, by alignment of mouse and human TCR, the same conformational differences appear, whereas CDR1 loops stay very similar. As in other TCRs, CDR3 domains are known to interact with the ligand – this flexibility in the unbound structure may reflect their capacity to recognize different ligands.

In order to understand the interaction between the CD1d–iGB3 complex and TCR, Zajonc et al. made a model where they positioned the TCR and CD1d in the exact same position as in the human structure with α-GalCer. However, this model reveals serious steric clashes between iGB3 and several TCR residues. This can be explained by the structural differences between iGB3 and α-GalCer (three versus one sugar). From these first results, they proposed two new models to obtain a better fit of iGB3 between the CD1d and TCR.

In the first, they described a "cavity" model where the mouse TCR presents a large and deep positive binding pocket. This cavity formed by the CDR3α and CDR3β loops would be able to fit the two "free" sugars of iGB3. In a second model, the three sugars of iGB3 are positioned between the CD1d and TCR. In this "squashed" model, the parallel orientation of CD1d along TCR as well as the interactions between CD1d and TCR described in the human structure are conserved. Each of these models allows us to probe the different ligand-binding constraints existing in the human structure; however, both of them have limitations. In the "cavity" model, the CDR3β flexibility permits TCR adaptability towards various ligands; however, it is commonly accepted that CDR3β is not involved in ligand recognition. On the other hand, the "squashed" model requires bond breaking/reforming to orient the polar group properly, which is energetically unfavorable.

Recent studies have begun to shed light on the role of the tetrasaccharide and sphingosine units found in many naturally occurring GSLs derived from Sphingomonads. Kinjo *et al.* made synthetic GSL-4A derivatives varying the nature of the tetrasaccharide, as well as generating GSLs with different lipid chains [116]. Since the initial studies of the effect of GSLs on NKT cells used material isolated from bacteria (probably a heterogeneous mixture) it was therefore important to go back and test homogeneous, synthetic versions of these antigens. In contrast to earlier studies, GSL-4A showed weak activity and the GSL-4B was inactive. Long *et al.* had also used synthetic GSL-1, GSL-2, GSL-3 and GSL-4 (as well as various lipid hybrids) to show that only the monosaccharide GSLs are potent antigens. Furthermore, they probed whether these higher order GSLs could be metabolized by various cell types. It may be that Sphingomonads and other bacteria have evolved a sugar coat to avoid detection by the immune system.

6.18
Conclusions

In both the LPS/endotoxin and GSL fields combinations of natural product chemistry, synthetic chemistry, structural biology, microbiology, immunology and genetics have begun to reveal the molecular interactions involved in the innate recognition between colonizing bacterial cells and their mammalian host. The exact details of what makes one microbe a beneficial symbiotic partner and another a life-threatening pathogen are beginning to emerge. These details are also opening our understanding of the nature of the inflammatory response and its links to a range of autoimmune diseases.

Acknowledgments

M.C.R. is funded by an EaStCHEM PhD studentship. D.J.C.'s laboratory is supported by the Biotechnology and Biological Sciences Research Council and the Engineering and Physical Sciences Research Council.

References

1. Beutler, B. and Rietschel, E.T. (2003) Innate immune sensing and its roots: the story of endotoxin. *Nature Reviews. Immunology*, **3**, 169–76.
2. Galloway, S. and Raetz, C.R.H. (1990) A mutant of *Escherichia coli* defective in the first step of endotoxin biosynthesis. *Journal of Biological Chemistry*, **265**, 6394–402.
3. Tanamoto, K., Zähringer, U., McKenzie, G., Galanos, C., Rietschel, E., Lüderitz, O. *et al.* (1984) Biological activities of synthetic lipid A analogs: pyrogenicity, lethal toxicity, anticomplement activity, and induction of gelation of Limulus amoebocyte lysate. *Infection and Immunity*, **44**, 421–6.
4. Kusumoto, S. and Fukase, K. (2006) Synthesis of endotoxic principle of bacterial lipopolysaccharide and its recognition by the innate immune systems of hosts. *Chemical Record*, **6**, 333–43.
5. Medzhitov, R. (2007) Recognition of microorganisms and activation of the immune response. *Nature*, **819**, 819.
6. Wright, S.D., Ramos, R.A., Tobias, P.S., Ulevitch, R.J. and Mathison, J.C. (1990) CD14, a receptor for complexes of lipopolysaccharide (LPS) and LPS binding protein. *Science*, **249**, 1431–3.
7. Miller, S.I., Ernst, R.K. and Bader, M.W. (2005) LPS, TLR4 and infectious disease diversity. *Nature Reviews. Microbiology*, **3**, 36–46.
8. Poltorak, A., He, X., Smirnova, I., Liu, M.Y., Van Huffel, C., Du, X. *et al.* (1998) Defective LPS signaling in C3H/HeJ and C57BL/10ScCr mice: mutations in Tlr4 gene. *Science*, **282**, 2085–8.
9. Aderem, A. and Ulevitch, R.J. (2000) Toll-like receptors in the induction of the innate immune response. *Nature*, **406**, 782–7.
10. Medzhitov, R., Preston-Hurlburt, P. and Janeway, C.A. (1997) A human homologue of the Drosophila Toll protein signals activation of adaptive immunity. *Nature*, **388**, 394–7.
11. Rock, F.L., Hardiman, G., Timans, J.C., Kastelein, R.A. and Bazan, J.F. (1998) A family of human receptors structurally related to Drosophila Toll. *Proceedings of the National Academy of Sciences of the United States of America*, **95**, 588–93.
12. Gay, N.J. and Gangloff, M. (2007) Structure and function of Toll receptors and their ligands. *Annual Review of Biochemistry*, **76**, 141–65.
13. Gay, N.J., Gangloff, M. and Weber, A.N. (2006) Toll-like receptors as molecular switches. *Nature Reviews. Immunology*, **6**, 693–8.
14. O'Neill, L.A. and Bowie, A.G. (2007) The family of five: TIR-domain-containing adaptors in Toll-like receptor signaling. *Nature Reviews. Immunology*, **7**, 353–64.
15. Shizuo Akira, S., Uematsu, S. and Takeuchi, O. (2006) Pathogen recognition and innate immunity. *Cell*, **124**, 783–801.
16. Rice, T.W. and Bernard, G.R. (2005) Therapeutic intervention and targets for sepsis. *Annual Review of Medicine*, **56**, 225–48.
17. Heine, H., Rietschel, E.T. and Ulmer, A.J. (2001) The biology of endotoxin. *Molecular Biotechnology*, **19**, 279–96.
18. Johnson, D. (2008) Synthetic TLR4-active glycolipids as vaccine adjuvants and stand-alone immunotherapeutics. *Current Topics in Medicinal Chemistry*, **8**, 64–79.
19. Raetz, C.R.H. and Whitfield, C. (2002) Lipopolysaccharide endotoxins. *Annual Review of Biochemistry*, **71**, 635–700.
20. Trent, M.S., Stead, C.M., Tran, A.X. and Hankins, J.V. (2006) Diversity of endotoxin and its impact on pathogenesis. *Journal of Endotoxin Research*, **12**, 205–23.
21. Barb, A.W., Jiang, L., Raetz, C.R.H. and Zhou, P. (2007) Structure of the deacetylase LpxC bound to the antibiotic CHIR-090: time-dependent inhibition and specificity in ligand binding. *Proceedings of the National Academy of Sciences of the United States of America*, **104**, 18433–8.

22 Meredith, T.C., Aggarwal, P., Mamat, U., Lindner, B. and Woodard, R.W. (2006) Redefining the requisite lipopolysaccharide structure in *Escherichia coli*. *ACS Chemical Biology*, **1**, 33–42.

23 Wang, X., Ribeiro, A., Guan, Z., Abraham, S. and Raetz, C. (2007) Attenuated virulence of a Francisella mutant lacking the lipid A 4′-phosphatase. *Proceedings of the National Academy of Sciences of the United States of America*, **104**, 4136–41.

24 Munford, R.S. (2008) Sensing Gram-negative bacterial lipopolysaccharides: a human disease determinant?. *Infection and Immunity*, **76**, 454–65.

25 Peschel, A. and Sahl, H.G. (2006) The co-evolution of host cationic antimicrobial peptides and microbial resistance. *Nature Reviews. Microbiology*, **4**, 529–36.

26 Raetz, C.R.H., Reynolds, C.M., Trent, M.S. and Bishop, R.E. (2007) Lipid A modification systems in Gram-negative bacteria. *Annual Review of Biochemistry*, **76**, 295–329.

27 Prost, L.R. and Miller, S.I. (2008) The *Salmonellae* PhoQ sensor: mechanisms of detection of phagosome signals. *Cellular Microbiology*, **10**, 576–82.

28 De Soyza, A., Silipo, A., Lanzetta, R., Govan, J. and Molinaro, A. (2008) Review: chemical and biological features of Burkholderia cepacia complex lipopolysaccharides. *Innate Immunity*, **14**, 127–44.

29 Silipo, A., Molinaro, A., Ieranò, T., De Soyza, A., Sturiale, L., Garozzo, D. *et al.* (2006) The complete structure and pro-inflammatory activity of the lipooligosaccharide of the highly epidemic and virulent Gram-negative bacterium Burkholderia cenocepacia ET-12 (strain J2315). *Chemistry – A European Journal*, **13**, 3501–11.

30 Shimomura, H., Matsuura, M., Saito, S., Hirai, Y., Isshiki, Y. and Kawahara, K. (2003) Unusual interaction of a lipopolysaccharide isolated from Burkholderia cepacia with polymyxin B. *Infection and Immunity*, **71**, 5225–30.

31 Ortega, X.P., Cardona, S.T., Brown, A.R., Loutet, S.A., Flannagan, R.S., Campopiano, D.J. *et al.* (2007) A putative gene cluster for aminoarabinose biosynthesis is essential for Burkholderia cenocepacia viability. *Journal of Bacteriology*, **189**, 3639–44.

32 Schmelzer, K., Fahy, E., Subramaniam, S. and Dennis, E. (2007) The lipid maps initiative in lipidomics. *Methods in Enzymology*, **432**, 171–83.

33 Raetz, C.R.H., Garrett, T.A., Reynolds, C.M., Shaw, W.A., Moore, J.D., Smith, D.C., Jr *et al.* (2006) Kdo2-Lipid A of Escherichia coli, a defined endotoxin that activates macrophages via TLR-4. *Journal of Lipid Research*, **47**, 1097–111.

34 Guy, B. (2007) The perfect mix: recent progress in adjuvant research. *Nature Reviews. Microbiology*, **5**, 505–17.

35 Stöver, A., Da Silva Correia, J., Evans, J., Cluff, C., Elliott, M., Jeffery, E. *et al.* (2003) Structure–activity relationship of synthetic toll-like receptor 4 agonists. *Journal of Biological Chemistry*, **279**, 4440–9.

36 North, S. and Butts, C. (2005) Vaccination with BLP25 liposome vaccine to treat non-small cell lung and prostate cancers. *Expert Review of Vaccines*, **4**, 249–57.

37 Zhang, Y., Gaekwad, J., Wolfert, M.A. and Boons, G.J. (2007) Modulation of innate immune responses with synthetic lipid A derivatives. *Journal of the American Chemical Society*, **129**, 5200–16.

38 Zhang, Y., Gaekwad, J., Wolfert, M.A. and Boons, G.J. (2008) Innate immune responses of synthetic lipid A derivatives of *Neisseria meningitidis*. *Chemistry – A European Journal*, **14**, 558–69.

39 Saitoh, S.I. and Miyake, K. (2006) Mechanism regulating cell surface expression and activation of Toll-like receptor 4. *Chemical Record*, **6**, 311–19.

40 Ohto, U., Fukase, K., Miyake, K. and Satow, Y. (2007) Crystal structures of human MD-2 and its complex with antiendotoxic lipid IVa. *Science*, **316**, 1632–4.

41 Sil, D., Shrestha, A., Kimbrell, M.R., Nguyen, T.B., Adisechan, A.K., Balakrishna, R. *et al.* (2007) Bound to

shock: protection from lethal endotoxemic shock by a novel, nontoxic, alkylpolyamine lipopolysaccharide sequestrant. *Antimicrobial Agents and Chemotherapy*, **51**, 2811–19.

42 David, S.A. (2001) Towards a rational development of anti-endotoxin agents: novel approaches to sequestration of bacterial endotoxins with small molecules. *Journal of Molecular Recognition*, **14**, 370–87.

43 Voss, S., Fischer, R., Jung, G., Wiesmüller, K.H. and Brock, R. (2007) A fluorescence-based synthetic LPS sensor. *Journal of the American Chemical Society*, **129**, 554–61.

44 Leevy, W.M., Lambert, T.N., Johnson, J.R., Morris, J. and Smith, B.D. (2008) Quantum dot probes for bacteria distinguish *Escherichia coli* mutants and permit *in vivo* imaging. *Chemical Communications*, **2008**, 2331–3.

45 Chaby, R. (2004) Lipopolysaccharide-binding molecules: transporters, blockers and sensors. *Cellular and Molecular Life Sciences*, **61**, 1697–713.

46 Beamer, L.J. (2003) Structure of human BPI (bactericidal/permeability-increasing protein) and implications for related proteins. *Biochemical Society Transactions*, **31**, 791–4.

47 Pristovsek, P., Simcic, S., Wraber, B. and Urleb, U. (2005) Structure of a synthetic fragment of the lipopolysaccharide (LPS) binding protein when bound to LPS and design of a peptidic LPS inhibitor. *Journal of Medicinal Chemistry*, **48**, 7911–14.

48 Kim, J., Lee, C., Jin, M., Lee, C., Paik, S., Lee, H. and Lee, J. (2005) Crystal structure of CD14 and its implications for lipopolysaccharide signaling. *Journal of Biological Chemistry*, **280**, 11347–51.

49 Ferguson, A.D., Hofmann, E., Coulton, J.W., Diederichs, K. and Welte, W. (1998) Siderophore-mediated iron transport: crystal structure of FhuA with bound lipopolysaccharide. *Science*, **282**, 2215–20.

50 Gangloff, M. and Gay, N.J. (2004) MD-2: the Toll "gatekeeper" in endotoxin signaling. *Trends in Biochemical Sciences*, **29**, 294–300.

51 Mata-Haro, V., Cekic, C., Martin, M., Chilton, P., Casella, C. and Mitchell, T. (2007) The vaccine adjuvant monophosphoryl lipid A as a TRIF-biased agonist of TLR4. *Science*, **316**, 1628–32.

52 Kim, H.M., Park, B.S., Kim, J.I., Kim, S.E., Lee, J., Oh, S.C. et al. (2007) Crystal structure of the TLR4-MD-2 complex with bound endotoxin antagonist Eritoran. *Cell*, **130**, 906–17.

53 Ward, A., Reyes, C.L., Yu, J., Roth, C.B. and Chang, G. (2007) Flexibility in the ABC transporter MsbA: alternating access with a twist. *Proceedings of the National Academy of Sciences of the United States of America*, **104**, 19005–10.

54 Suits, M.D., Sperandeo, P., Dehò, G., Polissi, A. and Jia, Z. (2008) Novel structure of the conserved Gram-negative lipopolysaccharide transport protein A and mutagenesis analysis. *Journal of Biological Chemistry*, **380**, 476–88.

55 Thudichum, J.L. (1884). *A Treatise on the Chemical Constitution of the Brain*, Bailliere, Tindall and Cox, London.

56 Carter, H.E., Haines, H.W., Ledyard, W.E. and Norris, W.P. (1947) Biochemistry of the sphingolipids. *Journal of Biological Chemistry*, **169**, 77–82.

57 Simons, K. and Ikonen, E. (1997) Functional lipid rafts in cell membranes. *Nature*, **387**, 569–72.

58 Futerman, A.H. and Hannun, Y.A. (2004) The complex life of simple sphingolipids. *EMBO Reports*, **5**, 777–82.

59 Hannun, Y.A. and Obeid, L.M. (2008) Principles of bioactive lipid signaling: lessons from sphingolipids. *Nature Reviews. Molecular Cell Biology*, **9**, 139–50.

60 Merrill, A.H. (2002) De Novo sphingolipid biosynthesis: a necessary, but dangerous pathway? *Journal of Biological Chemistry*, **277**, 25843–6.

61 Chalfant, C.E. and Spiegel, S. (2005) Sphingosine 1-phosphate and ceramide 1-phosphate: expanding roles in cell signaling. *Journal of Cell Science*, **118**, 4605–412.

62 Spiegel, S. and Milstien, S. (2003) Sphingosine-1-phosphate: an enigmatic signaling lipid. *Molecular and Cellular Biology*, **4**, 397–407.

63 Dickson, R.C., Sumanasekera, C. and Lester, R.L. (2006) Functions and metabolism of sphingolipids in *Saccharomyces cerevisiae*. *Progress in Lipid Research*, **45**, 447–65.

64 Lynch, D.V. and Dunn, T.M. (2004) An introduction to plant sphingolipids and a review of recent advances in understanding their metabolism and function. *New Phytologist*, **161**, 677–702.

65 Kawahara, K., Seydel, U., Matsuura, M., Danbara, H., Rietschel, E.T. and Zahringer, U. (1991) Chemical structure of glycosphingolipids isolated from *Sphingomonas paucimobilis*. *FEBS Letters*, **292**, 107–10.

66 White, D.C., Sutton, S.D. and Ringelberg, D.B. (1996) The genus *Sphingomonas*: physiology and ecology. *Current Opinion in Biotechnology*, **7**, 301–6.

67 Ikushiro, H., Hayashi, H. and Kagamiyama, H. (2001) A water-soluble homodimeric serine palmitoyl-transferase from *Sphingomonas paucimobilis* EY2395T strain. Purification, characterization, cloning, and overproduction. *Journal of Biological Chemistry*, **276**, 18249–56.

68 Hanada, K. (2003) Serine palmitoyl-transferase, a key enzyme of sphingolipid metabolism. *Biochimica et Biophysica Acta*, **1632**, 16–30.

69 Alexeev, D., Alexeeva, M., Baxter, R.L., Campopiano, D.J., Webster, S.P. and Sawyer, L. (1998) The crystal structure of 8-amino-7-oxononanoate synthase: a bacterial PLP-dependent, acyl-CoA-condensing enzyme. *Journal of Molecular Biology*, **284**, 401–19.

70 Astner, A., Schulze, J.O., van den Heuvel, J., Jahn, D., Schubert, W.D. and Heinz, D.W. (2005) Crystal structure of 5-aminolevulinate synthase, the first enzyme of heme biosynthesis, and its link to XLSA in humans. *EMBO Journal*, **24**, 3166–77.

71 Ferreira, G.C. and Gong, J. (1995) 5-Aminolevulinate synthase and the first step of heme biosynthesis. *Journal of Bioenergetics and Biomembranes*, **27**, 151–9.

72 Jordan, P.M. and Shemin, D. (1972) 5-Aminolevulinic acid synthase, in *The Enzymes*, Vol. VII, 3rd edn (ed. P.D. Boyer), Academic Press, New York, pp. 339–56.

73 Schmidt, A., Sivaraman, J., Li, Y., Larocque, R., Barbosa, J.A.R.G., Smith, C. et al. (2001) Three-dimensional structure of 2-amino-3-ketobutyrate CoA ligase from *Escherichia coli* complexed with a PLP-substrate intermediate: inferred reaction mechanism. *Biochemistry*, **40**, 5151–60.

74 Webster, S.P., Alexeev, D., Campopiano, D.J., Watt, R.M., Alexeeva, M., Sawyer, L. and Baxter, R.L. (2000) Mechanism of 8-amino-7-oxononanoate synthase: spectroscopic, kinetic, and crystallographic studies. *Biochemistry*, **39**, 516–28.

75 Eliot, A.C. and Kirsch, J.F. (2004) Pyridoxal phosphate enzymes: mechanistic, structural, and evolutionary considerations. *Annual Review of Biochemistry*, **73**, 383–415.

76 Alexeev, D., Baxter, R.L., Campopiano, D.J., Kerbarh, O., Sawyer, L., Tomczyk, N. et al. (2006) Suicide inhibition of α-oxamine synthases: structures of the covalent adducts of 8-amino-7-oxononanoate synthase with trifluoroalanine. *Organic and Biomolecular Chemistry*, **4**, 1209–12.

77 Hunter, G.A. and Ferreira, G.C. (1999) Lysine-313 of 5-aminolevulinate synthase acts as a general base during formation of the quinonoid reaction intermediates. *Biochemistry*, **38**, 3711–18.

78 Kerbarh, O., Campopiano, D.J. and Baxter, R.L. (2006) Mechanism of α-oxoamine synthases: identification of the intermediate Claisen product in the 8-amino-7-oxononanoate synthase reaction. *Chemical Communications*, 60–2.

79 Zaman, Z., Jordan, P.M. and Akhtar, M. (1973) Mechanism and stereochemistry of the 5-aminolaevulinate synthetase

reaction. *Biochemical Journal*, **135**, 257–63.
80 Ikushiro, H., Fujii, S., Shiraiwa, Y. and Hayashi, H. (2008) Acceleration of the substrate Cα deprotonation by an analogue of the second substrate palmitoyl-CoA in serine palmitoyltransferase. *Journal of Biological Chemistry*, **283**, 7542–53.
81 Hornemann, T., Richard, S., Rütti, M.F., Wei, Y. and von Eckardstein, A. (2006) Cloning and initial characterization of a new subunit for mammalian serine-palmitoyltransferase. *Journal of Biological Chemistry*, **281**, 37275–81.
82 Wilson, W.H., Schroeder, D.C., Allen, M.J., Holden, M.T., Parkhill, J., Barrell, B.G. et al. (2005) Complete genome sequence and lytic phase transcription profile of a Coccolithovirus. *Science*, **309**, 1090–2.
83 Han, G., Gable, K., Yan, L., Allen, M.J., Wilson, W.H., Moitra, P. et al. (2006) Expression of a novel marine viral single-chain serine palmitoyltransferase and construction of yeast and mammalian single-chain chimera. *Journal of Biological Chemistry*, **281**, 39935–42.
84 Yard, B.A., Carter, L.G., Johnson, K.A., Overton, I.M., Dorward, M., Liu, H. et al. (2007) The structure of serine palmitoyltransferase; gateway to sphingolipid biosynthesis. *Journal of Molecular Biology*, **370**, 870–86.
85 Bejaoui, K., Wu, C., Scheffler, M.D., Haan, G., Ashby, P., Wu, L. et al. (2001) SPTLC1 is mutated in hereditary sensory neuropathy, type 1. *Nature Genetics*, **27**, 261–2.
86 Dawkins, J.L., Hulme, D.J., Brahmbhatt, S.B., Auer-Grumbach, M. and Nicholson, G.A. (2001) Mutations in SPTLC1, encoding serine palmitoyltransferase, long chain base subunit-1, cause hereditary sensory neuropathy type I. *Nature Genetics*, **27**, 309–12.
87 Thevissen, K., Francois, I.E., Aerts, A.M. and Cammue, B.P. (2005) Fungal sphingolipids as targets for the development of selective antifungal therapeutics. *Current Drug Targets*, **6**, 923–8.
88 Kronenberg, M. and Gapin, L. (2002) The unconventional lifestyle of NKT cells. *Nature Reviews. Immunology*, **2**, 557–68.
89 Bendelac, A., Savage, P.B. and Teyton, L. (2007) The biology of NKT cells. *Annual Review of Immunology*, **25**, 297–336.
90 Akimoto, K., Natori, T. and Morita, M. (1993) Synthesis and stereochemistry of Agelasphin-9b. *Tetrahedron Letters*, **34**, 5593–6.
91 Natori, T., Koezuka, Y. and Higa, T. (1993) Agelasphins, novel α-galactosylceramides from the marine sponge Agelas mauritianus. *Tetrahedron Letters*, **34**, 5591–2.
92 Kobayashi, E., Motoki, K., Uchida, T., Fukushima, H. and Koezuka, Y. (1995) KRN7000, a novel immunomodulator, and its antitumor activities. *Oncology Research*, **7**, 529–34.
93 Kawano, T., Cui, J., Koezuka, Y., Toura, I., Kaneko, Y., Motoki, K. et al. (1997) CD1d-restricted and TCR-mediated activation of $V_\alpha 14$ NKT cells by glycosylceramides. *Science*, **278**, 1626–9.
94 Park, J.J., Lee, J.H., Ghosh, S.C., Bricard, G., Venkataswamy, M.M., Porcelli, S.A. and Chung, S.K. (2008) Synthesis of all stereoisomers of KRN7000, the CD1d-binding NKT cell ligand. *Bioorganic and Medicinal Chemistry Letters*, **18**, 3906–9.
95 Stanic, A.K., De Silva, A.D., Park, J.J., Sriram, V., Ichikawa, S., Hirabyashi, Y. et al. (2003) Defective presentation of the CD1d1-restricted natural $V_\alpha 14 J_\alpha 18$ NKT lymphocyte antigen caused by β-D-glucosylceramide synthase deficiency. *Proceedings of the National Academy of Sciences of the United States of America*, **100**, 1849–54.
96 Zhou, D., Mattner, J., Cantu, C., 3rd, Schrantz, N., Yin, N., Gao, Y. et al. (2004) Lysosomal glycosphingolipid recognition by NKT cells. *Science*, **306**, 1786–9.
97 Fischer, K., Scotet, E., Niemeyer, M., Koebernick, H., Zerrahn, J., Maillet, S. et al. (2004) Mycobacterial phosphatidylinositol mannoside is a natural

98 Savage, P.B., Teyton, L. and Bendelac, A. (2006) Glycolipids for natural killer T cells. *Chemical Society Reviews*, **35**, 771–9.

99 Sidobre, S., Hammond, K.J., Benazet-Sidobre, L., Maltsev, S.D., Richardson, S.K., Ndonye, R.M. et al. (2004) The T cell antigen receptor expressed by $V_\alpha 14i$ NKT cells has a unique mode of glycosphingolipid antigen recognition. *Proceedings of the National Academy of Sciences of the United States of America*, **101**, 12254–9.

100 Miyamoto, K., Miyake, S. and Yamamura, T. (2001) A synthetic glycolipid prevents autoimmune encephalomyelitis by inducing $T_H 2$ bias of natural killer T cells. *Nature*, **413**, 531–4.

101 Wu, D., Xing, G.W., Poles, M.A., Horowitz, A., Kinjo, Y., Sullivan, B. et al. (2005) Bacterial glycolipids and analogs as antigens for CD1d-restricted NKT cells. *Proceedings of the National Academy of Sciences of the United States of America*, **102**, 1351–6.

102 Schmieg, J., Yang, G., Franck, R.W. and Tsuji, M. (2003) Superior protection against malaria and melanoma metastases by a C-glycoside analogue of the natural killer T cell ligand α-galactosylceramide. *Journal of Experimental Medicine*, **198**, 1631–41.

103 Yang, G., Schmieg, J., Tsuji, M. and Franck, R.W. (2004) The C-glycoside analogue of the immunostimulant α-galactosylceramide (KRN7000): synthesis and striking enhancement of activity. *Angewandte Chemie (International Edition in English)*, **43**, 3818–22.

104 Goff, R.D., Gao, Y., Mattner, J., Zhou, D., Yin, N., Cantu, C., 3rd et al. (2004) Effects of lipid chain lengths in α-galactosylceramides on cytokine release by natural killer T cells. *Journal of the American Chemical Society*, **126**, 13602–3.

105 Ndonye, R.M., Izmirian, D.P., Dunn, M.F., Yu, K.O., Porcelli, S.A., Khurana, A. et al. (2005) Synthesis and evaluation of sphinganine analogues of KRN7000 and OCH. *Journal of Organic Chemistry*, **70**, 10260–70.

106 Yu, K.O., Im, J.S., Molano, A., Dutronc, Y., Illarionov, P.A., Forestier, C. et al. (2005) Modulation of CD1d-restricted NKT cell responses by using N-acyl variants of α-galactosylceramides. *Proceedings of the National Academy of Sciences of the United States of America*, **102**, 3383–8.

107 Oki, S., Chiba, A., Yamamura, T. and Miyake, S. (2004) The clinical implication and molecular mechanism of preferential IL-4 production by modified glycolipid-stimulated NKT cells. *Journal of Clinical Investigation*, **113**, 1631–40.

108 Giabbai, B., Sidobre, S., Crispin, M.D., Sanchez-Ruiz, Y., Bachi, A., Kronenberg, M. et al. (2005) Crystal structure of mouse CD1d bound to the self ligand phosphatidylcholine: a molecular basis for NKT cell activation. *Journal of Immunology*, **175**, 977–84.

109 Koch, M., Stronge, V.S., Shepherd, D., Gadola, S.D., Mathew, B., Ritter, G. et al. (2005) The crystal structure of human CD1d with and without α-galactosylceramide. *Nature Immunology*, **6**, 819–26.

110 Zajonc, D.M., Maricic, I., Wu, D., Halder, R., Roy, K., Wong, C.H. et al. (2005) Structural basis for CD1d presentation of a sulfatide derived from myelin and its implications for autoimmunity. *Journal of Experimental Medicine*, **202**, 1517–26.

111 Borg, N.A., Wun, K.S., Kjer-Nielsen, L., Wilce, M.C., Pellicci, D.G., Koh, R. et al. (2007) CD1d-lipid-antigen recognition by the semi-invariant NKT T-cell receptor. *Nature*, **448**, 44–9.

112 Zajonc, D.M., Savage, P.B., Bendelac, A., Wilson, I.A. and Teyton, L. (2008) Crystal structures of mouse CD1d–iGb3 complex and its cognate $V_\alpha 14$ T cell receptor suggest a model for dual

recognition of foreign and self glycolipids. *Journal of Molecular Biology*, **377**, 1104–16.

113 Burdin, N., Brossay, L., Degano, M., Iijima, H., Gui, M., Wilson, I.A. and Kronenberg, M. (2000) Structural requirements for antigen presentation by mouse CD1. *Proceedings of the National Academy of Sciences of the United States of America*, **97**, 10156–61.

114 Kamada, N., Iijima, H., Kimura, K., Harada, M., Shimizu, E., Motohashi, S. *et al.* (2001) Crucial amino acid residues of mouse CD1d for glycolipid ligand presentation to $V_\alpha 14$ NKT cells. *International Immunology*, **13**, 853–61.

115 Sidobre, S., Naidenko, O.V., Sim, B.C., Gascoigne, N.R., Garcia, K.C. and Kronenberg, M. (2002) The $V_\alpha 14$ NKT cell TCR exhibits high-affinity binding to a glycolipid/CD1d complex. *Journal of Immunology*, **169**, 1340–8.

116 Kinjo, Y., Pei, B., Bufali, S., Raju, R., Richardson, S.K., Imamura, M. *et al.* (2008) Natural sphingomonas glycolipids vary greatly in their ability to activate natural killer T cells. *Chemistry and Biology*, **15**, 654–64.

7
Supramolecular Assemblies of Polydiacetylenes for Biomolecular Sensing: Colorimetric and "Turn-On" Fluorescence Approaches

Guangyu Ma and Quan Jason Cheng

7.1
Introduction

There exists a lasting interest in the scientific community to develop biosensors for the detection of hazardous chemicals, disease-causing bacteria/viruses, cancer markers, and other molecules of environmental, food safety and healthcare significance. These sensors are compact, integrated, easy to operate robust, and can produce results in a straightforward fashion. Traditional analytical methods and immunoassays have been used for these measurements for a long period of time. However, the lengthy procedure and skill-demanding operation have made the traditional approaches less desirable as they usually fail to satisfy the compelling needs for speed and convenience.

In recent years, functional polymers, especially conjugated polymers (CPs), have attracted tremendous interest as "smart" sensing materials because of their unique electrical and optical properties [1–6]. A common feature of these polymers is their alternative single-bond–double (or triple)-bond backbone. The conjugated $\pi-\pi$ stacking system allows for electron delocalization along the backbone, which forms the basis for conductivity of these polymers. In addition, it offers attractive optical properties that are very desirable for molecular sensing.

Much of the early work on conjugated materials focused on conductivity properties [7] and therefore several conductivity-based sensing devices have been developed [7, 8]. Applications have also extended to electroactive coatings for electron transport studies between the electrode and the redox species [9], which led to electrochemical sensors [8, 10]. In the past decade, there has been increasing interest in the exploration of the optical properties of the CP materials [2], including their use in the areas of light-emitting diodes and sensing devices [4, 11–17]. Given that optical signals are well understood and straightforward, CP-based optical sensors have in fact been developed into an important branch in the biosensor family. Among all CPs studied for sensing applications, polypyrrole, polythiophene, polyaniline and polydiacetylene (PDA) are the most extensively investigated [3]. Swager *et al.* have examined a series of conjugated polymers as

Cellular and Biomolecular Recognition: Synthetic and Non-Biological Molecules. Edited by Raz Jelinek
Copyright © 2009 WILEY-VCH Verlag GmbH & Co. KGaA, Weinheim
ISBN: 978-3-527-32265-7

possible sensing materials for a variety of molecular targets [2, 3, 18–20]. Conductivity, electrochemical and optical properties were all employed in these sensor designs. This group was also among the first to develop the amplified fluorescence quenching sensing mechanism for CP-based chemical sensors, in which the excited states (excitons) propagating along the conjugated backbone are quenched when encountering an energy trap at an occupied receptor site, leading to diminished fluorescence [21–23]. Leclerc et al. developed water-soluble cationic polythiophene derivatives as sensing materials and have carried out a series of excellent works in biomolecular sensing from DNA oligonucleotides to thrombin [1, 16, 24–26]. Target recognition events are recognized via either colorimetric or fluorescent signals. Another series of excellent work has been presented by Bazan et al. who developed a series of DNA detection strategies utilizing fluorescence (or Förster) resonance energy transfer (FRET) between the conjugated polymers and fluorescent dyes [27–31].

Compared to other commonly used functional CPs, PDA stands out as a unique material that has been initially employed for the litmus-style colorimetric sensing. PDA is a conjugated polymer with alternating ene–yne conjugated backbone and can be obtained by photopolymerization of diacetylene monomers. The spatially aligned monomeric diacetylene lipids, under ultraviolet (UV) irradiation, undergo a photopolymerization process via a 1,4-addition mechanism and form π-conjugated polymer chains that give the material a sharp colored appearance [32, 33]. PDA exhibits a distinctive chromatic transition upon stimulation by temperature increase, mechanical stress or chemical solvents [34]. Charych and Bednaski were among the first to demonstrate the colorimetric biosensing of influenza virus [35]. Since then, PDA colorimetric sensors have been developed for cholera toxin [36], *Escherichia coli* [37], epitope-binding antibody [38], catecholamines [38] and and lipopolysaccharides [39], among others.

The biosensor research with CPs involves many different strategies and detection mechanisms. For the development of a "smart" sensor, a specific design must be implemented in order to achieve the required selectivity, sensitivity and signal transduction. Although tremendous research efforts have been invested in the area, applications of these sensors are still limited. Much work is still needed for fabricating new materials and creating novel designs, while the selectivity, sensitivity, simplicity and reproducibility need to be further improved. Our group has been interested in innovating PDA-based biosensors through establishment of new materials and detection schemes. Specifically, we have expanded the application of PDA colorimetric detection and developed the "turn-on" fluorescence detection scheme using PDA materials. These new systems, in our opinion, open new avenues for advancing biosensor technology that targets biomolecular interactions and sensing. In this chapter we will review some of the latest work in the development of PDA-based biosensors.

7.2 Vesicular PDA Sensors for Colorimetric Signaling of Bacterial Pore-Forming Toxin

Cytolytic pore-forming toxins (PFTs) comprise about 25% of all known bacterial protein toxins and are regarded as important for the virulence of many disease-causing bacteria [40]. One of the significant PFTs is streptolysin O (SLO) from *Streptococcus pyrogenes,* a cholesterol-dependent cytolysin that has served as a major virulence factor in streptococcus infections of soft tissue such as strep throat, impetigo and necrotizing fascitis [41]. Upon binding to the cell membrane, SLO can generate transmembrane pores of sizes ranging from a few nanometers up to 35 nm [40].

We took advantage of this unique pore-forming capability of the toxin and demonstrated a new sensing mechanism using PDA supramolecular assemblies. It is hoped that when SLO toxin binds to cholesterol molecules incorporated in the PDA vesicles and starts to form pores on the vesicle membrane, the toxin activity amplifies the perturbation to the conjugated backbone and thus generates significant colorimetric signals.

The vesicular SLO sensor was constructed with three lipid constituents: a glycine-terminated diacetylene monomer (Gly-PCDA), 1,2-bis(10,12-tricosadiynoyl)-sn-glycero-3-phosphocholine (PC-DIYNE) and cholesterol (CHO) (Figure 7.1). PC-DIYNE is membrane-mimicking component, and was used to adjust the stability and phase transition property of the vesicles. CHO serves as a bait molecule for SLO toxin since the first step in pore formation is believed to be its binding to CHO [41]. The working composition of the vesicle sensors was obtained by carrying out a series of screening experiments that optimize both the vesicle composition and experimental conditions. The key to inducing a color change by a membrane-targeting toxin appears to be the balance between the degree of polymer conjugation and the membrane fluidity.

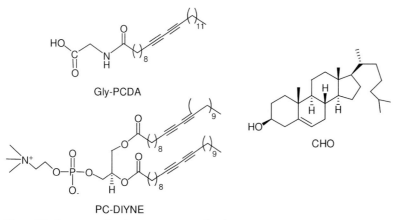

Figure 7.1 Structure of lipid constituents used in the fabrication of PDA vesicular biosensor for SLO toxin detection.

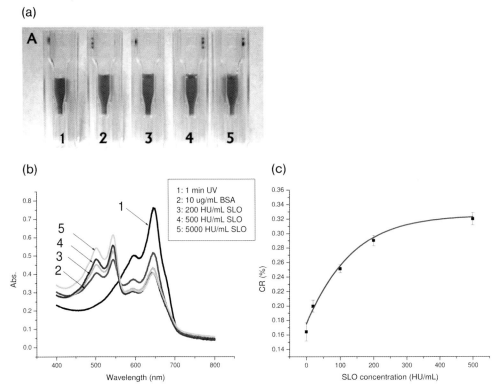

Figure 7.2 (a) Color change and (b) corresponding absorption spectra of PDA vesicle sensors in response to SLO toxin. (1) Original vesicles, (2) BSA control, (3) 200 HU/ml SLO, (4) 500 HU/ml SLO and (5) 5000 HU/ ml SLO. The incubation time was 30 min and the incubation temperature was 32 °C. (c) Calibration plot of CR versus SLO concentration. (From [42]).

It was demonstrated that the vesicle sensor consisting of 71% Gly-PCDA, 25% CHO and 4% PC-DIYNE (molar ratio) gave the strongest color change [42]. Incorporation of PC-DIYNE lipid into the vesicles appears to greatly improve the sensing response to SLO. Figure 7.2 shows the colorimetric response of the vesicle sensor against SLO toxin. Addition of 200 HU/ml SLO toxin (corresponds roughly to 2.5 nM), followed by incubation for 30 min, turned the solution from blue to red (Figure 7.2a, 3). Addition of more SLO toxin (Figure 7.2a, 4) resulted in greater color change, verifying the sensing response while also indicating the high end of the response range. The control experiment with excess bovine serum albumin (BSA; 10 mg/ml) only showed a very minor chromatic transition to blue/purple color (Figure 7.2a, 2) due to thermal effect.

The absorbance responses as a function of SLO concentration are shown in Figure 7.2b. Colorimetric response (CR) [35], the percent change before and after toxin incubation in the maximum absorption at 649 nm normalized with the total absorption at both 649 and 545 nm, is used to quantify the colorimetric signals generated when SLO toxin binds to the vesicles. The calibration plot of CR versus SLO concentrations is shown in Figure 7.2c. A correlation relationship was obtained in the range of 20–500 HU/ml (250 pM to 6.3 nM) of SLO toxin.

Transmission electron microscopy micrographs confirmed the pore formation of the PDA vesicles. After incubation with SLO toxin, pores can be clearly seen and the sizes were estimated to be around 30 nm, which agrees well with literature [43]. Additional characterization was performed with dynamic light scattering, which monitored the size change of vesicles with SLO incubation. It was found that incubation of the vesicles with SLO toxin gave a size measurement of 380 nm, yielding a 25 nm increase, which agrees well with previous results of pore formation on phosphatidylcholine (PC) vesicles [44].

This work outlines a new mechanism for colorimetric detection of proteins with PDA materials. Compared to receptor–ligand's bind-then-trigger mechanism, the biological activity of the protein toxin (i.e., pore formation) was used to further amplify the disturbance to the conjugated backbone, thus generating a strong signal. Nonspecific interaction is no longer an important interference factor and no synthesis is required to attach the receptor to the PDA backbone, which dramatically reduced the effort in sensor fabrication.

7.3
Fabrication of "Turn-On" Fluorescence Vesicle Sensors with PDAs

In CP-based biosensing, there is a considerable interest in the development of fluorescent materials that render direct fluorometric measurements of analytes by means of detecting changes of fluorescence intensity, wavelength or lifetime [2, 3, 17, 45]. Colorimetric detection techniques normally focus on simplicity, portability along with fast detection, while fluorometric detection is more accurate in quantitation and has a potential to reach lower detection limit. Most of the current PDA sensors use colorimetric detection because PDA is known to have low quantum yield, especially in the blue form, and thus not sufficient for fluorescence sensor applications. We set out to extend the PDA sensing with fluorescence signal by combining the conjugated chain of PDA with incorporated fluorophores.

For fluorescent sensors, a highly desirable and more sensitive approach is the "turn-on" mode of detection, in which the measured signal increases with target molecule concentration [3, 27, 46–48]. We first developed a very simple approach for the fabrication of a "turn-on" fluorescence sensor by using PDA vesicles. The sensor consists of a conjugated PDA moiety and a bleach-resistant, hydrophobic dipyrromethene boron difluoride (BODIPY) fluorescence dye, prepared by mixing Gly-PCDA and BODIPY 558/568 dye (BO558) in the molar ratio of 100:0.5. The excitation and emission spectra of the BODIPY dye are shown in Figure 7.3. This dye is highly photo-stable, making it suitable for this work since the polymerization of the diacetylene vesicles involves UV irradiation at 254 nm. After

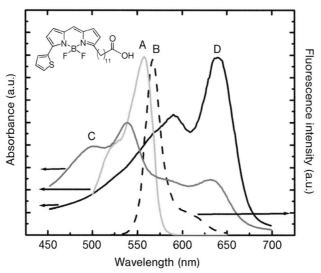

Figure 7.3 Absorption and emission spectra of PDA and BODIPY dye BO558. Curves A, C and D are absorption spectra of BO558, red-form PDA and blue-form PDA, respectively; curve B is the emission spectrum of BO558. Insert: Chemical structure of BO558.

photopolymerization, the solution exhibits a blue/purple mixed color of the blue PDA and the pink BODIPY dye.

These vesicles show interesting fluorescence "turn-on" properties in aqueous solution in response to changing pH. Figure 7.4 shows the fluorescence response upon polymerization as well as to addition of acid and base. Photopolymerization of the diacetylene lipids leads to a substantial quenching of the BODIPY dye in the vesicles. Then the fluorescence of the vesicles "turns on" at high pH and "turns off" at low pH. It is important to point out that this process is totally reversible, demonstrating a unique "on–off" switching behavior.

Comparison experiments on BO558-containing vesicles prepared from the saturated lipid L-α-phosphatidylglycerol and on free BO558 dye dissolved in organic solvent indicate that the quench–recover effect is not due to the photo-bleaching of the BO558 dye nor to that the dye itself is sensitive to pH changes. The mechanism is believed to be a FRET phenomenon between BO558 and the PDA backbone. Two types of energy transfer may be responsible for the fluorescence quenching in these vesicles: dynamic and static [49]. The dynamic quenching relies on a distance-dependent process (e.g., Förster energy transfer), where the quenching probability depends strongly on distance ($1/R^6$, where R is the intermolecular distance). Under this mechanism, when pH increases, the average distance between BO558 and the PDA segments increases, leading to reduced quenching. For a static mechanism, on the other hand, some fraction of dye is associated with a specific site in the PDA vesicles, which completely quenches its fluorescence.

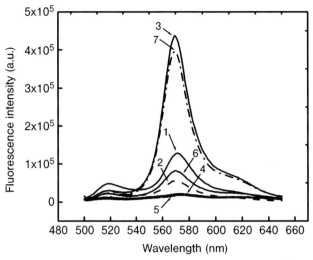

Figure 7.4 Fluorescence spectra of the Gly-PDA/BO558 vesicle system in response to acid–base addition. Curves 3–7 show subsequent addition of either acid or base into the testing solution. (1) Before UV irradiation; (2) after 1 min UV irradiation; (3) addition of 20 μl 0.1 M NaOH; (4) addition of 20 μl 0.1 M HCl; (5) addition of 20 μl 0.1 M HCl; (6) addition of 20 μl 0.1 M NaOH; (7) addition of 20 μl 0.1 M NaOH. (From [6]).

In order to distinguish between these two mechanisms, time-resolved fluorescence was used for a more detailed investigation. Figure 7.5 shows the normalized fluorescence decays at three different pH values (pH 7, 8.9 and 10.0). At all pH values, there is a fast, early time decay on the order of 100 ps, which is due to the fluorescence decay of the red PDA. This is followed by a much slower decay, whose spectral shape is identical to that of BO558. The decay time of the long-lived emission changes from 6.2 ns at pH 10.0 to 5.3 ns at pH 8.9, a decrease of only 15%, while the amplitude of the BODIPY steady-state fluorescence decreases by about 60%. The fact that the fluorescence decay time hardly changes while the fluorescence intensity declines dramatically would appear to rule out a conventional dynamic quenching process, either due to Förster energy transfer or due to some other rate-driven process. Furthermore, dynamic light scattering shows the size of the Gly-PDA vesicles increased by about 20 nm after treatment with NaOH solution, due to electrostatic repulsion of the deprotonated carboxylic acid head groups that lead to swelling of the vesicles. The increase in vesicle size suggests an increase in intermolecular distance as well, but the 6% increase in average distance R would only correspond to a 32% change in quenching rate, assuming a $1/R^6$ Förster mechanism. Again, this relatively small change is inconsistent with the much larger quenching observed in the steady-state spectra.

We thus proposed that the "on–off" fluorescence property is instead due to a static mechanism based on a pH-dependent association of BO558 with the PDA vesicle framework. This static quenching process would not affect the lifetime, but the fluorescence would be either on or off depending on whether the BODIPY

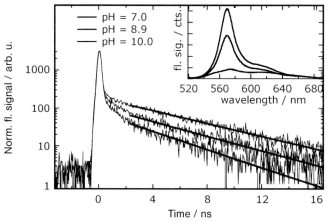

Figure 7.5 Time-resolved fluorescence decays of the GlyPDA/BO558 vesicles at pH 10.0 (top), pH 8.9 (middle) and pH 7.0 (bottom). The straight lines are single-exponential fits to the long-lived component of the decays. The corresponding steady-state fluorescence spectra are shown in the inset. (From [6]).

is bound or unbound. Both BO558 and the functionalized PCDA possess carboxylic acid groups, which will be completely deprotonated at pH 10. Coulombic repulsion thus prevents association of the two molecules, in the same way for the swelling of the vesicles. Rather than changing the average intermolecular distance, increase in pH shifts the equilibrium from bound to unbound dye, leading to break down of the quenching and occurrence of the fluorescence signal. Figure 7.6 is a cartoon illustration of the bound ("off") and unbound ("on") states. At high pH, the increased head group size cannot be accommodated by the original packing. Since the polymer chain prevents free expansion (as would be the case in unpolymerizable vesicles), the carboxylate head group would take a staggered arrangement to minimize the impact [50]. As the dye molecules are not covalently linked to the polymer backbone they would most likely be pushed towards the vesicle periphery, generating an "unbound" state that shows high fluorescence. At low pH, the staggered packing is reversed, restoring the quenching properties.

To demonstrate the sensing function of the PDA-BODIPY vesicles, three organic amines (triethylamine, diisopropylamine and benzylamine) were tested [51]. The results are shown in Figure 7.7. The steady-state fluorescence intensity responds linearly to the amine concentration in the range of 10 to 0.2 mM for all the amines investigated. The detection sensitivity (slope) appears to be associated with the basicity of the compounds. The three amine compounds can be preliminarily differentiated by the slope of the calibration curves (pK_as of triethylamine, diisopropylamine and benzylamine are 10.80, 10.76 and 9.40, respectively). Using the 3σ cut-off, the detection limit was determined to be 3.0 µM for triethylamine, 4.4 µM for diisopropylamine and 5.7 µM for benzylamine.

7.3 Fabrication of "Turn-On" Fluorescence Vesicle Sensors with PDAs | 185

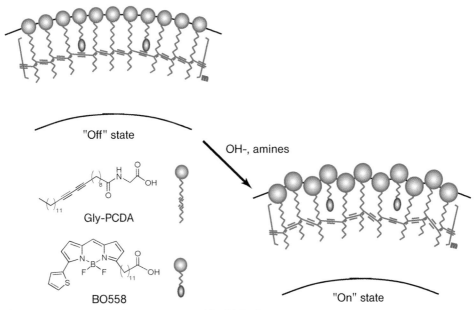

Figure 7.6 Structure of lipid constituents used for fabrication of the vesicle sensor and a schematic illustration of the fluorescence "turn-on" mechanism. Only one leaflet of the vesicle membrane is shown for better demonstration of the dye position.

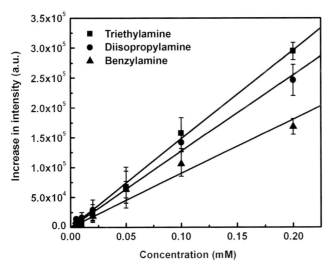

Figure 7.7 Calibration curves of the Gly-PDA/BO558 vesicle sensors for the detection of triethylamine, diisopropylamine and benzylamine in the concentration range of 5 to 200 μM. (From [51]).

7.4
"Mix-and-Detect" Type of "Turn-On" Fluorescence Sensor for Bacterial Toxin

Following the discovery of the "fluorescence switch" phenomenon based on the FRET mechanism between PDA and BO558 dye, we set out to explore the application of the system for detection of biological analytes. As discussed previously, DNA detection has been well demonstrated by using conjugated polymers and fluorometric detection schemes, while the detection of proteins with these methods remains challenging and has received far less success.

Using the FRET mechanism with PDA and BO-GM1 (a ganglioside), we developed a "mix-and-detect" type of fluorescent sensor for direct detection of bacterial cholera toxin. The principle of fluorescence sensing and the structure of BO-GM1 are shown in Figure 7.8. Similar to BO558, the hydrophobic BO-GM1 dye shows a strong tendency toward insertion into PDA vesicles in aqueous solution and the binding of a large molecule onto the lipid head group prohibits the insertion.

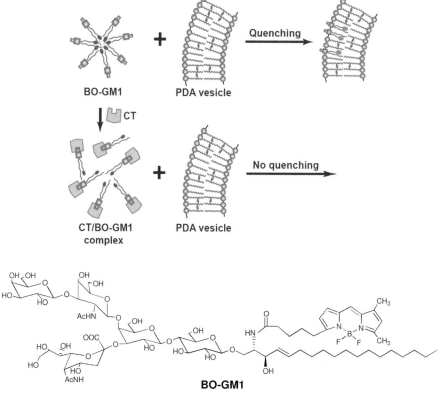

Figure 7.8 Schematic illustration of the sensing principle for the PDA vesicle based fluorescence sensor (top) and the structure of BO-GM1 (bottom).

In this design, charge-free BODIPY dye is responsible for signaling as its energy transfer with PDA moiety has been demonstrated in the previous study [6], whereas GM1 functions as the receptor for cholera toxin (CT) with picomolar binding affinity [52]. Parameters that affect quenching efficiency such as degree of polymerization and ratio of PDA to dye molecules have been investigated to optimize the sensing performance.

It was observed when the PDA vesicle solution was mixed with the BO-GM1 solution in phosphate-buffered saline buffer (pH 7.3), which is the medium for the CT test, sufficient quenching of the fluorescence signal was observed (above 60%) compared to that in deionized water. The low solubility and nonpolar ceramide tail in BO-GM1 is believed to promote insertion of the dye molecule into the membrane when the vesicles are present. The stable anchorage of the dye in the hydrophobic interior of the vesicle membrane and the resulting close proximity of the dye molecules to the conjugated chains enables energy transfer from dye to the polymer backbone, yielding the observed quenching effect. When CT was first mixed into the BO-GM1 solution, however, the quenching by PDA vesicles was essentially blocked. The observed fluorescence signal is found to be linearly dependent on the amount of CT present in solution. If saturated PC was used to prepare vesicles to replace PDA, no quenching effect was observed for BO-GM1, highlighting the need of a PDA backbone in the manipulation of the quenching effect [53].

We then compared the fluorescence quenching of BO558 to BO-GM1 in the CT test, the former was extensively studied in a previous work. A higher degree of quenching was observed with BO558 and Figure 7.9 shows the test results. Despite

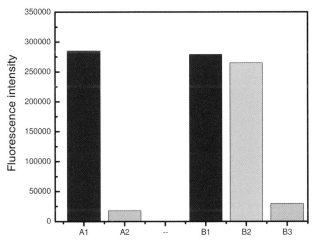

Figure 7.9 Fluorescence quenching of BO558 lipid dye by PDA vesicles. A1 and B1 are the initial intensity of BO558 in the Tris buffer (5.3 μM). A2 is the intensity with addition of PDA vesicles (1 mM stock solution). B2 shows addition of CT (0.1 mg/ml). B3 shows the addition of PDA vesicles into BO558 + CT solution.

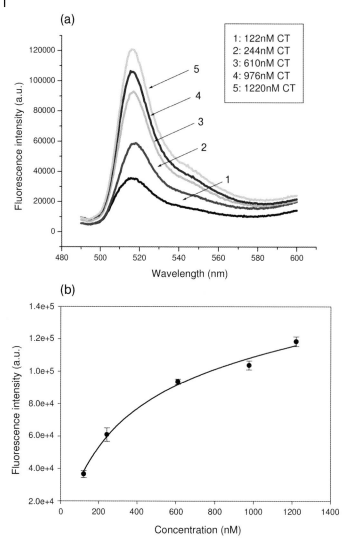

Figure 7.10 Fluorescence spectra of the vesicle sensor in response to different concentrations of CT (a) and the calibration curve for CT detection (b). (From [53]).

the strong quenching for BO558, addition of CT could not inhibit the quenching as seen with BO-GM1. This clearly demonstrates that the specific CT/GM1 interaction has a blocking effect on molecular insertion of the dyes into the vesicle membranes and therefore generates the sensing signal.

Figure 7.10 shows the fluorescence spectra as a function of CT concentration and the corresponding calibration curve. The fluorescence intensity increases with CT concentration, demonstrating a "turn-on" fluorescence detection characteristic, a desirable scheme as compared to the amplified quenching. The detection can reach 120 nM, which is comparable to the values observed using nonlabeling FRET with analyte analogs [54]. The "mix-and-detect" procedure is straightforward and simple. In addition to almost effortless fabrication, the batch-to-batch variations were found to be very small as the initial intensity of the BO-GM1 solution can be carefully controlled. In comparison to other FRET sensors, this design is unique in that the detection is based on molecular insertion into vesicle membrane, which potentially eliminates interference from nonspecific protein/backbone interactions.

7.5 Conclusions

In this chapter we summarize some of the latest work in the exploration of new bio/chemical sensing strategies based on functional conjugated polymers and our own approaches in the development of optical biosensors using functional PDAs. It is clear to us that the field is far from mature and tremendous work awaits before commercial success can be realized. Nevertheless, we believe the future of optical CP sensors is bright and it offers a unique design scheme for developing label-free, one-step sensitive devices for both medical and environmental applications. The innovative "smart" sensing techniques using functional polymers have already been demonstrated in the detection of metal ions, small organic molecules and certain bioanalytes. Unlike traditional analytical devices and immunoassays, these sensors require minimal experimental efforts and little professional skills, and many of them are simply "mix and detect" type. Optical signaling with color and fluorescence change is very desirable for applications because the signals are self-explanatory and compatible with one's intuition. In general, colorimetric detection is ideal for a low-cost, litmus paper style of sensing, while fluorometric detection is more sensitive, thus making it more suitable for the industrial setting where sensitive, quantitative and high-throughput assays are a prerequisite for sensing schemes.

It needs to be pointed out, however, that CP-based "biosensors" that can detect true biomolecules such as proteins, DNAs and polysaccharides are still very limited as compared to chemical sensors. The difficulty lies in that the impact to the signaling system from biorecognition events usually is much weaker than electrostatic attraction/repulsion, bond formation or other interactions of a chemical nature. CPs exhibit a unique optical property for signal amplification because of the electron configuration in the conjugated backbone. However, methods that take full advantage of this property for biomolecule detection are still undeveloped to this point. The work we present here focused on the effort in biomolecule detection with PDAs, in particular the detection of bacterial protein toxins. We hope that this chapter will inspire more interest in the use of PDAs for sensor design and stimulate fabrication of more exciting supramolecular assemblies as sensing platforms that meet the compelling challenges in biosensor technology.

Acknowledgments

The authors acknowledge the financial support from the US National Science Foundation (CHE-0719224).

References

1. Leclerc, M. (1999) Optical and electrochemical transducers based on functionalized conjugated polymers. *Advanced Materials*, **11** (18), 1491–8.
2. Thomas, S.W., Joly, G.D. and Swager, T.M. (2007) Chemical sensors based on amplifying fluorescent conjugated polymers. *Chemical Reviews*, **107** (4), 1339–86.
3. McQuade, D.T., Pullen, A.E. and Swager, T.M. (2000) Conjugated polymer-based chemical sensors. *Chemical Reviews*, **100** (7), 2537–74.
4. Leclerc, M. (2000) Smart materials based on functionalized conjugated polymers. *Canadian Chemical News*, **52** (6), 22–4.
5. Gaylord, B.S., Heeger, A.J. and Bazan, G.C. (2003) DNA hybridization detection with water-soluble conjugated polymers and chromophore-labeled single-stranded DNA. *Journal of the American Chemical Society*, **125**, 896–900.
6. Ma, G., Muller, A.M., Bardeen, C.J. and Cheng, Q. (2006) Self-assembly combined with photopolymerization for the fabrication of fluorescence "turn-on" vesicle sensors with reversible "on-off" switching properties. *Advanced Materials*, **18** (1), 55–60.
7. Malhotra, B.D., Kumar, N. and Chandra, S. (1986) Recent studies of heterocyclic and aromatic conducting polymers. *Progress in Polymer Science*, **12** (3), 179–218.
8. Nishizawa, M., Matsue, T. and Uchida, I. (1992) Penicillin sensor based on a microarray electrode coated with pH-responsive polypyrrole. *Analytical Chemistry*, **64**, 2642–4.
9. Sailor, M.J. and Curtis, C.L. (1994) Conducting polymer connections for molecular devices. *Advanced Materials*, **6** (9), 688–92.
10. Emge, A. and Bauerle, P. (1999) Molecular recognition properties of nucleobase-functionalized polythiophenes. *Synthetic Metals*, **102** (1–3), 1370–3.
11. Patil, A.O., Heeger, A.J. and Wudl, F. (1988) Optical properties of conducting polymers. *Chemical Reviews*, **88** (1), 183–200.
12. Leclerc, M. (2001) Polyfluorenes: twenty years of progress. *Journal of Polymer Science Part A–Polymer Chemistry*, **39** (17), 2867–73.
13. Shim, H.-K. and Jin, J.-I. (2002) Light-emitting characteristics of conjugated polymers. *Advances in Polymer Science*, **158** (Polymers for Photonics Applications 1), 193–243.
14. Comoretto, D. and Lanzani, G. (2003) Optical and spectroscopic properties of conjugated polymers. *Springer Series in Materials Science*, **60** (Organic Photovoltaics), 57–90.
15. Moliton, A. and Hiorns, R.C. (2004) Review of electronic and optical properties of semiconducting p-conjugated polymers: applications in optoelectronics. *Polymer International*, **53** (10), 1397–412.
16. Ho, H.A. and Leclerc, M. (2003) New colorimetric and fluorometric chemosensor based on a cationic polythiophene derivative for iodide-specific detection. *Journal of the American Chemical Society*, **125** (15), 4412–13.
17. Kim, J.M., Ji, E.K., Woo, S.M., Lee, H.W. and Ahn, D.J. (2003) Immobilized polydiacetylene vesicles on solid substrates for use as chemosensors. *Advanced Materials*, **15** (13), 1118–21.
18. Yang, J.S. and Swager, T.M. (1998) Fluorescent porous polymer films as

TNT chemosensors: electronic and structural effects. *Journal of the American Chemical Society*, **120** (46), 11864–73.

19 Zhang, S.W. and Swager, T.M. (2003) Fluorescent detection of chemical warfare agents: functional group specific ratiometric chemosensors. *Journal of the American Chemical Society*, **125** (12), 3420–1.

20 Shioya, T. and Swager, T.M. (2002) A reversible resistivity-based nitric oxide sensor. *Chemical Communications*, **13**, 1364–5.

21 Kuroda, K. and Swager, T.M. (2003) Self-amplifying sensory materials: energy migration in polymer semiconductors. *Macromolecular Symposia*, **201**, 127–34.

22 Zhou, Q. and Swager, T.M. (1995) Method for enhancing the sensitivity of fluorescent chemosensors: energy migration in conjugated polymers. *Journal of the American Chemical Society*, **117**, 7017–18.

23 Zhou, Q. and Swager, T.M. (1995) Fluorescent chemosensors based on energy migration in conjugated polymers: the molecular wire approach to increased sensitivity. *Journal of the American Chemical Society*, **117**, 12593–602.

24 Ho, H.A., Boissinot, M., Bergeron, M.G., Corbeil, G., Doré, K., Boudreau, D. and Leclerc, M. (2002) Colorimetric and fluorometric detection of nucleic acids using cationic polythiophene derivatives. *Angewandte Chemie (International Edition in English)*, **41**, 1548–51.

25 Doré, K., Dubus, S., Ho, H.A., Le'vesque, I., Brunette, M., Corbeil, G., Boissinot, M., Boivin, G., Bergeron, M.G., Boudreau, D. and Leclerc, M. (2004) Fluorescent polymeric transducer for the rapid, simple, and specific detection of nucleic acids at the zeptomole level. *Journal of the American Chemical Society*, **126**, 4240–4.

26 Ho, H.A. and Leclerc, M. (2004) Optical sensors based on hybrid aptamer/conjugated polymer complexes. *Journal of the American Chemical Society*, **126**, 1384–7.

27 Liu, B. and Bazan, G.C. (2004) Homogeneous fluorescence-based DNA detection with water-soluble conjugated polymers. *Chemistry of Materials*, **16** (23), 4467–76.

28 Wang, S., Gaylord, B.S. and Bazan, G.C. (2004) Fluorescein provides a resonance gate for FRET from conjugated polymers to DNA intercalated dyes. *Journal of the American Chemical Society*, **126** (17), 5446–51.

29 Gaylord, B.S., Massie, M.R., Feinstein, S.C. and Bazan, G.C. (2005) SNP detection using peptide nucleic acid probes and conjugated polymers: applications in neurodegenerative disease identification. *Proceedings of the National Academy of Sciences of the United States of America*, **102** (1), 34–9.

30 Liu, B. and Bazan, G.C. (2005) Methods for strand-specific DNA detection with cationic conjugated polymers suitable for incorporation into DNA chips and microarrays. *Proceedings of the National Academy of Sciences of the United States of America*, **102** (3), 589–93.

31 Hong, J.W., Hemme, W.L., Keller, G.E., Rinke, M.T. and Bazan, G.C. (2006) Conjugated-polymer/DNA interpolyelectrolyte complexes for accurate DNA concentration determination. *Advanced Materials*, **18** (7), 878–82.

32 Ringsdorf, H., Schlarb, B. and Venzmer, J. (1988) Molecular architecture and function of polymeric oriented systems: models for the study of organization, surface recognition and dynamics of biomembranes. *Angewandte Chemie (International Edition in English)*, **27**, 114–58.

33 O'Brien, D., Whitesides, T. and Klingbiel, R. (1981) The photopolymerization of lipid diacetylenes in bimolecular layer membranes. *Journal of Polymer Science, Polymer Letters Edition*, **19** (3), 95–101.

34 Okada, S., Peng, S., Spevak, W. and Charych, D. (1998) Color and chromism of polydiacetylene vesicles. *Accounts of Chemical Research*, **31** (5), 229–39.

35 Charych, D.H., Nagy, J.O., Spevak, W. and Bednarski, M.D. (1993) Direct colorimetric detection of a

receptor–ligand interaction by a polymerized bilayer assembly. *Science*, **261** (5121), 585–8.

36 Charych, D., Cheng, Q., Reichert, A., Kuziemko, G., Stroh, M., Nagy, J.O., Spevak, W. and Stevens, R.C. (1996) A "litmus test" for molecular recognition using artificial membranes. *Chemistry and Biology*, **3** (2), 113–20.

37 Ma, Z., Li, J., Liu, M., Cao, J., Zou, Z., Tu, J. and Jiang, L. (1998) Colorimetric detection of *Escherichia coli* by polydiacetylene vesicles functionalized with glycolipid. *Journal of the American Chemical Society*, **120** (48), 12678–9.

38 Kolusheva, S., Kafri, R., Katz, M. and Jelinek, R. (2001) Rapid colorimetric detection of antibody–epitope recognition at a biomimetic membrane interface. *Journal of the American Chemical Society*, **123** (3), 417–22.

39 Rangin, M. and Basu, A. (2004) Lipopolysaccharide identification with functionalized polydiacetylene liposome sensors. *Journal of the American Chemical Society*, **126** (16), 5038–9.

40 Alouf, J.E. and Palmer, M.W. (1999) *The Comprehensive Sourcebook of Bacterial Protein Toxins*, 2nd edn, Academic Press, London.

41 Bhakdi, S., Bayley, H., Valeva, A., Walev, I., Walker, B., Weller, U., Kehoe, M. and Palmer, M. (1996) Staphylococcal α-toxin, streptolysin-O, and *Escherichia coli* hemolysin: prototypes of pore-forming bacterial cytolysins. *Archives of Microbiology*, **165** (2), 73–9.

42 Ma, G.Y. and Cheng, Q. (2005) Vesicular polydiacetylene sensor for colorimetric signaling of bacterial pore-forming toxin. *Langmuir*, **21** (14), 6123–6.

43 Sekiya, K., Danbara, H., Yase, K. and Futaesaku, Y. (1996) Electron microscopic evaluation of a two-step theory of pore formation by streptolysin O. *Journal of Bacteriology*, **178** (23), 6998–7002.

44 Xu, D. and Cheng, Q. (2002) Surface-bound lipid vesicles encapsulating redox species for amperometric biosensing of pore-forming bacterial toxins. *Journal of the American Chemical Society*, **124** (48), 14314–15.

45 Wiskur, S.L., Ait-Haddou, H., Lavigne, J.J. and Anslyn, E.V. (2001) Teaching old indicators new tricks. *Accounts of Chemical Research*, **34** (12), 963–72.

46 Fan, L.J., Zhang, Y. and Jones, W.E. (2005) Design and synthesis of fluorescence "turn-on" chemosensors based on photoinduced electron transfer in conjugated polymers. *Macromolecules*, **38** (7), 2844–9.

47 Uchiyama, S., Kawai, N., de Silva, A.P. and Iwai, K. (2004) Fluorescent polymeric AND logic gate with temperature and pH as inputs. *Journal of the American Chemical Society*, **126** (10), 3032–3.

48 Tong, H., Wang, L.X., Jing, X.B. and Wang, F.S. (2003) "Turn-on" conjugated polymer fluorescent chemosensor for fluoride ion. *Macromolecules*, **36** (8), 2584–6.

49 Lakowicz, J.R. (1999) *Principles of Fluorescence Spectroscopy*, 2nd edn, Kluwer Academic, New York.

50 Cheng, Q. and Stevens, R.C. (1998) Charge-induced chromatic transition of amino acid-derivatized polydiacetylene liposomes. *Langmuir*, **14** (8), 1974–6.

51 Ma, G.Y. and Cheng, Q. (2005) A nanoscale vesicular polydiacetylene sensor for organic amines by fluorescence recovery. *Talanta*, **67** (3), 514–19.

52 Kuziemko, G.M., Stroh, M. and Stevens, R.C. (1996) Cholera toxin binding affinity and specificity for gangliosides determined by surface plasmon resonance. *Biochemistry*, **35**, 6375–84.

53 Ma, G.Y. and Cheng, Q. (2006) Manipulating FRET with polymeric vesicles: development of a "mix-and-detect" type fluorescence sensor for bacterial toxin. *Langmuir*, **22** (16), 6743–5.

54 Medintz, I.L., Anderson, G.P., Lassman, M.E., Goldman, E.R., Bettencourt, L.A. and Mauro, J.M. (2004) General strategy for biosensor design and construction employing multifunctional surface-tethered components. *Analytical Chemistry*, **76** (19), 5620–9.

8
Multivalent Synthetic Receptors for Proteins

Jolanta Polkowska, Peter Talbiersky, and Thomas Schrader

8.1
Aminopyrazoles for β-Sheet Capping in Protein Misfolding Events

One of the fundamental secondary structures found in proteins is called a "β-sheet" [1]. It is stabilized by an extended conformation of several peptide strands oriented parallel to each other. Hydrogen bonds are then formed between each carbonyl oxygen of one strand and the backbone amide NH group in an adjacent strand and *vice versa*. If the N-terminal residue of each strand "points" into the same direction the sheet is considered parallel; antiparallel sheets have their N-termini "pointing" in opposite directions (Figure 8.1).

β-Sheets fulfill important structural tasks in proteins, *inter alia* imparting rigidity or creating a binding pocket shielded from solvent. However, they also constitute an acute problem for the organism if they form spontaneously and circumvent the body's strict control mechanisms. Due to their mutual saturation of all hydrogen bond donors and acceptors with nonpolar side-chains extended to the solvent, these

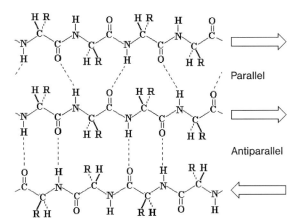

Figure 8.1 Schematic representing parallel and antiparallel β-sheet arrangement.

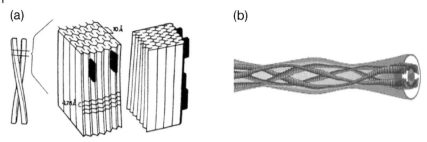

Figure 8.2 Typical amyloid fibrils (Alzheimer's disease with intercalated Congo red (a)) are 10 nm long and consist of twisted interdigitated β-sheets, called cross-β-sheets (b) [2, 3].

protein substructures have a high tendency to precipitate from aqueous solution, and subsequently form large aggregates. In many cases that have gained notoriety; uncontrolled β-sheet folding is associated with a clinically distinct disorder, leading to a whole class of so-called "protein misfolding diseases", comprising Alzheimer's, Creutzfeldt–Jakob as well as Parkinson's disease, to name just a few (Figure 8.2).

Alzheimer's disease is a progressive brain disorder named after the German physician Alois Alzheimer, who first described it in great detail in 1906 [4]. Today, more than 20 million people are affected and their number is growing rapidly. The histopathological characteristics of Alzheimer's disease are protein deposits in the patients' brains (amyloid plaques). Although recent results revealed the final stage of protein aggregation to be rather benign, intermediates on their way towards deposited plaques – especially soluble medium-size oligomers – were found to be extremely neurotoxic. Alzheimer's disease symptoms generally include a dramatic progressive loss of brain functions, caused both by protein aggregation and accompanying inflammatory processes, ultimately resulting in neuronal death. Typical symptoms encompass inattention, impaired memory, disoriented behavior and even impaired speech [5]. Sometimes, Alzheimer's disease patients are not able to perform even simple operations on their own and need constant supervision. More specifically, histology describes Alzheimer's disease by the presence of two misfolded protein components: extracellular amyloid deposits (Aβ peptide) and intracellular neurofibrillary tangles (Tau protein) [6, 7]. The major – and importantly also the first occurring – protein constituent of amyloid deposits in Alzheimer's disease is the Aβ peptide [4]. This is a peptide of 40 or 42 amino acids derived from proteolytic cleavage of the amyloid precursor protein by two enzymes, called β- and γ-secretase. Soluble Aβ above a critical threshold concentration spontaneously polymerizes to form oligomers with a regular cross-β-sheet structure, which fold to neurotoxic fibrils and finally accumulate around neurons. To date, only palliative therapies are available for this disease – with promising results in clinical trials for passive immunization with Aβ-specific antibodies. State-of-the-art medical treatment still focuses on symptom relief and is able to retard the loss of cognitive functions by approximately 1 year (acetylcholine esterase inhibitors, anti-inflammatory drugs). From a therapeutic point of view, however, it is imperative to elucidate the poorly understood mechanism of aggregation. Synthetic ligands with built-in β-sheet affinity may be useful as diagnostic tools and perhaps even for Alzheimer's disease therapy itself. They thus represent one aim of intense ongoing research. Approximately a decade ago, many laboratories began a world-wide search for small ligands against protein-folding diseases. Although various *in vitro* active compounds have been identified from library screening, their structures seem to be unrelated and very few of them demonstrate some kind of rational design. Those that do are peptides themselves and closely resemble a β-sheet motif. Some of them strongly retard fibril or subsequent plaque formation [8–11].

The first report in 1996 of a small peptide able to block Aβ aggregation featured the peptide fragment KLVFF taken directly from the central region of Aβ [12]. Soto *et al.* subsequently developed peptides with the LPFFD sequence that could be applied as β-sheet breakers against Aβ (which means that these compounds were able to dissolve existing β-sheets in a test tube) [13]. In a successful rational approach, Hugh and Meredith systematically modified every position of the KLVFF sequence by introduction of N-methylated amino acids or by elimination of the peptide's hydrogen bond donor capacity by replacing amides with esters [14–16]. Willbold's group had earlier solved the solution structure of Aβ and could demonstrate that it is mainly unordered with two α-helical regions, as opposed to the regular β-sheet found in solid-state nuclear magnetic resonance (NMR) spectroscopic experiments [17].

It was in this period of time when our group discovered that aminopyrazoles are complementary in their DAD hydrogen bond pattern to ADA dipeptides and first investigated multiple hydrogen bond formation in 1:1 complexes in organic solution [18]. This new recognition motif is almost perfectly complementary to the dipeptide fragment of a naturally occurring β-sheet with respect to atom distances and torsion angles (Figure 8.3).

In organic solution, even a single aminopyrazole is able to lock the conformation of freely rotatable glycine-containing dipeptide in an almost perfect β-sheet conformation, by formation of three contiguous hydrogen bonds with the top face of the peptidic guest [19]. Multiplication of these effects by oligomerization of the

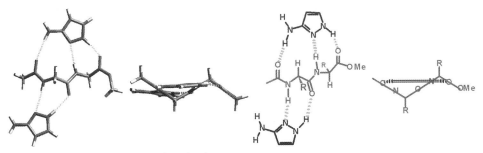

Figure 8.3 Dipeptide recognition with acylated aminopyrazoles employing every available hydrogen bond donor and acceptor.

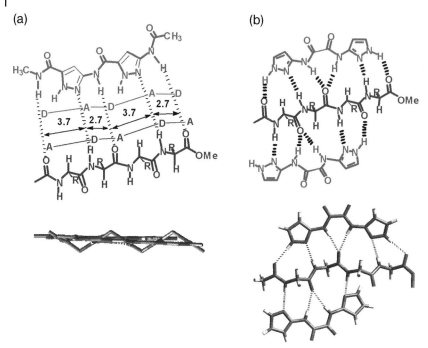

Figure 8.4 (a) A 1:1 complex between dimeric aminopyrazole and a tetrapeptide. (b) A 1:1 complex between the oxaloyl-bridged dimer and a tetrapeptide. At the bottom calculated structures are shown (MacroModel 7.0, Amber*, water).

aminopyrazole core structure produced much more stable complexes with larger peptides, which were also observed in buffered aqueous solution. To date, several successive generations of ever-improved and optimized aminopyrazole ligands have been developed in our laboratories. The second generation already contained two aminopyrazole moieties, connected by a rigid dicarboxylic acid linker to minimize entropic penalties. The smallest member with an oxaloyl bridge was already biologically active against prion protein aggregation [20]. Consequently, a carboxylate function was incorporated into the aminopyrazole generating a non-proteinogenic amino acid that could be subjected to iterative peptide synthesis (Figure 8.4).

The most potent trimeric structure is accessible in a modular manner by combining, for example, the dimeric acid **1** and the monomeric amine **2** or the other way round (Figure 8.5).

Since the underlying 3-aminopyrazole-5-carboxylic acid heterocycle had hitherto been unknown, a straightforward synthetic route was developed for mixed hybrid compounds comprising aminopyrazoles and proteinogenic amino acids. It starts from the respective nitropyrazolecarboxylic acid derivative, which was first transformed into a *p*-methoxybenzyl (PMB)-protected carboxylic acid. Reduction of the

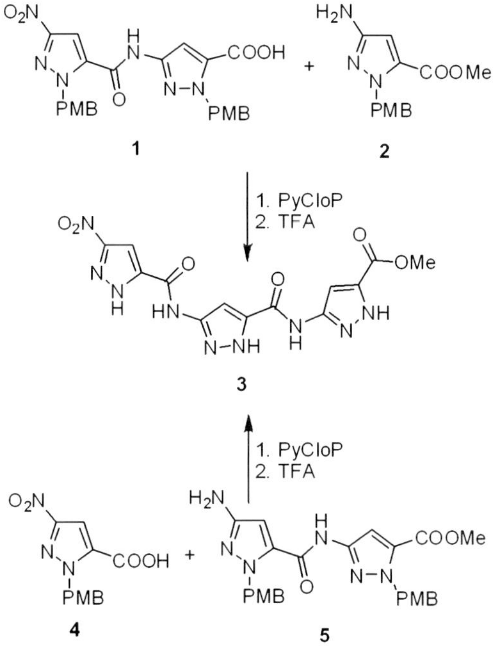

Figure 8.5 Modular synthesis of trimeric lead structure **3**. Note the PMB protection on the pyrazole nucleus that requires hot trifluoroacetic acid for deprotection.

nitro group liberates the nucleophilic amino group, which is subsequently coupled with PyCloP (chlorotripyrrolidinophosphonium hexafluorophosphate) either to the heterocyclic or any proteinogenic amino acid. Final protecting group cleavage occurs simultaneously under relatively harsh acidic conditions (2–5 h trifluoroacetic acid at 70 °C), but preserves the stereochemical integrity of all chiral units inside the molecule.

Some of the new hybrid compounds were investigated in greater detail with respect to their complex formation with small fragments of Aβ itself (Figure 8.6). The KLVFF sequence (17–21) from Aβ's centerpiece is assumed to be a nucleation site for the whole peptide. Direct 1:1 complex formation was established with aminopyrazoles **1–3** and characterized by downfield shifts of hydrogen-bonded NH protons as well as distinct upfield shifts of some aromatic protons pointing to arene–arene interactions. Karplus analyses revealed a change of the critical torsion angles in the model peptide KKLVFF in the direction of a perfect β-sheet. NMR titrations afforded dissociation constants for the fragment complexes in the low millimolar regime (K_d = 0.5–2 mM). Affinities of various hybrid ligands towards the model peptide KKLVFF strongly depend on their sequence and composition from natural and artificial amino acids [18].

A solid-phase protocol was developed using unmodified Wang resin or employing the Sieber amide linker for the construction of N-terminal peptide amides.

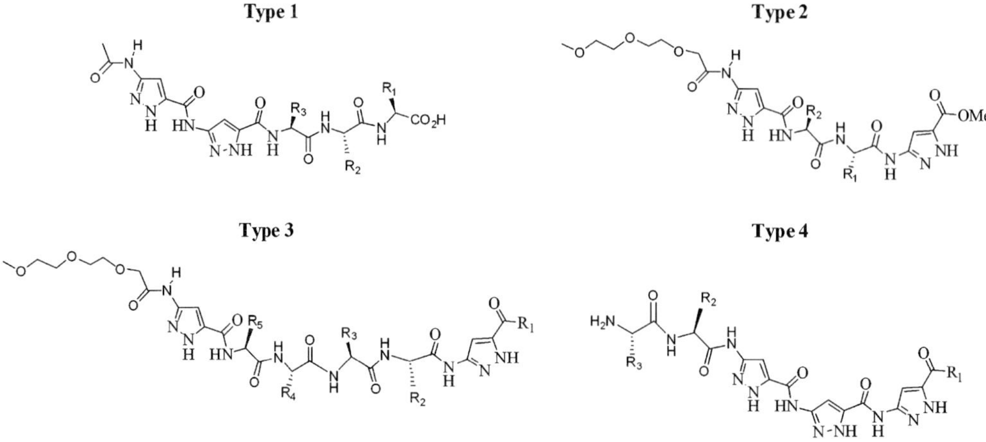

Figure 8.6 Type 1–4 aminopyrazole-peptide hybrid compounds that were synthesized by the above-detailed iterative protocol and subsequently tested for KKLVFF affinity.

To this end, the pyrazole was orthogonally protected with Fmoc (NH$_2$), PMB (nucleus) and ester groups (carboxylate). Proteinogenic and pyrazole amino acids can be combined at will, but only retain complementarity towards a β-sheet if two heterocycles are linked by an even number of α-amino acids. Thus, small libraries of ligands for octapeptides were synthesized and screened against characteristic truncated versions of Aβ. Surprisingly, the trimer unit was superior to all mixed hybrid versions, presumably due to its rigidity and propensity to undergo additional arene–arene interactions with aromatic amino acids in aqueous buffer (Figure 8.7).

In order to determine the best binders among various classes of aminopyrazole hybrid peptides, we also prepared them immobilized on an acid-stable polyacrylamide resin. Subsequent equilibration with a fluorescent target peptide (Dansyl-oligoethyleneglycol-spacer-AKLVFF) produced a distinct fluorescence quenching. Binding isotherms could be analyzed with standard nonlinear regression methods and translated into dissociation constants in the millimolar regime. Systematic variations of aminopyrazole composition on the resin and solvents revealed that again the trimeric aminopyrazole **10** was superior to all hybrid compounds irrespective of their sequence. Interestingly, aqueous buffered solution favored direct complex formation over organic solvents. This is a strong indication that aminopyrazoles recognize their peptide counterparts by a combination of hydrogen bonds and dispersive interactions (Figures 8.8 and 8.9) [21].

In a close cooperation with the Riesner group (Düsseldorf), a number of biophysical *in vitro* experiments were performed, which confirmed that especially the trimeric aminopyrazoles with or without proteinogenic amino acids in their backbone efficiently prevent the formation of large aggregates both from Aβ (1–40) as well as from Aβ (1–42) [22]. Aggregation was studied and quantified by various

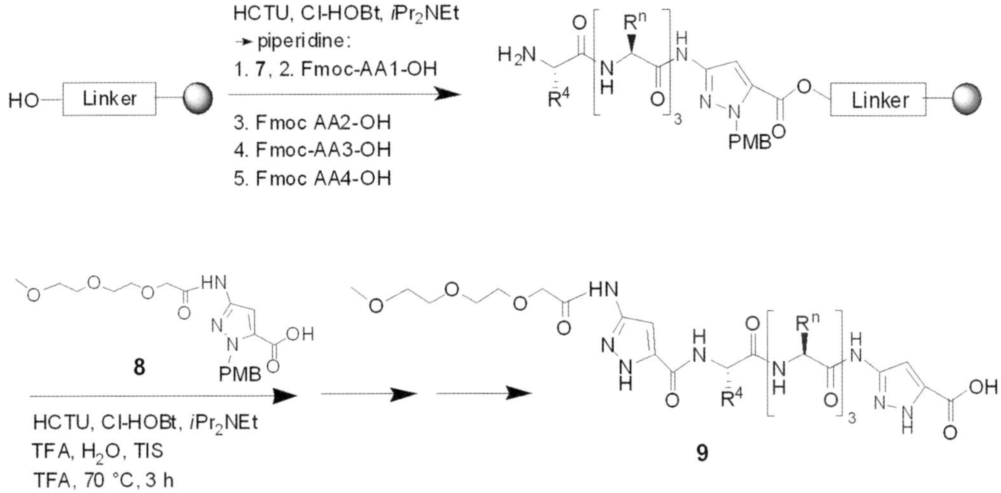

8.1 Aminopyrazoles for β-Sheet Capping in Protein Misfolding Events

Figure 8.7 Automated synthesis of type 3 hybrid compounds starting from the new PMB- and Fmoc-protected aminopyrazole carboxlyate **7**.

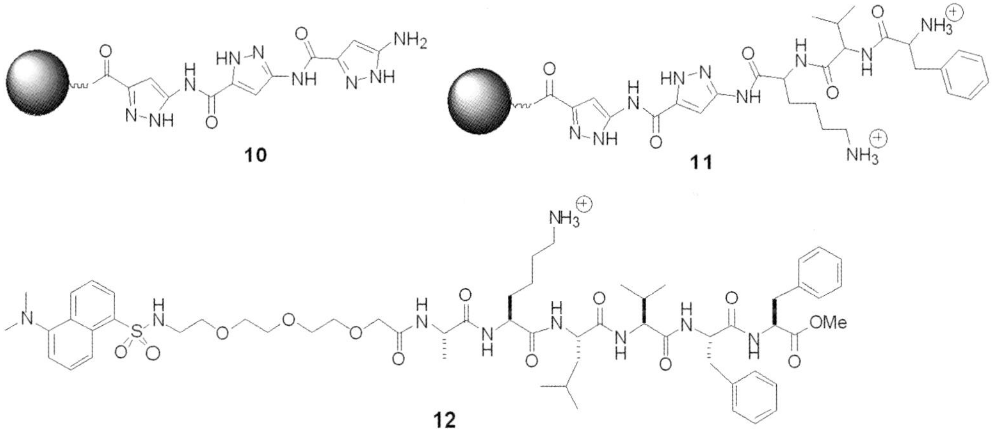

Figure 8.8 Immobilized aminopyrazole hybrid peptides **10/11** and fluorescent target peptide dansyl-oligoethyleneglycol-spacer-AKLVFF **12**.

methods, such as thioflavin T assays (total amount of aggregates), a new protocol to determine the Aβ aggregation kinetics employing electrospray ionization (ESI) mass spectroscopy, fluorescence correlation spectroscopy (FCS) for the quantification of numbers and sizes of Aβ aggregates, ultracentrifugation in combination with a sedimentation protocol for the kinetic analysis of size distributions, and density gradient ultracentrifugation.

Thioflavin intercalates into fibrillar peptide structures and starts to fluoresce quite intensely due to its reorientation within the fibrillar framework. The strong

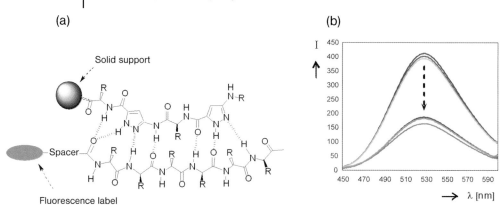

Figure 8.9 (a) Schematic illustration of solid-phase screening assay on peptide affinity. (b) Fluorescence quenching on complex formation between labeled peptide and immobilized aminopyrazole oligomer.

fluorescence observed in the absence of aggregation inhibitors with Aβ (1–40 and 1–42) was greatly reduced after addition of the best aminopyrazole dimers and trimers, without altering the typical lag phase (Aβ 1–40).

Introducing the first volatile neutral buffer for peptide aggregation (tetramethylammonium acetate, pH 6.8), the well-known aggregation inhibitor melatonin could be detected in its 1:1 complex with monomeric Aβ(1–40) [23, 24]; no complex formation occurred at 250 μM with the oxaloyl dimer or the trimeric aminopyrazole. Likewise, their aggregation profile monitored by the time-dependent decrease of the corresponding ESI signal for monomeric Aβ was indistinguishable from that of untreated Alzheimer's peptide. Obviously, aminopyrazoles do not prevent initial Aβ oligomerization, but rather bind to misfolded oligomers residing already in the β-sheet conformation.

Density gradient centrifugation indicated that Aβ alone (33 μM) was almost entirely aggregated 10 min after incubation, whereas the presence of a trimeric aminopyrazole retarded this event by more than 5 days. After 1 day no larger aggregates had still formed in the solution treated with trimeric aminopyrazole. Small Aβ oligomers prevailed, ranging from dimers to hexamers.

FCS allows us to measure the diffusion time of a fluorescent particle through the confocal volume of the laser optics. This method offers the possibility to monitor the aggregation process in the time range of minutes to hours; only a very small amount of peptide is required, imitating the nanomolar concentration regime in the brain of healthy patients. Aggregation assays were performed in the absence and presence of various classes of aminopyrazole derivatives. Again, the oxaloyl dimer and the aminopyrazole trimer proved superior to all other tested compounds, with the trimer being the best candidate. Quite recently, it was discovered that the trimeric lead structure is also able to disassemble existing fibrils and to revert the corresponding equilibrium strongly to the monomeric side (Figure 8.10).

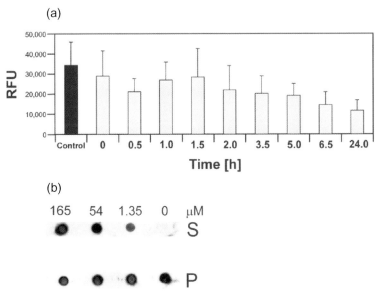

Figure 8.10 (a) Time-dependent disaggregation of preformed Aβ (1–42) fibrils by the trimeric aminopyrazole test compound shown by thioflavin T fluorescence measurements. (b) Disaggregation of preformed Aβ (1–42) fibrils by the test compound as shown with differential ultracentrifugation. After careful removal of the supernatant, supernatant (S) and pellet (P) were separately analyzed by a dot-blot procedure with an Aβ-specific antibody.

In order to examine if the promising potential of aminopyrazoles to prevent aggregation of isolated Aβ can be transferred to living cells, their antiaggregatory capacity was examined in various cell culture stress and lesion assays. Primary neuronal cultures were prepared from 8-day-old white Leghorn chicken embryo telencephalons. Surprisingly, the oxaloyl dimer and the trimeric aminopyrazole display neuroprotective potential in a 2% stress assay. In particular, the trimer is capable of counteracting neurodegeneration as a result of the partial withdrawal of important growth factors. More importantly, both compounds were also active against cell lesions with 33-μM preformed Aβ fibrils (1–42). Interestingly, the above-mentioned hybrid compounds with peptidic linkers between two aminopyrazoles were also active against monomeric Aβ and effectively prevented in the following test phase (8 days) the severe cell viability decrease as a consequence of Aβ aggregation as monitored with MTT [3-(4,5-dimethylthiazol-2-yl)-2,5-diphenyltetrazolium bromide] assays. Obviously, the internal flexible peptide linkers allow the hybrid compounds to adjust their hydrogen bond donor and acceptor groups to the curved peptide backbone of monomeric Aβ or its small oligomers.

Currently the trimer structure is adorned with small additional binding and labeling units, in order to increase its water solubility and affinity towards Aβ as well as render it specific towards the pathogenic Alzheimer's peptide.

8.2
New Mechanisms of Enzyme Inhibition by Molecular Clips and Tweezers

Enzymes are ubiquitous biocatalysts that control each metabolic pathway and accelerate a wide range of chemical transformations. Their activity can be influenced by so-called effectors: while *activators* raise the enzyme's activity, *inhibitors* lower it and thus inhibit the catalyzed process. Irreversible inhibition is brought about by compounds that react covalently with the enzyme and change its chemical structure (e.g., by modification of key amino acid residues which are essential for enzymatic activity). The result is infinite inhibition. Reversible inhibition, on the other hand, requires noncovalent interactions between inhibitor and enzyme that can be disrupted by displacement by a stronger binder; this ideally restores the enzyme's original activity [25–29].

Today, new artificial enzyme inhibitors can be mainly identified following two avenues, which are either based on combinatorial chemistry or on rational drug design. While the combinatorial approach screens huge libraries of compounds against a target enzyme for useful inhibitors, rational drug design uses the three-dimensional structure of the target enzyme's active site to predict which molecule could serve as a potential inhibitor and what structural changes to the inhibitor are necessary to generate a sufficient affinity and specificity [31–33]. Unfortunately the great combinatorial effort of pharmaceutical companies in the 1990s in the search for new drugs from compound libraries (including natural products) has produced only a very limited number of new lead structures.

New artificial enzyme inhibitors are either targeting a specific *human enzyme* with the aim of correcting a pathological condition. Alternatively, they inhibit enzymes needed for the survival of *pathogens*. An important example of the first case is the group of angiotensin-converting enzyme (ACE) inhibitors (e.g., captoprile or enalaprile) that are used in the treatment of high blood pressure (hypertension) and congestive heart failure by inhibiting the ACE (Figure 8.11).

As a vital part of the blood pressure regulation renin–angiotensin–aldosterone system cascade this enzyme converts the decapeptide angiotensin I to the vasoconstrictive octapeptide angiotensin II. This newly formed peptide then causes contraction of the muscles surrounding blood vessels and thereby increases blood pressure. However, in the presence of ACE inhibitors the production of angiotensin II remains low and the blood vessels enlarge or dilate leading to a significant drop in blood pressure. As a result, the function of a failing heart is greatly improved and the progression of kidney diseases resulting from hypertension or diabetes is stopped [29, 34–37]. The inhibitory effect of most ACE inhibitors relies on their close structural similarity to the bradykinin potentiating factor (BPF), which was extracted from the venom of the Brazilian lancehead snake *Bothrops jararaca* [38–41]. This nonapeptide showed the greatest ACE inhibition potency and hypotensive effect among natural products *in vivo*. Since BPF and its effective tripeptide sequence Try–Ala–Pro are quickly catabolized within the body, the structure of this molecule was modified several times in order to increase its metabolic lifetime. Its stability and inhibitory effect could both be markedly

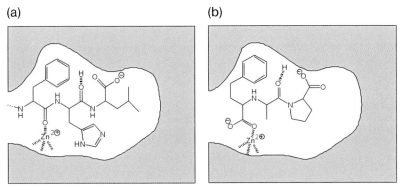

Figure 8.11 Schematic depicting the active site of ACE filled with the substrate angiotensin I (a) or the ACE inhibitor enalaprile (b). By imitating the substrate structure, enalaprile effectively blocks the active site of ACE and thereby competitively inhibits the enzymatic reaction.

improved by replacing the tripeptide sequence Try–Ala–Pro with the similar but much more stable sequence Phe–Ala–Pro and by introducing a succinic or glutaric acid structural element (Figure 8.12) [42]. All the therapeutically used ACE inhibitors, except captoprile and lisinoprile, also belong to the so-called family of prodrugs that are only activated within the body by the corresponding enzymes [43, 44].

Antibiotics such as penicillin and vancomycin are an example of the latter case: both compounds inhibit bacterial cell wall biosynthesis. More specifically, they block those enzymes responsible for the production and cross-linking of net-like peptidoglycan strands – the material of cell walls. If these essential enzymes are shut down, the bacterial cell wall loses its strength and the bacterium finally bursts [29, 45].

Classical design of enzyme inhibitors often imitates natural enzyme substrates as discussed above for ACE inhibitors. Even better is similarity to the transition state of an enzyme-catalyzed reaction. The latter case ensures that the inhibitor exploits the transition state stabilizing effect of the enzyme, resulting in optimum affinity and maximum retardation effect. Osletamivir, an antiviral drug, is an example of such a transition state inhibitor, mimicking the planar structure of the oxonium ion in the reaction of the viral enzyme neuraminidase [46]. A very important example for more conventional substrate mimics are protease inhibitors such as ritonavir which belongs to a class of antiretroviral drugs used for the successful treatment of human immunodeficiency virus (HIV) [47]. The structure of this protease inhibitor is again based on a peptide (Figure 8.13).

As it resembles the precursor protein that serves as the natural substrate for HIV protease, ritonavir acts as a potent competitive inhibitor. The well-known antitumor agent methotrexate is another important example of substrate mimicry. Methotrexate imitates the natural substrate folic acid and thus blocks the thymi-

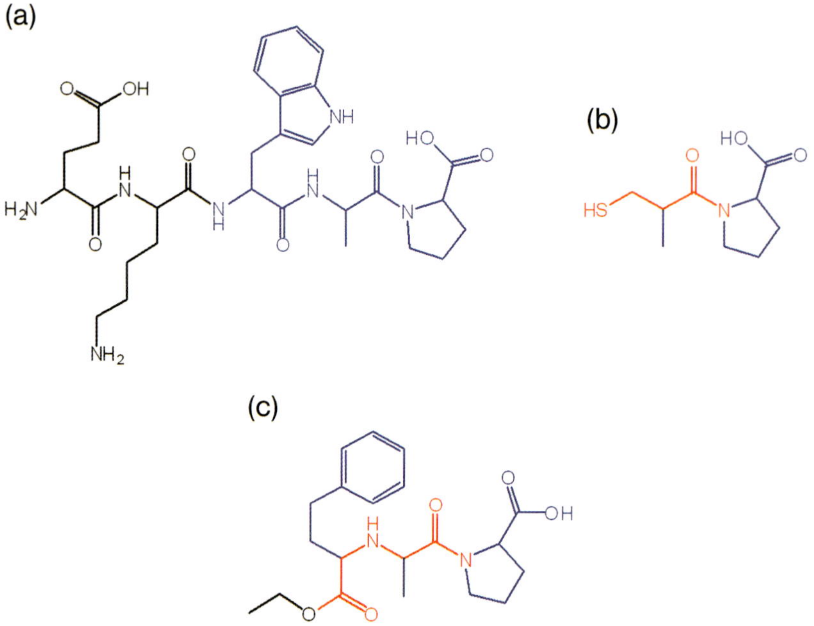

Figure 8.12 Structural analogy between the nonapeptide BPF (a) and the ACE inhibitors captoprile (b) and enalaprile (c). The inhibitory effect of BPF is based on the tripeptide sequence Try–Ala–Pro (blue). By imitating this tripeptide sequence (blue) and by introducing a succinic or glutaric acid analog (red), the stability and inhibitory effect of the ACE inhibitors captoprile and enalaprile could be enhanced in a remarkable way.

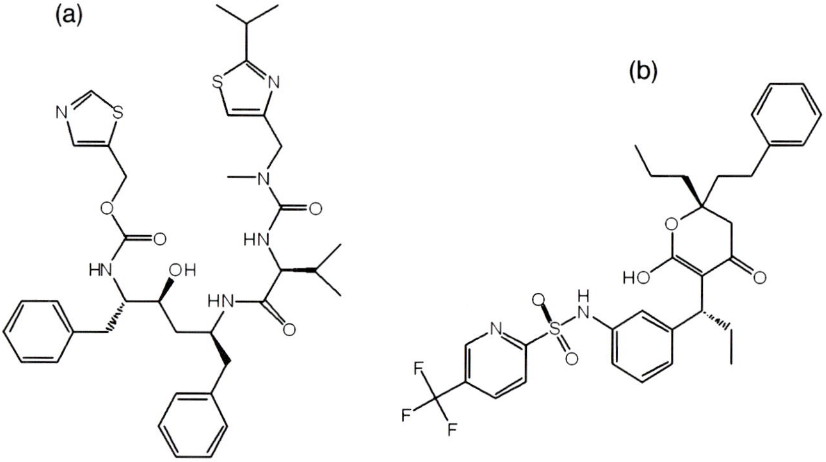

Figure 8.13 Structures of the HIV protease inhibitors titonavir (a) and tipranavir (b). While ritonavir clearly belongs to the peptide-based protease inhibitors, tipranavir represents an example of nonpeptidic protease inhibitors.

dine biosynthesis catalyzed by dihydrofolate reductase. This inhibition is especially toxic to rapidly growing cells so that methotrexate has been used worldwide against the most aggressive types of cancer. It has to be admitted that its side-effects are severe, so that more selective drugs are desired [48]. However, substrate similarity is not always the key to success. Tipranavir, another HIV protease, does not display any obvious structural similarities to the natural protein substrate. The advantages of such nonpeptidic inhibitors comprise, *inter alia*, their high chemical stability and indifference against peptidases (Figure 8.13) [49].

A well-known example for irreversible enzyme inhibition is the inhibition by so-called suicide inhibitors. These compounds seem to undergo a conventional catalysis that leads over a covalently bound intermediate. However, the intermediate cannot be processed any further towards the product, so that it is "stuck" or "commits suicide". Owing to their low reactivity outside the enzyme's active site combined with high specificity towards important metabolic enzymes, suicide inhibitors are generally used in cancer treatment. Their target enzymes are often glutamine amido-transferases, which are responsible for nucleotide biosynthesis and are hence effectively inhibited by glutamine analogs, such as azaserin and acivicin (Figure 8.14) [28, 50].

Since almost all inhibitors suffer from unwanted side-effects and drug resistance is an ever-increasing problem, new ways of enzyme inhibition would be highly welcome and bear immediate therapeutic promise. The Klärner and Schrader group have in the past years exploited the supramolecular action of molecular clips and tweezers. By means of peripheral phosphate groups these can be prepared in a water-soluble form and work especially efficient in physiological solution, since their binding depends on arene–cation as well as dispersive interactions, supported by the hydrophobic effect (see Figure 8.1) [51–54]. While the clip features an electron-rich cavity with parallel naphthalene side walls, the tweezer is more torus shaped, but likewise surrounds an almost circular cavity with highly negative electrostatic surface potential. This renders both receptors prone to receive either planar (clip) or linear (tweezer) cationic guests. The prototype biomolecule for the clip is NAD(P)$^+$, while the tweezer selectively embraces lysine derivatives. The tweezer and the clip can therefore act in two different ways as inhibitors for proteins. While the clip pulls the cofactor out of its cleft (cofactor capture), the tweezer forms a complex with single basic amino acids on the protein surface (lysine decoration).

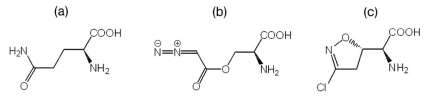

Figure 8.14 Structures of glutamine (a) and the suicide inhibitors azaserine (b) and acivicin (c). Both glutamine analogs inhibit the enzymatic reaction of glutamine amido-transferases.

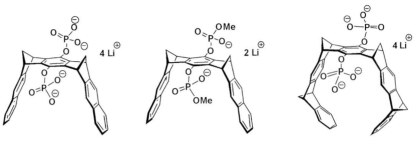

Figure 8.15 Molecular clip **13** and molecular tweezer **14** represent new types of enzyme inhibitors (from left to right: phosphate clip **13a**, phosphonate clip **13b**, phosphate tweezer **14**). While the phosphate- and phosphonate-substituted clips induce enzyme inhibition by cofactor capture, the phosphate-substituted tweezer achieves the same goal by lysine decoration.

These two binding events severely slow down enzymatic activity and can ultimately lead to a total enzyme shutdown. Evidently, both types of enzyme inhibition are intimately connected with their role as synthetic hosts and totally differ from classical mechanisms, for which many well-investigated examples exist. They will be presented and explained below (Figure 8.15).

Contrary to all the conventional enzyme inhibitors mentioned by way of introduction, the molecular clip and tweezer operate as hosts and not as ligands. In both cases inhibition is not triggered by imitation of the substrate or the transition state of an enzyme-catalyzed reaction, but by complexing either the cofactor or crucial amino acids on the protein surface. The clip with its highly electron-rich planar cavity acts as a good host for electron-poor (hetero)aromatic guests. The premier examples found among cell components are nicotinamide adenine dinucleotide (NAD$^+$) and also nicotinamide adenine dinucleotide phosphate (NADP$^+$), which serve as cofactors in many dehydrogenases like alcohol dehydrogenase (ADH) and glucose-6-phosphate dehydrogenase (G6PD). On the contrary, the torus-shaped tweezer strongly prefers slim guests with a terminal cation, above all the amino acid side-chain of lysine, which often occupies strategic places on protein surfaces (Figure 8.16).

Both novel types of enzyme inhibition were discovered and investigated in detail in the case study of the well-known NAD$^+$-dependent oxidoreductase ADH [55–59]. In spite of their divergent recognition profile both artificial water-soluble host molecules (**13** and **14**) achieve the same goal – to stop the enzymatic oxidation of ethanol by ADH. However, after extensive kinetic investigations it became apparent that the clip operates mainly in a competitive manner, since it pulls out the cofactor NAD$^+$ from the Rossman fold until the cofactor level is depleted below a critical threshold. This is contrasted by a noncompetitive inhibition mode for the tweezer, as evidenced by Lineweaver–Burk plots. The tweezer with its high lysine preference decorates the ADH surface, especially the entrance to the active site,

8.2 New Mechanisms of Enzyme Inhibition by Molecular Clips and Tweezers

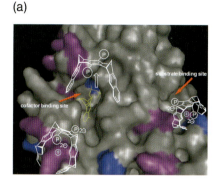
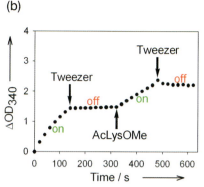

Figure 8.16 (a) Entry to the Rossman fold of ADH and divergent inhibition mechanisms of clip and tweezer. While the clip **13** pulls out the cofactor NAD⁺ (yellow), the tweezer **14** forms complexes with the basic amino acid residues (lysines: purple; arginines: blue) on the protein surface. (b) A successful ADH switch experiment. On addition of 0.6 equivalents tweezer (related to NAD⁺) to the normal ADH assay all enzyme activity is stopped. Subsequent addition of 3 equivalents of Ac-Lys-OMe restores around 40% of the original enzyme activity, which can finally be totally shut down again by further addition of 2.4 equivalents tweezer. Hence, the new noncovalent tweezer inhibition is reversible, and the equilibrium can be shifted back and forth by addition of appropriate complexing agents.

and thereby blocks access of the cofactor to the catalytic center. Circular dichroism spectroscopy at micromolar enzyme concentrations reveals a significant degree of denaturation after docking of both hosts, but only the latter case is reversible by addition of a guest excess. Enzyme switch experiments beautifully confirmed this effect: after a total enzyme shutdown only the tweezer **14** allows to switch on enzyme activity again by an appropriate additive such as Ac-Lys-OMe (Figure 8.16). Interestingly, all other amino acid derivatives except arginine fail to do this job, reflecting the tweezers recognition profile, which selects lysine and (to a lesser extent) arginine as guests for **14**. By stripping the side walls off the clip and tweezer, a substituted hydroquinone diphosphate moiety remains, which is biologically virtually inactive. Hence, ADH inhibition by both inhibitors must hinge on authentic host–guest interactions, most likely a combination of π-cation, dispersive and electrostatic interactions. Nonspecific electrostatic interactions between basic amino acid residues on the protein surface and the diphosphate substituents of the centerpiece can be ruled out. Another piece of evidence is the prevailing ADH inhibition even in the presence of 250 mM NaCl, which goes far beyond physiological salt loads. Pure Coulomb attraction would break down under these conditions. We further characterized the two new different inhibition mechanisms by IC_{50} values, gel electrophoresis and detailed enzyme kinetics. Interestingly, the clip features a relatively large IC_{50} value for ADH inhibition, exactly reflecting the high NAD⁺ concentration (1.5 mM; around 1.3 equivalents NAD⁺). The tweezer operates substoichiometrically at 0.09 equivalents of NAD⁺ (IC_{50} = 180 µM), reflecting the lysine and arginine density on the protein surface. Native gel electrophoresis on

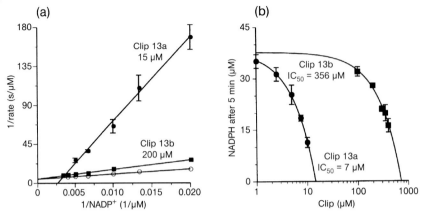

Figure 8.17 (a) Conventional competitive (**13b**) and new unknown (**13a**) Lineweaver–Burk plot. (b) Drastic increase in G6PD inhibition efficiency from phosphonate to phosphate clip.

agarose makes the complex between enzyme and tweezer visible to the naked eye, since it almost fails to migrate in the electric field due to charge neutralization between the tweezer phosphate anions and the protein lysine cations.

Another enzyme that was investigated with these new inhibitor tools was G6PD. This NADP$^+$-dependent oxidoreductase catalyses the oxidation of glucose-6-phosphate to 6-phosphogluconate, which represents the first step in the pentose phosphate pathway [28, 29]. Critically lowered G6PD levels represent the most common human enzyme deficiency in the world; it affects an estimated 400 million people and causes oxidative stress in red blood cells (erythrocytes) [30]. Surprisingly, the phosphate- and phosphonate-substituted clips **13a** and **13b** show completely different mechanisms of enzyme inhibition. While the bisphosphonate clip triggers enzyme inhibition by cofactor capture just as in ADH ($IC_{50} = 350\,\mu M$; around 1.8 equivalents NADP$^+$), the related bisphosphate clip points to a totally different but much more effective kind of inhibition mode. Although both clips possess comparable affinities towards the cofactor NADP$^+$ in the $100\,\mu M$ K_d range, the inhibitory effect for the bisphosphate was boosted into the far substoichiometric range ($IC_{50} = 7\,\mu M$; around 0.03 equivalents NADP$^+$). Obviously, inhibition cannot be related to competition with the Rossman fold for NADP$^+$ inclusion. Extensive enzyme kinetic investigations lead to a mixed substrate inhibition pathway and a new hitherto unknown Lineweaver–Burk plot where the native and inhibition curve intersect in the positive x/y quadrant (Figure 8.17). Ultrafiltration experiments and isothermal calorimetry (ITC) measurements indicate a high affinity of the clip towards the enzyme itself, in the micromolar range ($K_d \sim 3\,\mu M$). The 2:1 stoichiometry found for the clip/enzyme complex by ITC points to a second place at the bottom of the substrate-binding pocket beside the Rossman fold offering multiple favorable interactions with lysine, arginine and aromatic residues. Thus clip **13a** can occupy both cofactor and substrate binding site. Truncated versions without the side walls loose a substantial amount of their inhibitory power. It

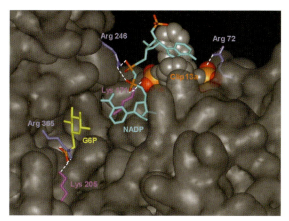

Figure 8.18 Putative formation of the ternary complex between G6PD, clip **13a** and NADP⁺ [77].

seems that clip **13a** operates as a guest to the enzyme and simultaneously as a host to the cofactor. A putative assumption is the formation of a ternary complex between enzyme, cofactor and clip. In this ternary complex the clip could act as a guest to the enzyme, most likely inside the Rossman fold, and also as a host for the cofactor (Figure 8.18). But this partial uncompetitive inhibition mechanism is not sufficient to account for the observed total enzyme shut down. The inhibitory efficiency of clip **13a** could therefore only be explained by the combination of these two inhibition mechanisms. If this picture of a ternary complex is correct, it should be possible to subsequently add clip and cofactor in both orders to G6PD and obtaining in both cases two favorable steps with a large enthalpy release corresponding to successive formation of both inhibitory complexes. This was indeed confirmed by careful ITC measurements, providing strong experimental evidence for the existence of this postulated ternary complex [77].

The two above-mentioned model cases of ADH and G6PD inhibition demonstrated that the guest specificity of artificial host molecules can be retained in physiological solution and in the biological environment of a protein-binding pocket. Provided the selectivity issue is solved, modified versions of the clip and tweezer may therefore serve in the future as starting points for therapeutic approaches relying on the new inhibition mechanisms. There is a plethora of NAD⁺-dependent enzymes that catalyze essential biochemical transformations, some of which may constitute novel targets for external enzyme shutdown in the treatment of pathological processes. Another promising medicinal application of phosphate clip **13a** as a pyridinium host already exists: it can be used for treatment, diagnosis and prevention of age-related macula degeneration. We found out recently that the lipophilic cation *N*-retinyl-*N*-retinylidene-ethanolamine (A2E) produced in the retina is strongly complexed inside the electron-rich cavity. This inclusion prevents A2E forming a problematic complex with cytochrome *c* oxidase which would otherwise trigger apoptosis due to increased oxidative stress. Cell culture experiments with retinal pigment epithelium cells reveal IC_{50} values for apoptosis prevention in the low micromolar regime (around 15 µM) [53].

Thus, we will pursue the goal of applying molecular clips and tweezers for selective and efficient interference with biological systems, in a way that opens new avenues for biodiagostics and medicinal treatment. The structural framework of these molecules allows facile functionalization with recognition and fluorescence labels, and should provide a means to tune these host/guest molecules selective for a given protein target.

8.3
Protein Surface Recognition by Tailor-Made Polymers

Natural processes are often mediated by direct protein–protein contacts. These not only play a key role in all metabolic processes, but also during cell cycle and transcription control as well as in signal transduction. Their eminent importance is further highlighted by the well-known dramatic consequences of abnormal protein folding and subsequent aggregation. Sickle cell anemia can be traced back to a single point mutation at the hemoglobin protein surface which leads under physiological conditions to non-native oligomerization. Permanent dimerization of receptor tyrosine kinases effects continuous signal transduction and triggers pathologic cell growth. On the other hand, diseases can evolve from a sudden loss of protein–protein interactions. Oxidation of a specific methionine residue in the α_1-protease inhibitor prevents its complex formation with various proteases and leads, *inter alia*, to elastase activation, causing respiratory distress [60].

The recognition of protein surfaces with artificial means remains a superb challenge for biological and medicinal chemistry. The specific disruption of protein surface interactions with other ligands or proteins offers numerous alternatives for interference by external drugs. Large surface areas and complex topologies have to be taken into account, requiring extended receptor molecules, which must still be able to adjust to the dynamic ensemble of water-exposed surface amino acid residues. Lessons from protein–protein interactions teach us that nature prefers contact areas of around 1500 Å^2, with an unusual amount of arginine and aromatic amino acid residues, involved in polar interactions (Coulomb, π-cation), complemented and enforced by hydrophobic attraction between approaching nonpolar domains or patches [61]. This combination is partly reflected in recent attempts to simplify the complex task by designed artificial protein binders. Contemporary approaches by chemists and biologists involve secondary structure mimetics [62], multiplication of single specific [63] or unspecific binding sites [64], evolutive optimization of biomacromolecular scaffolds [65] and molecular imprinting [66].

Often multiple copies of single weak binding motifs were used to increase affinities towards proteins: thus, linear anionic oligomers were reported to adopt heparin-like properties [67] or efficiently inhibit human leukocyte elastase ($K_i \leq$ 0.2 µM) [68]. By contrast, anionically functionalized amphiphilic nanoparticles (i.e., monolayer-protected gold clusters) use nonspecific interactions to efficiently inhibit chymotrypsin through electrostatic binding followed by protein denaturation [69]. Similarly, Kiessling et al. developed postsynthetically modified polymers in the form of multivalent mannose displays that nonspecifically inhibited hemagglutination [70]. Bioconjugates involving covalent attachment of smart polymers just outside the active site represent another elegant way to control protein activity. Employing azobenzene–NIPAM copolymers on an endoglucanase 12A mutant allowed photoinduced switching between a compact hydrophobic state with a closed active site and an extended soluble state providing free access to the active site [71]. Similarly, amphiphilic polymer scaffolds have been described that nonspecifically bind to chymotrypsin, inhibit its peptidase activity and modulate substrate specificity; very high ionic strengths again release the protein from the polymer [72]. However, most of these multivalent approaches require elaborated synthetic protocols while their single protein interactions remain nonspecific.

We have therefore recently embarked on a program that uses dendritic or polymeric skeletons to multiply binding events and decorates them with a set of binding sites selective for those amino acid residues most often found on protein surfaces. Our first attempt involved polypropyleneimine dendrimers whose terminal NH_2 groups were subjected to a multiple reductive amination with formyl-substituted m-xylylene bisphosphonates [73]. The resulting three generations of polyanionic dendrimers (4-mer, 8-mer, 16-mer) were studied with pulsed field-gradient longitudinal eddy-current delay techniques in order to secure that in neutral buffered solution no backfolding occurs in spite of their zwitterionic nature. Equilibration with buffered protein solutions was quantitatively monitored either with colored ultraviolet-active proteins (cytochrome c, ferritin) or with OregonGreen-labeled proteins; the best results, however, were obtained with fluorescein-labeled dendrimers. Their resulting dissociation constants strongly varied between the proteins ranging from low millimolar to low micromolar K_d values with prevailing 1:1 or 1:2 stoichiometries depending on protein sizes. Basic arginine-rich proteins were bound most tightly, as expected from the bisphosphonate recognition motif, which itself prefers arginine over lysine; however, only at an estimated 100 mM K_d for the single amino acid recognition event (Figure 8.19).

Next, we proceeded to polymeric backbones, which are much easier to synthesize, especially with respect to higher dendrimer generations. Methacrylamide was selected as the common comonomer unit, because it is relatively stable against spontaneous polymerization at sunlight and air, and the respective binding monomers can be purified by chromatography, characterized spectroscopically and stored almost indefinitely. In addition, any binding site with a terminal amino group can easily be coupled with methacryloyl chloride and the resulting polymer backbone carries multiple amide groups resembling a polypeptide. In our first series, we prepared polybisphosphonates by free radical polymerization initiated by 2,2′-azobisisobutyronitrile in 2,4-dinitrofluorobenzene, with (15) or without hydroxyl comonomers for increased solubility [74]. A methacrylamide-based dansyl monomer was added for fluorescence labeling. The polymers were selectively monodealkylated with LiBr in a quantitative polymer-analogous

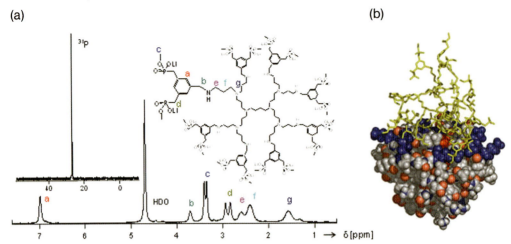

Figure 8.19 (a) ^{31}P- and ^{1}H-NMR spectrum of the crude octameric bisphosphonate (400 MHz, D$_2$O). (b) Minimized 1:1 complex between cytochrome c (C: gray; O: red; N: blue; lysines around the active site: blue) and the hexadecamer dendrimer (yellow) (SYBYL 6.9, MMFF94).

reaction. The resulting new materials displayed an excellent water solubility of above 100 mM. Fluorescence titrations immediately revealed the superior complexation ability of the polymers over the dendrimers. Dissociation constants were often in the micromolar regime, and reflected pI values, protein size, but also surface topologies. For comparison the best binders were electrostatically immobilized on a glass surface covered with a cationic polyethyleneimine polymer layer. Reflectrometric interference spectroscopy (RIfS) was used to monitor protein binding on the immobilized polymer and to quantify from titration experiments association thermodynamics as well as dissociation kinetics. Here, the association constants were generally one order of magnitude lower than in free aqueous solution; moreover, the strongest binders also showed the slowest dissociation and displayed a significant amount of irreversible binding, a prerequisite of indefinite immobilization for rebinding experiments (Figure 8.20).

The obvious consequence of the encouraging properties of the parent polymers was extension to other binding monomers, which should be specific for various amino acid residues on protein surfaces. Thus, a set of simple binding comonomers was synthesized, which allowed interaction with basic, acidic, aromatic and hydrophobic residues. These comonomers were combined in various ratios (including the dansyl monomer for fluorimetric detection) and subjected to free radical polymerization as well as final protecting group cleavage in polymer-analogous reactions [75]. The resulting copolymers were in some cases insoluble in water; this problem was overcome with copolymerization of a sugar comonomer based on free glucosamine. Polymer weights were in the range of 50 000 Da, featuring 50–100 binding monomers for the interaction with protein surface residues.

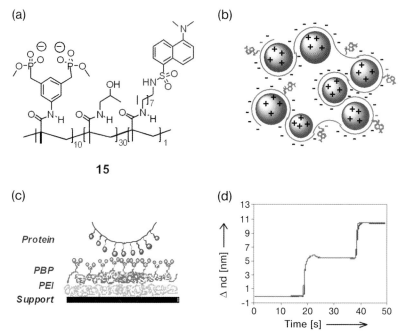

Figure 8.20 (a) Schematic structure of copolymer **15**. (b) Schematic depicting the postulated binding mode between **15** and cytochrome c. (c) Schematic showing the attachment of molecular polyethyleneimine (PEI) and polybisphosphonate (PBP) layers onto the signal transducer. (d) Corresponding stepwise increase in RIfS signal.

Screening of small polymer and protein libraries revealed some "hits", especially in the case of basic proteins with hydrophobic patches on their surfaces, as evidenced from the inspection of crystal structures from the Protein Data Bank (PDB). Thus, a copolymer from a 3:1:1 mixture of bisphosphonate, sugar and dodecyl comonomer was able to bind lysozyme more than 100 times more tightly (K_d = 25 nM) than cytochrome c, although these proteins are very similar in size and pI (Figure 8.21).

With this polymer a detailed investigation was carried out on an enzyme assay with lysozyme and the bacterial substrate *Micrococcus lysodeicticus* in 50 mM phosphate buffer [76]. The optimized polymer for lysozyme displayed a submicromolar IC$_{50}$ value (0.7 µM) for cell wall degradation monitored by the optical density decrease in the assay. Related copolymers lacking one specific component revealed that for efficient enzyme inhibition, all the three comonomers must be present in the polymer. Together with a textbook example of competitive inhibition (Lineweaver–Burk plot) this is a strong indication for noncovalent blocking of the active site by the sugar, reinforced by electrostatic attraction between the bisphosphonate binding sites and the arginine residues surrounding the binding site, and

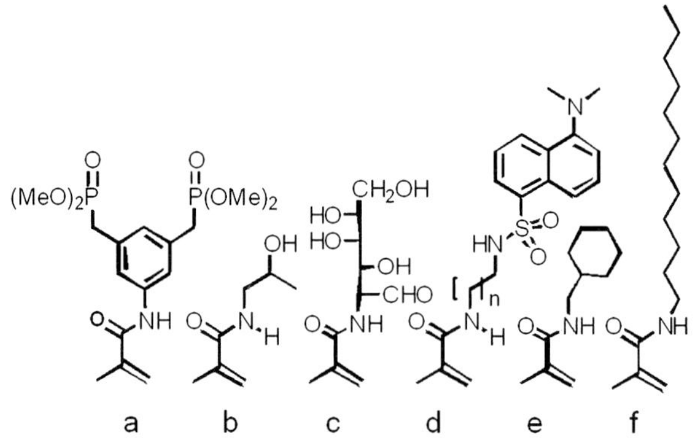

Polymer	Bisphosphonate a	Alcohol b, c	Dansyl d	Cyclohexyl e	Dodecyl f
16	1	–	0.1	–	–
17	1	3[a]	0.4	–	–
18	1	1[b]	1	–	–
19	3	1[b]	0.5	1	–
20	1	1[b]	0.3	–	1
21	3	1[b]	0.5	–	1

Figure 8.21 Modular set of methacrylamide-based comonomers ($n = 1$ or 7) and polymers **a–f** derived thereof; each line indicates the relative ratio of comonomers for a given copolymer. [a]Alcohol monomer from 2-hydroxypropylamine. [b]Alcohol monomer from 2-glucosamine.

further supported by dispersive and hydrophobic interactions with phenylalanine residues in their close vicinity. As the circular dichroism spectrum did not change during the inhibition process, we assume a nondenaturing mild coverage of the enzyme surface including the active site. Subsequent addition of polyarginine effectively restored greater than 90% of the original enzyme activity, so that a fully reversible enzyme inhibition was discovered, which can be used as a structural switch to turn off and on bacterial cell wall degradation by external agents (Figure 8.22).

We will in the future extend this principle of protein surface recognition with linear polymers and try to extend this methodology to molecular imprinting (i.e., polymerization in the presence of the protein template) in order to achieve the ultimate goal – artificial antibodies.

8.4
Conclusions and Outlook

The examples summarized in this chapter show how chemists have gained more and more insight into the structural factors governing highly complex biological recognition events. With this knowledge and the power of chemical synthesis, a

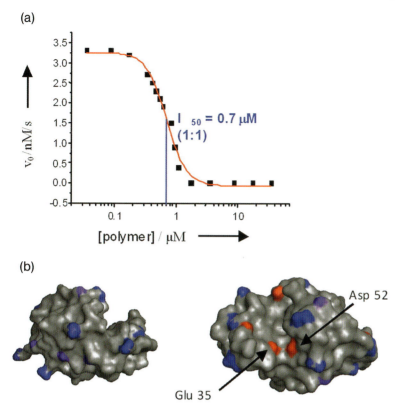

Figure 8.22 (a) Lysozyme activity determined for 0.7 µM enzyme treated with increasing amounts of polymer **21**. The polymer concentration at 50% maximum activity indicates the IC$_{50}$ value. (b) The lysozyme molecule depicted with its Connolly surface as projected from the side and from the top (PyMOL v0.98). Arginine residues are blue, lysines are purple. The catalytically active Asp52 and Glu35 in red are responsible for glycoside cleavage inside the cleft.

new door is now wide open to prepare – by design or with evolutionary methods – artificial receptor molecules that are not only able to interfere with biological systems in a predictable manner, but also self-assemble into novel artificial systems for signal transduction, self-replication and information processing.

References

1 Branden, C. and Tooze, J. (1999) *Introduction to Protein Structure*, 2nd edn, Garland Publishing, New York.
2 Glenner, N. (1980) *New England Journal of Medicine*, **302**, 1283–92.
3 Sacchettini, J.C. and Kelly, J.W. (2002) *Nature Reviews Drug Discovery*, **1**, 267.
4 Alzheimer, A. (1906) *Neurologisches Centralblatt*, **23**, 1129–36.
5 Selkoe, D.J. (1991) *Neuron*, **6**, 478–98.
6 Selkoe, D.J. (2004) *Annals of Internal Medicine*, **140**, 627–38.
7 Hardy, J. (2003) *Journal of Molecular Neuroscience*, **20**, 203–6.

8 Sato, C. (1998) *Journal of Biological Chemistry*, **273**, 7185–8.
9 Crooxs, E. (2000) *Journal of Structural Biology*, **130**, 247–58.
10 Kelly, J.W. (2003) *Science*, **299**, 713–16.
11 Yamada, M. (2002) *Journal of Neurochemistry*, **81**, 434–40.
12 Tjernberg, L.O. (1996) *Journal of Biological Chemistry*, **271**, 8545.
13 Sato, C. (1998) *Nature Medicine*, **4**, 822–6.
14 Watanabe, K. (2002) *Biochemical and Biophysical Research Communications*, **290**, 121–4.
15 Meredith, S.C. (2001) *Biochemistry*, **40**, 8237–45.
16 Meredith, S.C. (2002) *Biochemistry*, **42**, 475–85.
17 Willbold, D. (1995) *European Journal of Biochemistry*, **233**, 293–8.
18 Schrader, T. (2005) *Synthesis*, **12**, 1815–26.
19 Schrader, T. (1997) *Journal of the American Chemical Society*, **119**, 12061.
20 Schrader, T. (2005) *Journal of the American Chemical Society*, **127**, 3016–25.
21 Kinsman, R.G. (1997) *Tetrahedron*, **53**, 6977.
22 Schrader, T., Riesner, D., Rzepecki, P., Nagel-Steger, L., Wehner, M., Kirsten, C., Molt, O., Zadmard, R. and Aschermann, K. (2003) PTC INT Appl, DE 10221052 A1, 79.
23 Skribanek, Z., Balaspiri, L. and Mak, M. (2001) *Journal of Mass Spectrometry*, **36**, 1226–9.
24 Chen, X.G., Brining, S.K., Nguyen, V.Q. and Yergey, A.L. (1997) *FASEB Journal*, **11**, 817–23.
25 Bisswanger, H. (2000) *Enzymkinetik. Theorie und Methoden*, Wiley-VCH, Weinheim.
26 Schellenberger, A. (1989) *Enzymkatalyse: Einführung in die Chemie, Biochemie und Technologie der Enzyme*, Gustav Fischer, Jena.
27 Berg, J., Tymoczko, J. and Stryer, L. (2002) *Biochemistry*, Freeman, New York.
28 Löffler, G. and Petrides, P.E. (2003) *Biochemie und Pathobiochemie*, 7th edn, Springer, Berlin.
29 Nelson, D. and Cox, M. (2001) *Lehninger Biochemie*, 3rd edn, Springer, Berlin.
30 Filosa, S., Fico, A., Paglialunga, F., Balestrieri, M., Crooke, A., Verde, P., Abrescia, P., Bautista, J.M. and Martini, G. (2003) *Biochemical Journal*, **370**, 935–43.
31 Koppitz, M. and Eis, K. (2006) *Drug Discovery Today*, **11**, 561–8.
32 Scapin, G. (2006) *Current Pharmaceutical Design*, **12**, 2087–97.
33 Hunter, W.N. (1995) *Molecular Medicine Today*, **1**, 31–4.
34 Paul, M., Mehr, A.P. and Kreutz, R. (2006) *Physiological Reviews*, **86**, 747–803.
35 Atkinson, A.B. and Robertson, J.I. (1979) *Lancet*, ii, 863–9.
36 Skeggs, L.T., Jr, Kahn, J.R. and Shumway, N.P. (1956) *Journal of Experimental Medicine*, **103**, 295–9.
37 Patchett, A.A., Harris, E., Tristram, E.W., Wyvratt, M.J., Wu, M.T., Taub, D., Peterson, E.R., Ikeler, T.J., ten Broeke, J., Payne, L.G., Ondeyka, D.L., Thorsett, E.D., Greenlee, W.J., Lohr, N.S., Hoffsommer, R.D., Joshua, H., Ruyle, W.V., Rothrock, J.W., Aster, S.D., Maycock, A.L., Robinson, F.M., Hirschmann, R., Sweet, C.S., Ulm, E.H., Gross, D.M., Vassil, T.C. and Stone, C.A. (1980) *Nature*, **288**, 280–3.
38 Smith, C.G. and Vane, J.R. (2003) *FASEB Journal*, **17**, 788–9.
39 Ferreira, S.H. (1965) *British Journal of Pharmacology*, **24**, 163–9.
40 Rioli, V., Prezoto, B.C., Konno, K., Melo, R.L., Klitzke, C.F., Ferro, E.S., Ferreira-Lopes, M., Camargo, A.C.M. and Portaro, C.V. (2008) *FEBS Journal*, **275**, 2442–54.
41 Ferreira, S.H., Greene, L.H., Alabaster, V.A., Bakhle, Y.S. and Vane, J.R. (1970) *Nature*, **225**, 379–400.
42 Nemec, K. and Schubert-Zsilavecz, M. (2003) *Pharmazie in unserer Zeit*, Wiley-VCH, Weinheim.
43 Brown, N. and Vaughan, D. (1998) *Circulation*, **97**, 1411–20.
44 Henck, H.H.W. and Drieman, H.C. (1990) *Cardiovascular Drug Reviews*, **8**, 386–400.

45 Katz, A.H. and Caufield, C.E. (2003) *Current Pharmaceutical Design*, **9**, 857–66.
46 Lew, W., Chen, X. and Kim, C.U. (2000) *Current Medicinal Chemistry*, **7**, 663–72.
47 Hsu, J.T., Wang, H.C., Chen, G.W. and Shih, S.R. (2006) *Current Pharmaceutical Design*, **12**, 1301–14.
48 McGuire, J.J. (2003) *Current Pharmaceutical Design*, **9**, 2593–613.
49 Fischer, P.M. (2003) *Current Protein and Peptide Science*, **4**, 339–56.
50 Chittur, S.V., Klem, T.J., Shafer, C.M. and Davisson, V.J. (2001) *Biochemistry*, **40**, 876–87.
51 Jasper, C., Schrader, T., Panitzky, J. and Klärner, F.-G. (2002) *Angewandte Chemie (International Edition in English)*, **41**, 1355–8.
52 Schrader, T., Fokkens, M., Klärner, F.-G., Polkowska, J. and Bastkowski, F. (2005) *Journal of Organic Chemistry*, **70**, 10227–37.
53 Schrader, T., Klärner, F.-G., Fokkens, M., Zadmard, R., Polkowska, J., Bastkowski, F. and Jasper, C. (2006) Neue Wirkstoffe zu Therapie, Diagnostik und Prophylaxe der Makula-Degeneration, Patent DE 10 2004 056 822.7.
54 Fokkens, M., Schrader, T. and Klärner, F.-G. (2005) *Journal of the American Chemical Society*, **127**, 14415–21.
55 Dickinson, F.M. and Monger, G.P. (1973) *Biochemical Journal*, **131**, 261–70.
56 Schöpp, W. and Aurich, H. (1976) *Biochemical Journal*, **157**, 15–22.
57 Ganzhorn, A.J., Green, D.W., Hershey, A.D., Gould, R.M. and Plapp, B.V. (1987) *Journal of Biological Chemistry*, **262**, 3754–61.
58 Leskovac, V. and Trivic, S. (1994) *Biochemistry and Molecular Biology International*, **32**, 399–407.
59 Leskovac, V. et al. (2002) *FEMS Yeast Research*, **2**, 481–94.
60 Stites, W.E. (1997) *Chemical Reviews*, **97**, 1233.
61 (a) Larsen, T.A., Olsen, A.J. and Goodsell, D.S. (1998) *Structure*, **6**, 421–7; (b) Janin, J., Miller, S. and Chothia, C. (1988) *Journal of Molecular Biology*, **204**, 155–64; (c) Bogan, A.A., Thorn, K.S. (1998) *Journal of Molecular Biology*, **280**, 1–9; (d) Sheinerman, F.B., Norel, R. and Honig, B. (2000) *Current Opinion in Structural Biology*, **10**, 153–9.
62 (a) α-helices: Yin, H., Lee, G.I., Sedey, K.A., Rodriguez, J.M., Wang, H.G., Sebti, S.M. and Hamilton, A.D. (2005) *Journal of the American Chemical Society*, **127**, 5463; (b) β-turns: Smith, A.B., Wang, W.Y., Sprengeler, P.A. and Hirschmann, R. (2000) *Journal of the American Chemical Society*, **122**, 11037.
63 (a) Baldini, L., Wilson, A.J., Hong, J. and Hamilton, A.D. (2004) *Journal of the American Chemical Society*, **126**, 5656–7; (b) Merritt, E.A., Zhang, Z., Pickens, J.C., Ahn, M., Hol, W.G.J. and Fan, E. (2002) *Journal of the American Chemical Society*, **124**, 8818–24; (c) Ojida, A., Miyahara, Y., Kohira, T. and Hamachi, I. (2004) *Biopolymers*, **76**, 177–84; (d) Fazal, M.A., Roy, B.C., Sun, S., Mallik, S. and Rodgers, K.R. (2001) *Journal of the American Chemical Society*, **123**, 6283–90.
64 (a) Gold clusters: Fischer, N.O., Verma, A., Goodman, C.M., Simard, J.M. and Rotello, V.M. (2003) *Journal of the American Chemical Society*, **125**, 13387–91; (b) postsynthetically modified polymers: Strong, L.E. and Kiessling, L.L. (1999) *Journal of the American Chemical Society*, **121**, 6193–6.
65 (a) Protein grafting: Rutledge, S.E., Volkman, H.M. and Schepartz, A. (2003) *Journal of the American Chemical Society*, **125**, 14336; (b) RNA aptamers: Schneider, D.J., Feigon, J., Hostomsky, Z. and Gold, L. (1995) *Biochemistry*, **34**, 9599.
66 (a) Reconstructed liposomes: Santos, M., Roy, B.C., Goicoechea, H., Campiglia, A.D. and Mallik, S.J. (2004) *Journal of the American Chemical Society*, **126**, 10738–45; (b) Cu-II-imprinted silica: Mallik, S., Plunkett, S.D., Dhal, P.K., Robinson, R.D., Pack, D., Shenk, D. and Arnold, F.H. (1994) *New Journal of Chemistry*, **18**, 299; (c) saccharide/polymer coating: Shi, H.Q., Tsai, W.B., Garrison, M.D., Ferrari, S. and Ratner, B.D. (1999) *Nature*, **398**, 593.

67 Benezra, M., Vlodavsky, I., Yayon, A., Bar-Shavit, R., Regan, J., Chang, M. and Ben-Sasson, S. (1992) *Cancer Research*, **52**, 5656–62.

68 Regan, J., McGarry, D., Bruno, J., Green, D., Newman, J., Hsu, C.-Y., Kline, J., Barton, J., Travis, J., Choi, Y.M., Volz, F., Pauls, H., Harrison, R., Zilberstein, A., Ben-Sasson, A.A. and Chang, M. (1997) *Journal of Medicinal Chemistry*, **40**, 3408–22.

69 (a) Fischer, N.O., McIntosh, C.M., Simard, J.M. and Rotello, V.M. (2002) *Proceedings of the National Academy of Sciences of the United States of America*, **99**, 5018–23. (b) Fischer, N.O., Verma, A., Goodman, C.M., Simard, J.M. and Rotello, V.M. (2003) *Journal of the American Chemical Society*, **125**, 13387–91.

70 Strong, L.E. and Kiessling, L.L. (1999) *Journal of the American Chemical Society*, **121**, 6193–6.

71 Shimoboji, T., Larenas, E., Fowler, T., Kulkarni, S., Hoffman, A.S. and Stayton, P.S. (2002) *Proceedings of the National Academy of Sciences of the United States of America*, **99**, 16592–6.

72 Sandanaraj, B.S., Rao Vutukuri, D., Simard, J.M., Klaikherd, A., Hong, R., Rotello, V.M. and Thayumanavan, S. (2005) *Journal of the American Chemical Society*, **127**, 10693–8.

73 Arendt, M., Sun, W., Thomann, J., Xie, X. and Schrader, T. (2006) *Chemistry: An Asian Journal*, **1**, 544–54.

74 Renner, C., Piehler, J. and Schrader, T. (2006) *Journal of the American Chemical Society*, **128**, 620–8.

75 Koch, S., Renner, C., Xie, X. and Schrader, T. (2006) *Angewandte Chemie (International Edition in English)*, **45**, 6352–5.

76 Wenck, K. and Schrader, T. (2007) *Journal of the American Chemical Society*, **129**, 16015–9.

77 Kirsch, M., Talbiersky, P., Polkowska, J., Bastkowski, F., Schaller, T., de Groot, H., Klärner, F. and Schrader, T. (2009) *Angewandte Chemie*, **48**, 2886–90.

# 9
Analysis of Biological Interactions and Recognitions by Surface Plasmon Resonance

Sang Jun Sim and Cuong Cao

9.1
Introduction to Surface Plasmon Resonance Technology

Recently, we have witnessed that surface plasmon-based immunosensors are providing a highly promising analytical approach for quantification and qualification of chemical and biochemical substances in various applications. Since the first observation by Wood in 1902 [1], the physical phenomenon of surface plasmon resonance (SPR) has paved the way for practical applications in sensitive detectors and optical reflection measurements. In 1957, Rufus Ritchie reported a study on electron energy losses in thin films [2] and this was the first theoretical description of surface plasmons, which showed that plasmon modes can exist near the surface of metals. A complete explanation of the phenomenon was not possible until 1968, when Otto [3] and Kretschmann and Raether [4] reported the excitation of surface plasmons. However, the application of SPR in biomolecular interaction monitoring was first reported in the landmark paper of Liedberg *et al.* in the early 1980s [5]. Since then, thousands of scientific publications relying on SPR sensor technology have been reported for measurements related to numerous physical and chemical quantities, including humidity measurements [6], pressure [7], gas concentration [8], and interaction analysis of various chemical and biochemical compounds [5]. Among these applications, analysis of biomolecular recognitions based on SPR has recently become the center of research and is rapidly growing – it has gained considerable contributions in fundamental biological studies, food quality and safety analysis, medical diagnostics, drug delivery, and environmental monitoring.

SPR, an optical transduction based on the excitation of surface plasmons, is a widely used method for real-time biomolecular interaction analysis. The major advantages of SPR biosensors are the possibility of real-time analysis of biomolecular interactions without the need for labeled molecules and that the SPR biosensor requires only a small sample quantity. Furthermore, it is possible to reuse the sensor chip via surface regeneration and the analysis of biomolecular interactions is completed in a few minutes [9]. However, the major disadvantage

Cellular and Biomolecular Recognition: Synthetic and Non-Biological Molecules. Edited by Raz Jelinek
Copyright © 2009 WILEY-VCH Verlag GmbH & Co. KGaA, Weinheim
ISBN: 978-3-527-32265-7

is that it is difficult to determine an analyte at low concentration or with a low molecular mass. The detection limit is approximately 1–10 nM for a 20-kDa molecule and is even higher for smaller molecules [10]. This chapter gives an overview of the analysis of biological interactions and recognitions by means of SPR-based biosensors; current advanced achievements, shortcomings as well as current solutions will be summarized.

9.2
Working Principle of SPR

SPR is a charge-density oscillation that occurs at the interface of two media with dielectric constants of opposite signs (e.g., a metal and a dielectric). When a polarized light beam propagating in a medium of higher refractive index (e.g., glass) meets an interface at a medium of lower refractive index (e.g., air, water) at a defined angle called the angle of resonance [11, 12], the light is completely reflected at the interface and propagates back into the high-refractive-index medium (total internal reflection). If the surface of the glass is coated with a layer of a suitable conducting material such as a gold metal, this reflection is not total because the light beam leaks an electrical field intensity called an evanescent field wave into the metallic film. Consequently, oscillation of free electrons at the surface of the metal film will be excited and these oscillating waves are called electromagnetic surface plasmon waves, which propagate parallel along the conductive gold surface contacting with the low-refractive-index medium (sample medium). SPR will be produced when the light wavevector of the incident light matches the wavelength of the surface plasmons. The coupling of the incident light to the surface plasmons results in a considerable loss of energy owing to energy absorption of free electrons at an angle of incident light called the SPR angle and therefore it is accountable for a characteristic decrease in the intensity of the reflected light. Since the surface plasmon wave is on the boundary of the metal and the external medium (sample medium), the resonant frequency of these oscillations is very sensitive to any change of this boundary, and therefore a direct insight into changes of the refractive index of the sample medium at the surface interface is made possible by monitoring the intensity and the resonance angle of the reflected light, induced by the biospecific interactions of (bio)chemical molecules at the metal surface [13].

The surface plasma wave propagating at the interface between a semi-infinite dielectric and metal is represented by:

$$k_{spw} = \frac{\omega}{c}\sqrt{\frac{\varepsilon_m n_s^2}{\varepsilon_m + n_s^2}}, \tag{9.1}$$

where k_{spw} is the surface plasma wave propagation constant, ω/c is the propagation constant of light in vacuum (ω is the angular frequency and c is the velocity of light in vacuum), ε_m is the dielectric constant of the metal ($\varepsilon_m = \varepsilon_r + \varepsilon_i$, real and imaginary dielectric constants, respectively) and n_s is the refractive index of the dielectric [14]. It is seen in Equation 9.1 that SPR propagation can be supported

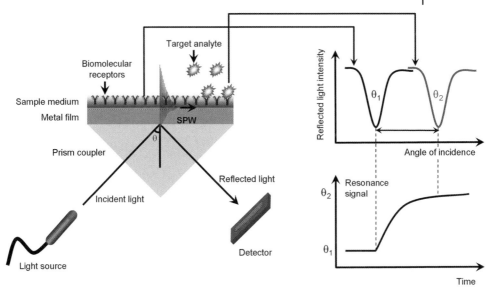

Figure 9.1 Schematic illustrating the prism coupler-based SPR biosensor configuration (left) and SPR resonance shift induced by binding of target analyte to the biomolecular receptors on the sensor surface (right). SPW = surface plasmon wave.

by the structure only if $\varepsilon_r < -n_s^2$, providing that the dielectric constant of the metal and the refractive index of the medium are of opposite sign [15]. Moreover, the SPR is characterized by two main factors: the intrinsic properties of the transduction metallic surface and the nature of the medium in contact with the transduction surface. A system of a gold film (50–100 nm) in contact with a water or buffer interface containing a target analyte is commonly utilized in SPR measurements [14]. Those two media have dielectric constants of opposite signs; gold has a negative dielectric function in the infrared and visible areas of the electromagnetic spectrum, while water has a positive dielectric function [16] (Figure 9.1).

SPR sensors belonging to the group of refractometric sensing devices; an SPR biosensor is mainly comprised of two main parts: the specific, sensitive layer and the SPR refractometer. Every alteration in the refractive index caused by specific biological interactions on the sensing layer will be optically converted by the SPR refractometer into binding data as sensorgrams (Figure 9.2). In a typical biosensing experiment, biomolecular recognition elements are immobilized on a sensor surface and their other binding partners are passed over this surface in solution. Almost any molecule (e.g., DNA, aptamer, protein, enzyme, etc.) can be immobilized on the sensing surface and monitored for binding to soluble analytes ranging from low-molecular-mass drugs to large, multicomponent virus particles. The specific biological interactions will lead to an increase in refractive index at the biointerface and, in turn, the change of refractive index causes a shift in the angle

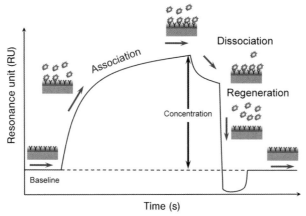

Figure 9.2 Schematic representation of a typical SPR-commercialized sensorgram. Spectroscopy of surface plasmon-based sensors is capable to supplying rich information on the concentration of target analyte, kinetics of the reaction, affinity of the biological recognition as well as specificity of the analytical measurement.

of incidence at which the SPR phenomenon occurs. Therefore, the detection of target molecules in a complicated mixture is possible. Such shifts are continuously monitored automatically and shown as graphs called sensorgrams. The angle of incidence required to create the SPR phenomenon (the SPR angle) is therefore altered and this change is measured as a response signal. It is believed that the signal generated in arbitrary resonance units (RUs) is approximately proportional to a change in mass (1 RU = 1 pg/ mm² for protein) [17]. The SPR sensorgrams are performed in real-time and without any labeling; they provide rich information on the kinetics, specificity, affinity and concentration levels of the target analyte from a complex sample (Figure 9.2).

9.3
Sensor Surface Chemistry and Its Fabrications

Although SPR transduction is a totally nonspecific phenomenon, it is impossible to discriminate different changes of various biological or chemical recognitions specifically by the optical phenomenon itself. While this may appear to be a limitation, it is really a powerful advantage since the principle of the SPR biosensor is based on the change in the refractive index on a thin gold film surface. The transducer surface is practically modified with various materials and the specificity of the SPR assay strongly depends upon selection of pairs of molecules that react only with each other (typically, the specific interactions of any pair of molecules can be antigen–antibody, protein–aptamer, DNA–DNA, enzyme–substrate, chelating agent–metal ion, etc.).

Biorecognition elements immobilized on the transducer surface have various entities that have been extensively utilized in affinity SPR biosensors. Antibodies are immune system-related proteins called immunoglobulins (Igs), which are the most frequently used biorecognition element. They are produced in the immunosystem of vertebrates in order to specifically react to a foreign substance (antigen) [18]. There are five different classes of antibodies: IgA, IgD, IgE, IgG, and IgM; each antibody "monomer" has a molecular weight of approximately 150 000 Da, and is composed of two identical heavy polypeptide chains and two identical light chains joined to form a "Y"-shaped molecule covalently bonded via interchain disulfide (S–S) linkages between cysteine residues. Moreover, these monomers are arranged in three discrete domains – two Fab fragments at the tips of the "Y" and one Fc at its pole. The enzyme papain can be used to cleave an Ig monomer into two Fab fragments and an Fc fragment [19, 20]. The enzyme pepsin cleaves below the hinge region, so an F(ab')$_2$ fragment and an Fc fragment are formed. Since it is the Fab fragments that bind to antigens, the size of an antibody can be reduced substantially by removing the Fc fragment. The antibodies used in most immunoassays are of two kinds: monoclonal and polyclonal antibodies. While monoclonal antibodies are produced from an individual clone of plasma cells and only react with a specific binding site on the antigen, polyclonal antibodies are derived from different lymphocyte cell lines and recognize different epitopes of the antigen. Therefore, monoclonal antibodies are generally more specific than polyclonal antibodies [18]. However, owing to the fact that polyclonal antibodies contain the entire antigen-specific antibody population, one antigen molecule can form a complex with several antibody molecules leading to stronger affinity towards their binding antigen partner. Recently, single-chain Fab fragments of an antibody comprising an intact antigen-binding site have been utilized in SPR or microcantilever-based immunosensors [21, 22]. The Fab fragments are not only able to react with antigens with the same affinity, but also provide advantages over intact Ig molecules, such as their minimized size, the possibility for directed and dense immobilization on interfaces, and their ease of production [22]. Peptides are also considered as recognition elements for SPR biosensors. While proteins are supposed to be difficult to construct because of the complexity and heterogeneity, as well as it being laborious to preserve their conformational folding after covalent binding to a surface [23], peptides become an attractive alternative owing to their lower complexity, easier production, less nonspecific binding and greater stability. However, the major disadvantage preventing peptides from being excellent recognition elements is that sometimes they lack high affinity and specificity against the target [24]. Recently, aptamers have emerged as another promising candidate for recognition of various biomolecules based on SPR biosensors, such as amino acids, drugs and proteins, with high affinity [25–29]. Aptamers are composed of DNA, RNA or peptide molecules that bind to many different target molecules such as small molecules, nucleic acids, proteins or living cells. The synthesis of aptamers is straightforward, cost-effective and reproducible; they can be created by a SELEX process (Systematic Evolution of Ligands by EXponential enrichment) related to absorbing, recovering and amplifying them from a large random sequence pool [30, 31]. It is believed that aptamers have several advantages over their corresponding antibodies in some aspects, such as stronger binding affinity to the target protein and stability [32].

The construction of suitable surface chemistry and the immobilization of biorecognition elements on the transducer surface are very important to SPR measurements. Many book chapters, review papers and notes have been extensively focused on these issues [18, 24, 33–35]. The biological components (capture molecules) are commonly immobilized on the surface of the transducer in order to interact with their target of interest. As a consequence, the biological recognition will be translated into an optical signal by changing the refractive index at the surface interface and the signal generated at the inorganic transducer surface will be proportionally related to the concentration of the analyte in the sample. Therefore, the surface chemistry and immobilization strategy for biomolecules are of paramount importance to preserve the biological activity as well as to assure the sensitivity and specificity of the SPR-based quantifications. Simple and rapid immobilization strategies that maintain biological activity by avoiding any conformation change, unfavorable orientation, chemical inactivation or denaturation of the biorecognition elements; prevent nonspecific binding of undesired molecules; and prevent steric hindrance effects caused by too high density of immobilized biomolecules on the transducer surface are strongly required and very crucial to performance of SPR-based analysis of biorecognitions.

Langmuir–Blodgett (LB) deposition and self-assembly are two widely used methods to construct molecular layers on the sensing (gold) surface. LB films are mechanically formed by one or more monolayers of an amphiphilic material in which the molecules are compressed to the desired organization and then transferred to the solid substrate with very accurate thickness (Figure 9.3a). Self-assembled monolayers (SAMs) are surfaces consisting of a single layer of surfactant molecules with surface-active head groups (thiols, sulfides, disulfides) spontaneously adsorbing on an appropriate substrate (gold, silver, platinum or copper). When a metal (e.g., gold) is immersed into solution containing thiol derivatives, sulfur atoms of disulfides (R-S–S-R), sulfides (R-S-R) and thiols (R-SH) coordinate very strongly onto the metal surface (Figure 9.3b). For specific use, the terminal end of the thiol-containing molecules has been functionalized with a number of active groups like -OH, $-NH_2$, -COOH, -COOR, biotin, etc., thereby producing surfaces with different potentials. Indeed, one of benefit of SAMs is that they provide desired surface functional groups to covalently immobilize biorecognition elements over the transducer surface as well as to control the hydrophobic/hydrophilic properties of the surface chemistry easily. Moreover, a mixed thiol monolayer enhances the surface coverage of immobilized biorecognition elements by decreasing the steric hindrance effect. Patel *et al.* constructed SAMs of pure 3-mercaptopropionic acid (3-MPA), pure 11-mercaptoundecanoic acid (11-MUA) and a mixture of these thiols, and concluded that the mixed SAMs had improved accessibility for protein binding due to reduced steric hindrance [36]. Lee *et al.* tested surfaces constructed by mixing different molar ratios between 3-MPA

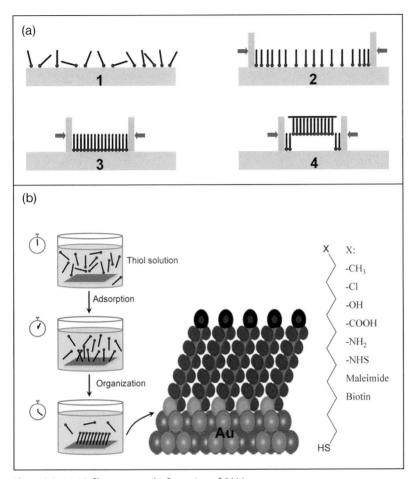

Figure 9.3 (a) LB film process, (b) formation of SAMs.

and 11-MUA (SAM 1 = 20:1, SAM 2 = 10:1, SAM 3 = 1:1 and SAM 4 = homogeneous 11-MUA) to obtain an optimum surface for the immobilization of capture protein [37]. They found that SAMs in a molar ratio of 10:1 (3-MPA:11-MUA) had the best streptavidin adsorption and immobilization of biotinylated glutamic acid decarboxylase (GAD) (Figure 9.4). Different mixed SAMs of oligo(ethylene glycol) (OEG) have also been recognized as ideal linkers to immobilize biomolecules and to prevent nonspecific adsorption of proteins [38–41]. The OEG-terminated SAMs on the bare gold surface are very effective in reducing nonspecific binding because of the fact that the OEG has the ability to render the surface biocompatibility. The SAM surface of OEG provides a template for water nucleation and protein resistance of OEG-terminated SAMs is a consequence of the stability of the interfacial water layer, which prevents direct contact between the surface and proteins [42, 43].

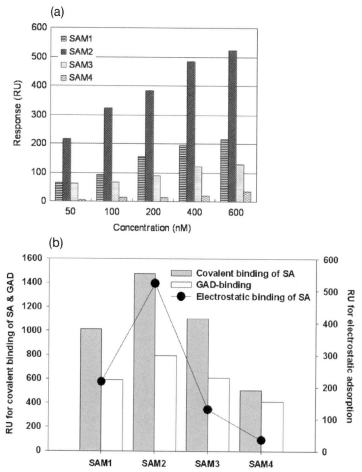

Figure 9.4 SPR signals for (a) adsorption of different concentrations of streptavidin on SAMs and (b) for the covalent bonding of streptavidin (200 μg/ml in 10 mM acetate buffer, pH 4.5), biotinylated GAD (1.2 μg/ml in HBSS) and a linear graph for the adsorption of 600 nM of the streptavidin. SA = streptavidin. (Reproduced with permission from [37], Copyright 2005 by Elsevier).

While most biomolecules can be covalently immobilized onto the gold surface by means of SAMs, noncovalent physical adsorption is another alternative method to fabricate SPR sensors. The physical adsorption exploiting hydrophobic, electrostatic interactions between the molecule of interest and the sensing surface does not require chemical modification of the biorecognition elements, thus the biological activity of the immobilized components can be maintained. Although this method is very simple and straightforward, it also possesses some limitations. For example, the biomolecules cannot be immobilized in a stable and well-ordered manner. Therefore, such random orientation and less stability will limit the regeneration of the sensor surface as well as specificity of the SPR-based assays, excluding them from various applications.

More specific and better-oriented immobilization of biomolecules is possibly achieved by high-affinity capture, where tagged molecules are noncovalently attached to immobilized capture molecules via biotin–streptavidin coupling, histidine–chelated metal ion interaction or DNA hybridization [24]. Protein A or protein G, which have a powerful ability to bind to proteins from many species (especially to the Fc region of Ig) [44], have been utilized as capture molecules for binding of antibodies on the SPR transducer surface with good orientation [45, 46]. Well-oriented immobilization of the recognition elements will lead to highly efficient immunoreactions and enhance detection system performance. Shankaran and Miura have fully covered a variety of immobilization methods for recognition elements; more details are given in their review paper [18] (Figure 9.5).

9.4
Important Factors Impacting on the Performance of SPR-Based Analyses of Biological Interactions on the Nonbiological Transducer Surface

As mentioned previously, SPR is a completely nonspecific phenomenon whereby the electrons in the metal transducer surface are resonant with the photon frequency of the incident light as a plasmon. Any change to the chemical environment within the electric field produced by the plasmon will be recorded as changes in the resonance wavelength or angle. Therefore, the transducing performance of SPR biosensors (sensitivity, specificity, reproducibility, etc.) is strongly impacted by nonfouling surfaces, recognition elements and detection formats.

9.4.1
Nonspecific Interactions

In any analysis of biological recognition, interaction between nontargeted components in crude environments (e.g., blood, cell lysate, urine, waste water, etc.) and the sensing materials could be a source of noise leading to aberrance of the final results. Proteins or DNA could be nonspecifically adsorbed onto the gold surface via hydrophobic and ionic interactions, and this could result in reduced sensitivity and specificity of the SPR sensor systems; limitations have prevented versatile applications of SPR biosensor chips from being implemented. Therefore, one of the most important missions of surface fabrication is to provide abundant functional groups for immobilization of biomolecular recognition elements while minimizing the nonspecific adsorption of other components. Several chemistries have been developed for the fabrication of a nonfouling surface employing covalent binding of gold to the free sulfhydryl groups of alkanethiols, aromatic thiols and other important functional molecules by interfacial ligand-exchange reactions [47, 48] or place-exchange reactions [49, 50]. It is proposed that the functional

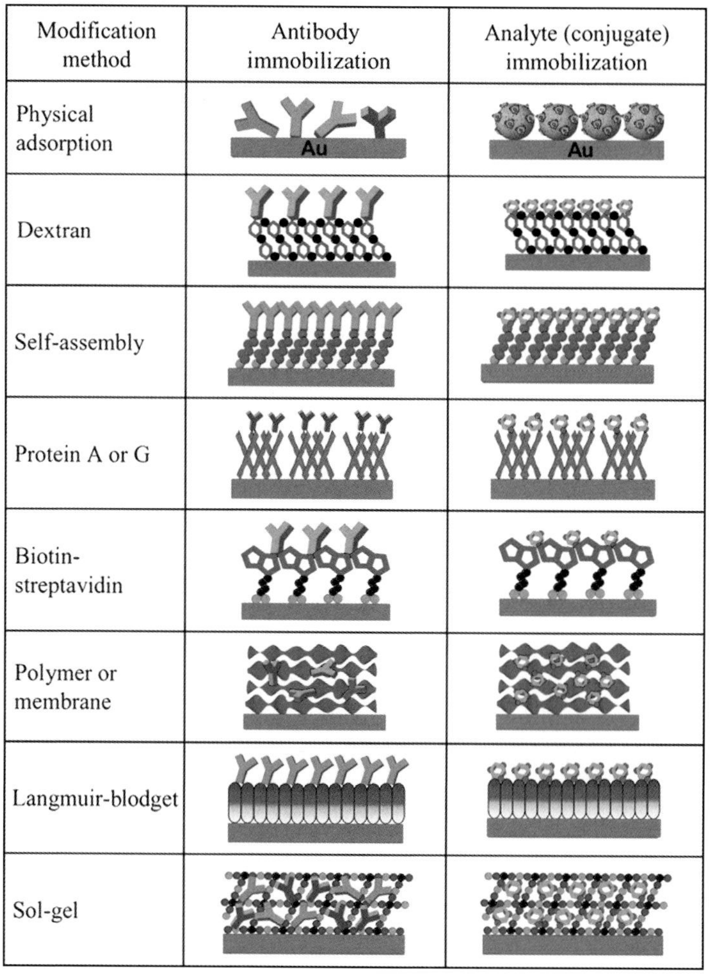

Modification method	Antibody immobilization	Analyte (conjugate) immobilization
Physical adsorption		
Dextran		
Self-assembly		
Protein A or G		
Biotin-streptavidin		
Polymer or membrane		
Langmuir-blodget		
Sol-gel		

Figure 9.5 Schematic overview of the formation of biorecognition elements fabricated by different modification methods. (Reproduced with permission from [18], Copyright 2007 by IOP Publishing).

groups of a low fouling transducer surface should possess several essential features, such as hydrophilicity, electrical neutrality, ability to accept a hydrogen bond as well as inability to donate a hydrogen bond [51].

SAMs of alkanethiols terminated in different functional groups are the first candidates to serve as a functionalized structure for further modification of the capture surface, as well creating a barrier to prevent proteins and other ligands from contacting with the metal. Alkanethiol molecules can be chemisorbed and physisorbed on the gold surface to form a monolayer, and then the recognition elements can be attached directly to the alkanethiol monolayer. The advantage of the alkanethiol SAMs is that they show a closely packed, highly ordered structure. However, within this structure, the terminal functional groups of the SAM sterically prevent the formation of covalent bonds with the immobilized target of interest, leading to a decrease in the immobilization efficiency. Therefore, to prevent the steric hindrance effect and to maximize the amount of immobilized protein, a mixed thiol surface (e.g., using various thiol mixtures of different molar ratios of alkanethiol-COOH and alkanethiol-OH) immobilizes protein more efficiently than homogenous alkanethiols (Figure 9.4) [36, 37, 52, 53].

Poly(ethylene glycol) (PEG) is a hydrophilic, electrically neutral polymer that has been widely utilized to fabricate a nonfouling surface for SPR biosensors [38, 54]. It is explained that PEG molecules resist nonspecific adsorption of protein because of their steric-entropy barrier characteristics and high degree of hydration [24, 55]. The gold surface having PEG tethered chains greatly improved the stability in aqueous milieu due to the steric repulsion effects of PEG and proved very effective in ameliorating the protein nonfouling character. Pasche et al. have studied the relationship between interfacial forces and protein resistance of PEG-grafted poly(L-lysine) adlayers on niobia surfaces; they found that the protein resistance of PEGylated surfaces correlates with a net repulsive force versus distance curve [56]. A mixture of PEG derivatives has also been exploited to fabricate a low fouling background surface with maximum binding sites for the immobilization of biological materials. For example, Yu et al. have developed a gold-coated SPR chip functionalized with mixed OEG SAMs for the detection of domoic acid, where the long-chain $HS(CH_2)_{11}(C_2H_5O)_6NH_2$ thiol serves as a functional site to react with carboxyl groups on domoic acid while the short-chain $HS(CH_2)_{11}(C_2H_5O)_4OH$ thiol serves as a protein nonfouling background. Control experiments using several nontargeted proteins [anti-domoic acid, anti-human chorionic gonadotropin and bovine serum albumin (BSA)] showed that there was no detectable nonspecific protein adsorption on the sensor surface immobilized with domoic acid [57]. In our group, we constructed a biochip by using various OEG mixtures of different molar ratios of $HS(CH_2)_{11}(OCH_2CH_2)_6OCH_2COOH$ and $HS(CH_2)_{11}(OCH_2CH_2)_3OH$ to reduce the nonspecific binding and steric hindrance effect. The SAMs were consequently biotinylated in order to facilitate the immobilization of streptavidin. Nonspecific binding between the surface materials and several serum proteins, such as BSA, IgG and fibrinogen, has been investigated; the results showed that the adsorption of BSA, IgG and fibrinogen was negligible. Using the chip surfaces, a prostate-specific antigen (PSA)–$α_1$-antichymotrypsin (AAT) complex in both Hank's balanced saline solution (HBSS) and human serum could be detected at 20.7 and 47.5 ng/ml, respectively [58].

Alternatively, surface polymers other than PEG have also been employed to produce a protein-resistant surface for SPR-based applications. Cellulose derivatives and dextran grafted to polystyrene have been shown to be nearly as effective as PEG at preventing protein adsorption [59]. SAMs of oligo(phosphorylcholine) (OPC) and phosphorylcholine (PC) have demonstrated that the material exhibits strong resistance to protein adsorption (fibrinogen, lysozyme, bovine serum albumin) and cell adhesion (bovine aortic endothelial cells) [60]. This work has reported that OPC SAMs resist protein adsorption as effectively as or better than PC SAMs formed from highly purified PC thiols. Polysaccharides cross-linked to poly(ethylenimine) bound noncovalently to polystyrene are also highly protein repellent; the amount of model proteins such as fibrinogen and IgG was significantly reduced when they were exposed to the polymer surface [61]. A dual-functional biocompatible material based on zwitterionic poly(carboxybetaine methacrylate) has been introduced to be not only highly resistant to protein adsorption/cell adhesion, but also abundant in functional groups convenient for the immobilization of biological ligands [62].

9.4.2
Recognition Elements

The recognition elements, which can be antibodies, nucleic acids, aptamers, enzymes or receptors, provide the main selective element, thus they are the major limitation of SPR biosensors. The quality of every measurement greatly depends on the selectivity, specificity and affinity of the capture element used. Normally, the biorecognition elements have been isolated from living systems, therefore the production process has several limitations, such as complexity, high cost and low quantity. However, thanks to recombinant technology, many biorecognition elements now available are synthesized at the laboratory scale. Antibodies have become the most extensively used recognition elements for detection purposes. Monoclonal antibody technology allows us to produce large quantities of antibody with high specificity and affinity to the target of interest. Recombinant antibodies consisting of genetically manipulated fused antigen-binding domains (Fab fragment) of common antibodies are now available. The Fab fragment shows the same binding affinity as the whole antibody molecule; however, the size can be reduced substantially [63, 64]. Aptamers, short and single-stranded oligonucleotides, can recognize their binding partner primarily by shape and conformation due to their self-annealing properties [65]. The ability to bind with high affinity and specificity to a broad range of target molecules (e.g., organic dyes, amino acids, biological cofactors, antibiotics, peptides, proteins and whole cells) makes aptamers suitable for applications based on molecular recognition as analytical, diagnostic and therapeutic tools [29, 66, 67].

9.4.3
Detection Formats

Various assay formats have been developed and applied for the analysis of biological interactions (Figure 9.6) [36, 68].

9.4.3.1 Direct Binding Assays (Figure 9.6a)
In direct binding assays, biorecognition elements are immobilized on the transducer surface. The analyte of interest interacts with the capture elements, producing a change in refractive index detected by the SPR sensor. The direct sensing

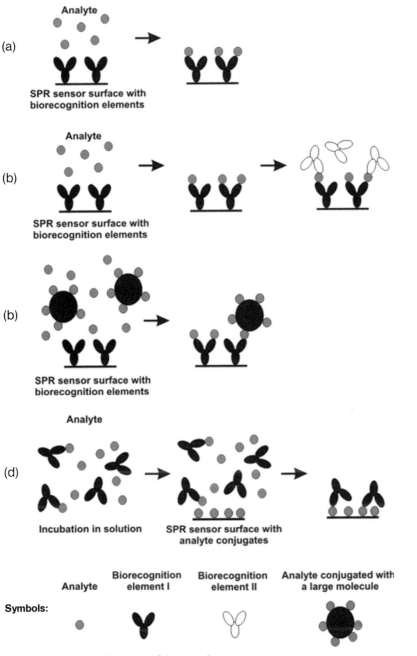

Figure 9.6 Schematic illustration of the main detection formats used in SPR biosenors: (a) direct binding assays, (b) sandwich detection assays, (c) competitive detection assays, (d) inhibition detection assays. (Reproduced with permission from [24], Copyright 2008 by American Chemical Society).

platform is only suitable for detection of large molecules having high molecular weight (above 10 kDa) or high concentrations of target analyte to produce a measurable SPR signal. However, this detection format is very straightforward and easy to perform; it is useful not only for concentration measurements, but also for affinity and kinetic analyses.

9.4.3.2 Sandwich Detection Assays (Figure 9.6b)

Sandwich detection assays are utilized to enhance the specificity and sensitivity of the direct binding assays. Figure 9.6b shows that a sandwich detection assay comprises two simultaneous recognition steps where the primary component immobilized on the transducer surface reacts with its specific molecule and then the secondary component is allowed to interact with the bound analyte. Therefore, the assay enables detection of large molecules (above 5000 Da) or enhancement of SPR sensitivity.

9.4.3.3 Competitive Detection Assays (Figure 9.6c)

The basis of the competitive detection format is competition between the analyte and its conjugated analog (normally conjugated to a high-molecular-weight carrier protein) in order to interact with a limited number of binding sites on the surface. A capture component binding to the analyte will be immobilized on the sensing surface, and when a constant amount of conjugated analyte is added to the sample, the SPR signal will be the total contributions from the target analyte and its conjugated analog, and this measure value will be inversely proportional to the analyte concentration in the sample.

9.4.3.4 Inhibition Detection Assays (Figure 9.6d)

In this sensing platform, the target analyte is immobilized onto the transducer surface instead of the biorecognition elements (e.g., antibodies). Practically, a known amount of biorecognition elements is first incubated with an unknown amount of target analyte in solution. Consequently, the reacted solution is allowed to flow over the SPR sensor surface. The free biorecognition element remaining from the first reaction then binds to the analyte molecules immobilized on the sensor surface. As a result, the SPR response measured for the free biorecognition element is inversely related to the concentration of analyte in the sample solution. This method is powerful for the detection of low-molecular-weight analytes; however, it may also be a useful alternative for macromolecules or cells whose dimensions are larger than the working range of SPR biosensors (normally about 300 nm), leading to poor penetration of cells within the evanescent field and low SPR signal generation. For example, Kang et al. have developed a SPR-based inhibition assay for the real-time detection of *Cryptosporidium parvum* oocysts. Secondary polyclonal anti-mouse IgM specific to *Cryptosporidium* was immobilized onto the SPR surface chip via heterogeneous SAMs. The inhibition assay consisted of the immunoreaction step between monoclonal anti-*C. parvum* oocyst (primary antibody) and oocysts, followed by the binding step of the unbound primary antibody onto the secondary antibody surface. They found that this method enhanced not only the immunoreaction yield of the oocysts by batch reaction, but also the accessibility of analytes to the chip surface by antibody–antibody interaction. The results showed that *Cryptosporidium* could be detected in the range of 10^2–10^6 oocysts/ml, and the SPR-based inhibition assay using the *Cryptosporidium* sensor chip has high application potential for the real-time analysis of *C. parvum* oocysts in laboratory and field water monitoring [69].

9.4.4
Several Approaches for Sensitivity Enhancements of the SPR Bioassays

As already shown, SPR has been applied to wide spectrum of applications. However, it is a fact that lack of a highly sensitive analytical method for detecting an analyte at low concentrations is a major impediment to SPR biosensor technology. Several approaches have been developed to improve the limitation in sensitivity and limit of detection (LOD) of SPR biosensors.

The sandwich detection format is the simplest and the most straightforward method to amplify the SPR signal [58]. We have applied this method for detection of PSA–ACT complex in both HBSS and human serum (Figure 9.7). The SPR sensor chips was constructed by using various OEG mixtures of different molar

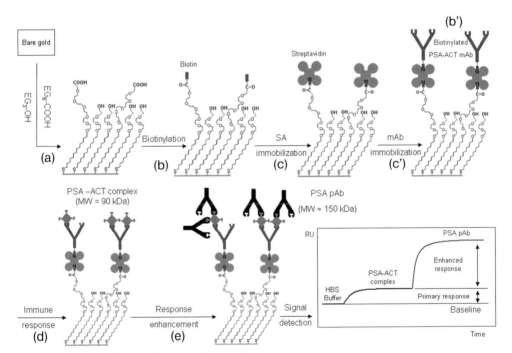

Figure 9.7 Schematic diagram illustrating the steps in sensor chip fabrication (e.g., molar ratio of EG_6-COOH:EG_3-OH = 1:2) and detection of PSA–ACT complex based on SPR. (Reproduced with permission from [58], Copyright 2006 by Elsevier).

ratios of $HS(CH_2)_{11}(OCH_2CH_2)_6OCH_2COOH$ and $HS(CH_2)_{11}(OCH_2CH_2)_3OH$. The SAMs were consequently biotinylated in order to facilitate the immobilization of streptavidin. Nonspecific binding between the surface materials and several serum proteins, such as BSA, IgG and fibrinogen, has been investigated; the results showed that the adsorption of BSA, IgG and fibrinogen was negligible [58]. However, the LOD could be simply enhanced by a sandwich strategy to improve the sensitivity and specificity of the immunoassay. An intact PSA polyclonal antibody was used as an amplifying agent in the strategy. As a result, PSA–ACT complex concentrations as low as 10.2 and 18.1 ng/ml were found in the HEPES-buffered saline (HBS) and human serum sample, respectively. The result indicates that this approach could satisfy our goal without modifying the secondary interactant.

Gold nanoparticles have been used in many reports [70–75] to improve the LOD or sensitivity of SPR due to not only their high mass, but also their ability to undergo SPR phenomenon leading to strengthening of the SPR signal [76]. Lyon *et al.* reported that immobilization of around 11-nm colloidal gold to an evaporated gold film results in a large shift in plasmon angle, a broadened plasmon resonance and an increase in minimum reflectance [70]. A typical immune reaction between the Fab fragment of human IgG molecules and sheep anti-human IgG molecules has been performed using a Kretschmann-configured SPR system, and it was found that colloidal gold could increase the sensitivity of the SPR instrument by a factor of about 300 [71]. Mitchell *et al.* reported techniques for nanoparticle enhancement of sensitivity in small-molecule immunoassays using progesterone as a model compound [74]. Their results demonstrated that a 13-fold signal enhancement and an improvement in LOD of more than two orders of magnitude could be achieved using 25-nm gold nanoparticles to amplify the SPR signal.

"Macromolecules" are good materials to increase the refractive index at the biointerface. Liposomes have also been utilized as a material for enhancement of SPR spectroscopy [77]. The authors rationalized that a considerable shift in resonance angle will considerably change when large liposome vesicles are bound to the transducer surface. The amplification has been applied to detection of interferon-γ with a LOD of 1 μg/ml. Without the liposome amplification, the lowest concentration of interferon-γ was about 100 pg/ml. Streptavidin-biotinylated antibody complex is another alternative to increase biomass for enhancement purposes [78]. Streptavidin and biotin can interact to form a cross-linking network; therefore, a complex of streptavidin-biotinylated antibody molecules is able to enhance the SPR signal due to its large molecular size. As a result, detection of human IgG protein could be enhanced about 200-fold by this amplification strategy and the detection level ranged from 0.005 to 10 μg/ml.

Recently, another strategy has emerged as a novel technique to amplify SPR sensitivity, exploiting a horseradish peroxidase (HRP)-catalyzed precipitation mechanism. HRP catalyzes the oxidation of some compounds such as 3,3′,5,5′-tetramethylbenzidine, 4-chloro-1-naphthol, 3,3′-diaminobenzidine tetrahydrochloride and 3-amino-9-ethyl carbazol to yield insoluble products. Alfonta *et al.* prepared liposomes labeled with biotin and HRP to amplify antigen–antibody or oligonucleotide–DNA biosensing processes by the precipitation of an insoluble product on the electrode [79]. Su et al. reported that the HRP-catalyzed oxidation of 4-chloro-1-naphthol could be applied to enhance the signals of cuvette-based SPR and quartz crystal microbalance (QCM) biosensors as well [80]. The deposition of the oxidized 4-chloro-1-naphthol-insoluble products leads to SPR angle shifts that are linear to the glucose concentration. They also found that the SPR sensitivities were greater than those of the QCM.

Based on the state of art of the enhancement strategies, a combination of sandwich immunoassay, enzyme precipitation and gold nanoparticle enhancement has been proposed. Gold nanoparticles and precipitation of an insoluble product formed by HRP-biocatalyzed oxidation of 3,3'-diaminobenzidine in the presence of H_2O_2 were used to enhance the signal obtained from the SPR biosensor [81]. The gold nanoparticles were synthesized and functionalized with $HS-OEG_3$-COOH by the self-assembling technique. Thereafter, the $HS-OEG_3$-COOH functionalized nanoparticles were covalently conjugated with HRP and anti-IgG antibody to form an enzyme–immunogold complex. Then, the gold–anti-IgG–HRP complex has been applied to enhance SPR immunoassays using a sensor

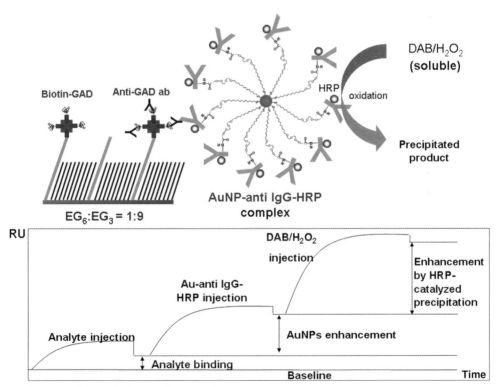

Figure 9.8 Schematic diagram illustrating the steps in anti-GAD antibody detection and enhancement strategy based on a SPR immunosensor. AuNP = gold nanoparticle; DAB = 3,3'-diaminobenzidine. (Reproduced with permission from [81], Copyright 2007 by Elsevier).

chip constructed of a 1:9 molar ratio of HS-OEG$_6$-COOH and HS-OEG$_3$-OH for detection of anti-GAD. As a result, the LOD was found to be as low as 0.03 ng/ml of anti-GAD antibody (or 200 fM). The overall scheme of this study is illustrated in Figure 9.8.

9.5
Localized SPR of Inorganic Nanoparticles for Analyses of Biological Interaction

Surface plasmons, which are collective charge oscillations that occur at the interface between conductors and dielectrics, have various forms. They are denoted as SPR for planar surfaces or localized SPR (LSPR) for nanometer-sized metallic structures. The LSPR of noble metal nanoparticles causes unique properties such as strong ultraviolet/visible absorption bands that are not present in the planar metal or brilliant colors observed for the particles in solution. It is this fascination with the optical properties that led to their many applications in the development of biosensors.

Nanometer-sized metallic structures are ionic clusters covered by free electronic clouds. When the nanometer-sized metallic structures are excited by light, the electric field of the incoming light will induce polarization of the free electrons relative to the ionic cluster. As a result, the collective electrons of the nanoparticles oscillate with the incident photon frequency, which leads to the so-called LSPR. The excitation of free electrons of the nanoparticles by an electric filed at an incident wavelength where the resonance occurs will result in strong light scattering or a strong ultraviolet/visible absorption band that is not present in the spectrum of the bulk metal (Figure 9.9) [82, 83].

To model the LSPR of spherical nanoparticles, Mie theory is the simplest calculation available for estimating the extinction of a metallic nanoparticle in a long wavelength with an electrostatic dipole limit by using [82–84]:

$$E(\lambda) = \left[\frac{24\pi N_A a^3 \varepsilon_m^{3/2}}{\lambda(\ln 10)} \frac{\varepsilon_i}{(\varepsilon_r + \chi\varepsilon_m)^2 + \varepsilon_i^2} \right] \quad (9.2)$$

where $E(\lambda)$ is the extinction (= absorption + Rayleigh scattering), N_A is the areal density of nanoparticles, a is the radius of the metallic nanoparticle, χ is the shape coefficient or the aspect ratio of the nanoparticle (equal to 2 for a sphere and up to values of 17 for a 5:1 aspect ratio nanoparticle), ε_m is the dielectric constant of the medium surrounding the metallic nanoparticle, λ is the wavelength of the absorbing radiation, ε_i is the imaginary dielectric constant of the metallic nanoparticle and ε_r is the real dielectric constant of the metallic nanoparticle. As indicated Equation 9.2, the LSPR spectrum and the location of the extinction maximum of noble metal nanoparticles will strongly depend on the nanoparticle's radius a, shape χ and material (ε_i and ε_r), and the nanoenvironment's dielectric constant (ε_m); and wavelength shifts in the extinction maximum of nanoparticles can be used to detect molecule-induced changes surrounding the

9.5 Localized SPR of Inorganic Nanoparticles for Analyses of Biological Interaction

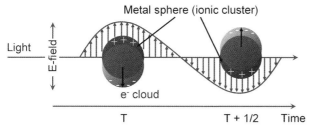

Figure 9.9 Schematic diagram illustrating the excitation of localized surface plasmons of a metal nanoparticle. The electric field of an incident light wave induces the polarization of the free conduction electrons with respect to the much heavier ionic core of a spherical metal nanoparticle. The charge difference only happens on the nanoparticle surfaces.

nanoparticles. Therefore, there are at least four different nanoparticle-based sensing mechanisms that enable the transduction of macromolecular or chemical-binding events into optical signals based on changes in the LSPR extinction or scattering intensity shifts in LSPR λ_{max}, or both. These mechanisms are (i) resonant Rayleigh scattering from nanoparticle labels [85–87], (ii) nanoparticle aggregation [88], (iii) charge transfer interactions at nanoparticle surfaces [89] and (iv) local refractive index changes [90, 91].

Nanoparticle aggregation induced by specific biomolecular interactions is one of the nanoparticle-based sensing mechanisms that enables the transduction of biological/chemical binding events into optical signals based on the wavelength shifts in LSPR λ_{max} [92]. The sensing phenomenon induces a color change of the colloidal solution from red (free particles) to blue (aggregated particles) and it has been developed as simple colorimetric assays for the detection of DNA hybridization. For example, Mirkin et al. developed a colorimetric DNA hybridization assay using the spectral properties of gold nanoparticles [93–95]. In their experiments, gold nanoparticles were coated with single-stranded DNA oligonucleotides that aggregated in the presence of target oligonucleotides with a matching sequence as indicated by a color change of the sample from red to purple. The method also has potential applications in immunoassays [96, 97]. For detection of antigen–antibody interaction, Thanh et al. have quantified the level of antibodies on the basis of the aggregation of gold nanoparticles [98]. Gold nanoparticles coated with protein A were used to determine the level of anti-protein A. The rate of aggregation of the protein A-coated gold nanoparticles in the presence of anti-protein A was monitored by measuring the absorption of the gold colloid suspension at 620 nm. Their result showed that a LOD of 1 µg/ml of anti-protein A was observed. Recently, several research groups have begun to explore alternative strategies for the development of optical biosensors and chemosensors based on the tunable LSPR properties of noble metal nanoparticles materials such as silver and gold [87–99]. These extraordinary optical properties of the nanostructures mean that they can be applied as materials for surface-enhanced spectroscopy [100, 101], plasmonic devices [102] and sensors [103–105]. It is now well established that optical excitation of the LSPR of silver and gold nanoparticles results in absorptions with extremely large molar extinction coefficients of around 3×10^{11} $M^{-1}cm^{-1}$, resonant Rayleigh scattering with an efficiency equivalent to that of 10^6 fluorophors and strong enhancement of the local electromagnetic fields near the nanoparticle surface [87, 106]. Furthermore, the peak extinction or resonant Rayleigh scattering wavelength, λ_{max}, intensity, and line width of these LSPR spectra are strongly dependent on their size [107, 108], shape [109–112], interparticle spacing [114], and local dielectric environment [109, 113]. More interestingly, Van Duyne's group has shown that the LSPR spectrum of noble metal nanoparticles is very sensitive to adsorbate-induced changes in the dielectric constant of the surrounding nanoenvironment, which provides several improvements over existing array- or cluster-based techniques; the resonant Rayleigh scattering pattern of a single silver nanoparticle can be strongly differentiated when about 60 000 molecules (or 100 zmol) of 1-hexadecanthiol or less than 100 streptavidin molecules are adsorbed on its surface [87]. Raschke et al. have developed a method for biomolecular recognition using light scattering of a single gold nanoparticle functionalized with biotin [99]. Additional of streptavidin and subsequent specific binding events alter the dielectric environment of the nanoparticle, resulting in a spectral shift of the particle plasmon resonance. Spectral shifts as low as 2 meV could be detected as they used single nanoparticles as sensing counterparts. Using a detection method based on resonant Rayleigh light scattering of single nanoparticles has several advantages. (i) it is a nonlabeling detection; no toxic, erosive, radioactive or invasive materials are used, making them ideal platforms for *in vivo* quantification of chemical species and monitoring of dynamic processes inside biological cells. (ii) The LOD will be dramatically reduced leading to very high sensitivity being achieved. (iii) The volume of reactants is greatly reduced to reduce the cost of detection. (iv) Fast detection can be performed owing to low volume and low analyte molecule detection. (v) It provides a possibility to miniaturize the instrumentation. (vi) Single nanoparticle sensing platforms offer further advantages because they are readily implemented in multiplex detection schemes. By controlling the size, shape and chemical modification of individual nanoparticles, several sensing platforms can be fabricated in which each unique nanoparticle can be distinguished from others on the basis of the spectral location of its LSPR λ_{max}. Several of these unique nanoparticles can then be incorporated into one device, allowing for the rapid, simultaneous, label-free detection of thousands of different chemical or biological species.

References

1. Wood, R.W. (1902) On a remarkable case of uneven distribution of light in a diffraction grating spectrum. *Philosophical Magazine*, **4**, 396–402.
2. Ritchie, R.H. (1957) Plasma loses by fast electrons in thin films. *Physical Review*, **106**, 874–81.
3. Otto, A. (1968) Excitation of surface plasma waves in silver by the method of frustrated total reflection. *Zeitschrift für Physik (Leipzig)*, **216**, 398–410.
4. Kretschmann, E. and Raether, H. (1968) Radiative decay of non-radiative surface plasmons excited by light. *Zeitschrift für Naturforschung A*, **23**, 2135–6.
5. Liedberg, B., Nylander, C. and Lundstrom, I. (1983) Surface plasmon resonance for gas detection and biosensing. *Sensors and Actuators*, **4**, 299–304.
6. Homola, J., Schwotzer, G., Lehman, H., Willsch, R., Ecke, W. and Bartelt, H. (1995) A new optical fiber sensor for humidity measurement. *EOS Annual Meeting Digest Series*, **2A**, 245–8.
7. Schilling, A., Yava, O., Bischof, J., Boneberg, J. and Leiderer, P. (1996) Absolute pressure measurements on a nanosecond time scale using surface plasmons. *Applied Physics Letters*, **69**, 4159–61.
8. Nylander, C., Liedberg, B. and Lind, T. (1982) Gas detection by means of surface plasmon resonance. *Sensors and Actuators*, **3**, 79–88.
9. Homola, J., Yee, S.S. and Gauglitz, G. (1999) Surface plasmon resonance sensors: review. *Sensors and Actuators B*, **54**, 3–15.
10. Gomes, P. and Andreu, D. (2002) Direct kinetic assay of interactions between small peptides and immobilized antibodies using a surface plasmon resonance biosensor. *Journal of Immunological Methods*, **259**, 217–30.
11. Mirabella, F.M. and Harrick, N.J. (1985) *Internal Reflection Spectroscopy: Review and Supplement*, Harrick Scientific Products, Pleasantville, NY.
12. De Mello, A.J. (1996) *Surface Analytical Techniques for Probing Biomaterial Processes*, CRC Press, Boca Raton, FL.
13. Luppa, P.B., Sokoll, L.J. and Chan, D.W. (2001) Immunosensors – principles and applications to clinical chemistry. *Clinica Chimica Acta*, **314**, 1–26.
14. Raether, H. (1988) *Surface Plasmons on Smooth and Rough Surfaces and on Gratings*, Springer, Berlin.
15. Ordal, M.A., Long, L.L., Bell, R.J., Bell, S.E., Bell, R.R., Alexander, R.W., Ward, J. and Ward, C.A. (1983) Optical properties of metals Al, Co, Cu, Au, Fe, Pb, Ni, Pd, Pt, Ag, Ti, and W in the infrared and far infrared. *Applied Optics*, **11**, 1099–119.
16. Johnson, P.B. and Christy, R.W. (1972) Optical constants of the noble metals. *Physical Review B*, **6**, 4370–9.
17. Sternberg, E., Person, B., Roos, H. and Urbaniczky, C. (1991) Quantitative determination of surface concentration of protein with surface plasmon resonance using radiolabeled proteins. *Journal of Colloid and Interface Science*, **143**, 513–26.
18. Shankaran, D.R. and Miura, N. (2007) Trends in interfacial design for surface plasmon resonance based immunoassays. *Journal of Physics D: Applied Physics*, **40**, 7187–200.
19. Ng, P.C. and Osawa, Y. (1997) Preparation and characterization of the Fab and F(ab')$_2$ fragments of an aromatase activity-suppressing monoclonal antibody. *Steroids*, **62**, 776–81.
20. Harris, L.J., Larson, S.B., Hasel, K.W., Day, J., Greenwod, A. and McPherson, A. (1992) The three-dimensional structure of an intact monoclonal antibody for canine lymphoma. *Nature*, **360**, 369–72.
21. Plückthun, A. (1990) Antibodies from Escherichia coli. *Nature*, **347**, 497–8.
22. Backmann, N., Zahnd, C., Huber, F., Bietsch, A., Plückthun, A., Lang, H.P., Güntherodt, H.J., Hegner, M. and Gerber, C. (2005) Microcantilever-based immunosensor for protein detection

using single-chain antibody fragments (scFv). *Proceedings of the National Academy of Sciences of the United States of America*, **102**, 14587–92.

23 Cherif, B., Roget, A., Villiers, C.L., Calemczuk, R., Leroy, V., Marche, P.N., Livache, T. and Villiers, M.B. (2006) Clinically related protein–peptide interactions monitored in real time on novel peptide chips by surface plasmon resonance imaging. *Clinical Chemistry*, **52**, 255–62.

24 Homola, J. (2008) Surface plasmon resonance sensors for detection of chemical and biological species. *Chemical Reviews*, **108**, 462–93.

25 Ellington, A.D. and Szostak, J.W. (1990) In vitro selection of RNA molecules that bind specific ligands. *Nature*, **346**, 818–22.

26 Tuerk, C. and Gold, L. (1990) Systematic evolution of ligands by exponential enrichment: RNA ligands to bacteriophage T4 DNA polymerase. *Science*, **249**, 505–10.

27 Wilson, D.S. and Szostak, J.W. (1999) In vitro selection of functional nucleic acids. *Annual Review of Biochemistry*, **68**, 611–47.

28 Tombelli, S., Minunni, M. and Mascini, M. (2005) Analytical applications of aptamers. *Biosensors and Bioelectronics*, **20**, 2424–34.

29 Tombelli, S., Minunni, M., Luzi, E. and Mascini, M. (2005) Aptamer-based biosensors for the detection of HIV-1 Tat protein. *Bioelectrochemistry*, **67**, 135–41.

30 Cho, E.J., Collett, J.R., Szafranska, A.E. and Ellington, A.D. (2006) Optimization of aptamer microarray technology for multiple protein targets. *Analytica Chimica Acta*, **564**, 82–90.

31 Musheev, M.U. and Krylov, S.N. (2006) Selection of aptamers by systematic evolution of ligands by exponential enrichment: addressing the polymerase chain reaction issue. *Analytica Chimica Acta*, **564**, 91–6.

32 Jayasena, S.D. (1999) Aptamers: an emerging class of molecules that rival antibodies in diagnostics. *Clinical Chemistry*, **45**, 1628–50.

33 Gordon, L. (2005) *Biosensors and Modern Biospecific Analytical Techniques*, Elsevier, Amsterdam.

34 Gizeli, E. and Lowe, C.R. (2002) *Biomolecular Sensors*, Taylor & Francis, New York.

35 Kambhampati, D. (2003) *Protein Microarray Technology*, Wiley-VCH, Weinheim.

36 Patel, N., Davies, M.C., Heaton, R.J., Roberts, C.J., Tendler, S.J.B. and Williams, P.M. (1998) A scanning probe microscopy study of the physisorption and chemisorption of protein molecules onto carboxylate terminated self-assembled monolayers. *Applied Physics A: Materials Science and Processing*, **66**, S569–74.

37 Lee, J.W., Sim, S.J., Cho, S.M. and Lee, J. (2005) Characterization of a self-assembled monolayer of thiol on a gold surface and the fabrication of a biosensor chip based on surface plasmon resonance for detecting anti-GAD antibody. *Biosensors and Bioelectronics*, **20**, 1422–7.

38 Chapman, R.G., Ostuni, E., Yan, L. and Whitesides, G.M. (2000) Preparation of mixed self-assembled monolayers (SAMs) that resist adsorption of proteins using the reaction of amines with a SAM that presents interchain carboxylic anhydride groups. *Langmuir*, **16**, 6927–36.

39 Benesch, J., Svedhem, S., Svensson, S.C.T., Valiokas, R., Liedberg, B. and Tengvall, P. (2001) Protein adsorption to oligo (ethylene glycol) self assembled monolayers with amide linkage experiments with fibrinogen, heparinised plasma and serum. *Journal of Biomaterials Science. Polymer Edition*, **12**, 581–97.

40 Chen, H., Zhang, Z., Chen, Y., Brook, M.A. and Sheardown, H. (2005) Protein repellant silicone surfaces by covalent immobilization of poly (ethylene oxide). *Biomaterials*, **26**, 2391–9.

41 Frederix, F., Bonroy, K., Reekmans, G., Laureyn, W., Campitelli, A., Abramov, M.A., Dehaen, W. and Maes, G. (2004) Reduced nonspecific absorption on

covalently immobilized protein surfaces using poly (ethylene glycol) containing blocking agents. *Journal of Biochemical and Biophysical Methods*, **58**, 67–74.

42 Silin, V., Weetall, H. and Vanderah, D.J. (1997) SPR studies of the nonspecific adsorption kinetics of human IgG and BSA on gold surfaces modified by self-assembled monolayers (SAMs). *Journal of Colloid and Interface Science*, **185**, 94–103.

43 Wang, R.L.C., Kreuzer, H.J. and Grunze, M. (1997) Molecular conformation and solvation of oligo(ethylene glycol)-terminated self-assembled monolayers and their resistance to protein adsorption. *Journal of Physical Chemistry B*, **101**, 9767–73.

44 Harlow, E. and Lane, D. (1999) *Using Antibodies: A Laboratory Manual*, Cold Spring Harbor Laboratory, Cold Spring Harbor, NY.

45 Lee, W., Lee, D.B., Oh, B.K., Lee, W.H. and Cho, J.W. (2004) Nanoscale fabrication of protein A on self-assembled monolayer and its application to surface plasmon resonance immunosensor. *Enzyme and Microbial Technology*, **35**, 678–82.

46 Soh, N., Tokuda, T., Watanabe, T., Mishima, K., Imato, T., Masadome, T., Asano, Y., Okutani, S., Niwa, O. and Brown, S. (2003) A surface plasmon resonance immunosensor for detecting a dioxin precursor using a gold binding polypeptide. *Talanta*, **60**, 733–45.

47 Brust, M., Walker, M., Bethell, D., Schiffrin, D.J. and Whyman, R. (1994) Synthesis of thiol-derivatised gold nanoparticles in a two-phase liquid–liquid system. *Journal of the Chemical Society. Chemical Communications*, **7**, 801–2.

48 Zou, X., Bao, H., Guo, H., Zhang, L., Qi, L., Jiang, J., Niu, L. and Dong, S. (2006) Mercaptoethane sulfonate protected, water-soluble gold and silver nanoparticles: syntheses, characterization and their building multilayer films with polyaniline via ion–dipole interactions. *Journal of Colloid and Interface Science*, **295**, 401–8.

49 Hostetler, M.J., Green, S.J., Stokes, J.J. and Murray, R.W. (1996) Monolayers in three dimensions: synthesis and electrochemistry of ω-functionalized alkanethiolate-stabilized gold cluster compounds. *Journal of the American Chemical Society*, **118**, 4212–13.

50 Wang, T., Zhang, D., Xu, W., Li, S. and Zhu, D. (2002) New approach to the assembly of gold nanoparticles: formation of stable gold nanoparticle ensemble with chainlike structures by chemical oxidation in solution. *Langmuir*, **18**, 8655–9.

51 Holmlin, R.E., Chen, X., Chapman, R.G., Takayama, S. and Whitesides, G.M. (2001) Zwitterionic SAMs that resist nonspecific adsorption of protein from aqueous buffer. *Langmuir*, **17**, 2841–50.

52 Spinke, J., Liley, M., Schmitt, F.J., Guder, H.J., Angermaier, L. and Knoll, W. (1993) Molecular recognition at self-assembled monolayers: optimization of surface functionalization. *Journal of Chemical Physics*, **99**, 7012–19.

53 Bain, C.D., Troughton, E.B., Tao, Y.T., Evall, J., Whitesides, G.M. and Nuzzo, R.G. (1989) Formation of monolayer films by the spontaneous assembly of organic thiols from solution onto gold. *Journal of the American Chemical Society*, **111**, 321–35.

54 Harris, J.M. and Zalipsky, S. (1997) *Poly(ethylene Glycol): Chemistry and Biological Applications*, American Chemical Society, Washington, DC.

55 Blattler, T.M., Pasche, S., Textor, M. and Griesser, H.J. (2006) High salt stability and protein resistance of poly(l-lysine)-g-poly(ethylene glycol) copolymers covalently immobilized via aldehyde plasma polymer interlayers on inorganic and polymeric substrates. *Langmuir*, **22**, 5760–9.

56 Pasche, S., Textor, M., Meagher, L., Spencer, N.D. and Griesser, H.J. (2005) Relationship between interfacial forces measured by colloid-probe AFM and protein resistance of PEG-grafted PLL adlayers on niobia surfaces. *Langmuir*, **21**, 6508–20.

57 Yu, Q., Chen, S., Taylor, A.D., Homola, J., Hock, B. and Jiang, S. (2005) Detection of low-molecular-weight domoic acid using surface plasmon resonance sensor. *Sensors and Actuators B: Chemical*, **107**, 193–201.

58 Cao, C., Kim, J.P., Kim, B.W., Chae, H., Yoon, H.C., Yang, S.S. and Sim, S.J. (2006) A strategy for sensitivity and specificity enhancements in prostate specific antigen–α1-antichymotrypsin detection based on surface plasmon resonance. *Biosensors and Bioelectronics*, **21**, 2106–13.

59 Österberg, E., Bergstrom, K., Holmberg, K., Riggs, J.A., Van Alstine, J.M., Schuman, T.P., Burns, N.L. and Harris, J.M. (1993) Comparison of polysaccharide and poly(ethylene glycol) coatings for reduction of protein adsorption on polystyrene surfaces. *Colloids and Surfaces A*, **77**, 159–69.

60 Chen, S., Liu, L. and Jiang, S. (2006) Strong resistance of oligo(phosphorylcholine) self-assemblies to protein adsorption. *Langmuir*, **22**, 2418–21.

61 Brink, C., Österberg, E., Holmberg, K. and Tiberg, F. (1992) Using poly(ethylene imine) to graft poly(ethylene glycol) or polysaccharide to polystyrene. *Colloids and Surfaces*, **66**, 149–56.

62 Zhang, Z., Chen, S. and Jiang, S. (2006) Dual-functional biomimetic materials: nonfouling poly(carboxybetaine) with active functional groups for protein immobilization. *Biomacromolecules*, **7**, 3311–15.

63 Canziani, G.A., Klakamp, S. and Myszka, D.G. (2004) Kinetic screening of antibodies from crude hybridoma samples using Biacore. *Analytical Biochemistry*, **325**, 301–7.

64 Steukers, M., Schaus, J.M., van Gool, R., Hoyoux, A., Richalet, P., Sexton, D.J., Nixon, A.E. and Vanhove, M. (2006) Rapid kinetic-based screening of human Fab fragments. *Journal of Immunological Methods*, **310**, 126–35.

65 Lim, D.V., Simpson, J.M., Kearns, E.A. and Kramer, M.F. (2005) Current and developing technologies for monitoring agents of bioterrorism and biowarfare. *Clinical Microbiology Reviews*, **18**, 583–607.

66 Kim, D.K., Kerman, K., Hiep, H.M., Saito, M., Yamamura, S., Takamura, Y., Kwon, Y.S. and Tamiya, E. (2008) Label-free optical detection of aptamer–protein interactions using gold-capped oxide nanostructures. *Analytical Biochemistry*, **379**, 1–7.

67 Tang, Q., Su, X. and Loh, K.P. (2007) Surface plasmon resonance spectroscopy study of interfacial binding of thrombin to antithrombin DNA aptamers. *Journal of Colloid and Interface Science*, **315**, 99–106.

68 Shankaran, D.R., Gobi, K.V. and Miura, N. (2007) Recent advancements in surface plasmon resonance immunosensors for detection of small molecules of biomedical, food and environmental interest. *Sensors and Actuators B: Chemical*, **121**, 158–77.

69 Kang, C.D., Cao, C., Lee, J., Choi, I.S., Kim, B.W. and Sim, S.J. (2008) Surface plasmon resonance-based inhibition assay for real-time detection of Cryptosporidium parvum oocyst. *Water Research*, **42**, 1693–9.

70 Lyon, L.A., Musick, M.D. and Natan, M.J. (1998) Colloidal Au-enhanced surface plasmon resonance immunosensing. *Analytical Chemistry*, **70**, 5177–83.

71 Gu, J.H., Lü, H., Chen, Y.W., Liu, L.Y., Wang, P., Ma, J.M. and Lu, Z.H. (1998) Enhancement of the sensitivity of surface plasmon resonance biosensor with colloidal gold labeling technique. *Supramolecular Science*, **5**, 695–8.

72 Zayats, M., Pogorelova, S.P., Kharitonov, A.B., Lioubashevki, O., Katz, E. and Willner, I. (2003) Au-nanoparticle-enhanced surface plasmon resonance sensing of biocatalytic transformations. *Chemistry – A European Journal*, **9**, 6108–14.

73 Huang, L., Reekmans, G., Saerens, D., Friedt, J.M., Frederix, F., Fracis, L., Muyldermans, S., Campitelli, A. and Hoof, C.V. (2005) Prostate-specific antigen immunosensing based on mixed self-assembled monolayers, camel antibodies and colloidal gold enhanced

sandwich assays. *Biosensors and Bioelectronics*, **21**, 483–90.

74 Mitchell, J.S., Wu, Y., Cook, C.J. and Main, L. (2005) Sensitivity enhancement of surface plasmon resonance biosensing of small molecules. *Analytical Biochemistry*, **343**, 125–35.

75 Pieper-Fürst, U., Stöcklein, W.F.M. and Warsinke, A. (2005) Gold nanoparticle-enhanced surface plasmon resonance measurement with a highly sensitive quantification for human tissue inhibitor of metalloproteinases-2. *Analytica Chimica Acta*, **550**, 69–76.

76 Lyon, L.A., Musick, M.D., Smith, P.C., Reiss, B.D., Pena, D.J. and Natan, M.J. (1999) Surface plasmon resonance of colloidal Au-modified gold films. *Sensors and Actuators B: Chemical*, **54**, 118–24.

77 Wink, T., Zuilen, S.J., Bult, A. and Bennekom, W.P. (1998) Liposome-mediated enhancement of the sensitivity in immunoassays of proteins and peptides in surface plasmon resonance spectrometry. *Analytical Chemistry*, **70**, 827–32.

78 Pei, R., Yang, X. and Wang, E. (2001) Enhanced surface plasmon resonance immunosensing using a streptavidin-biotinylated protein complex. *Analyst*, **126**, 4–6.

79 Alfonta, L., Singh, A.K. and Willner, I. (2001) Liposomes labeled with biotin and horseradish peroxidase: a probe for the enhanced amplification of antigen–antibody or oligonucleotide–DNA sensing processes by the precipitation of an insoluble product on electrodes. *Analytical Chemistry*, **73**, 91–102.

80 Su, X. and O'Shea, S. (2001) Determination of monoenzyme- and bienzyme-stimulated precipitation by a cuvette-based surface plasmon resonance instrument. *Analytical Biochemistry*, **299**, 241–6.

81 Cao, C. and Sim, S.J. (2007) Signal enhancement of surface plasmon resonance immunoassay using enzyme precipitation-functionalized gold nanoparticles: a femto molar level measurement of anti-glutamic acid decarboxylase antibody. *Biosensors and Bioelectronics*, **22**, 1874–80.

82 Haes, A.J. and Van Duyne, R.P. (2004) A unified view of propagating and localized surface plasmon resonance biosensors. *Analytical and Bioanalytical Chemistry*, **379**, 920–30.

83 Hayes, C.L., Van Duyne, R.P. (2001) Nanosphere lithography: a versatile nanofabrication tool for studies of size-dependent nanoparticle optics. *Journal of Physical Chemistry B*, **105**, 5599–611.

84 Kreibig, U. and Vollmer, M. (1995) *Cluster Material*, Springer, Berlin.

85 Yguerabide, J. and Yguerabide, E.E. (1998) Light-scattering submicroscopic particles as highly fluorescent analogs and their use as tracer labels in clinical and biological applications: II. Experimental characterization. *Analytical Biochemistry*, **262**, 157–76.

86 Taton, T.A., Mirkin, C.A. and Letsinger, R.L. (2000) Scanometric DNA array detection with nanoparticle probes. *Science*, **289**, 1757–60.

87 McFarland, A.D., Van Duyne, R.P. (2003) Single silver nanoparticles as real-time optical sensors with zeptomole sensitivity. *Nano Letters*, **3**, 1057–62.

88 Storhoff, J.J., Lazarides, A.A., Mucic, R.C., Mirkin, C.A., Letsinger, R.L. and Schatz, G.C. (2000) What controls the optical properties of DNA-linked gold nanoparticle assemblies? *Journal of the American Chemical Society*, **122**, 4640–50.

89 Malinsky, M.D., Kelly, K.L., Schatz, G.C. and Van Duyne, R.P. (2001) Chain length dependence and sensing capabilities of the localized surface plasmon resonance of silver nanoparticles chemically modified with alkanethiol self-assembled monolayers. *Journal of the American Chemical Society*, **123**, 1471–82.

90 Nath, N. and Chilkoti, A. (2002) A colorimetric gold nanoparticle sensor to interrogate biomolecular interactions in real time on a surface. *Analytical Chemistry*, **74**, 504–9.

91 Eck, D., Helm, C.A., Wagner, N.J. and Vaynberg, K.A. (2001) Plasmon

resonance measurements of the adsorption and adsorption kinetics of a biopolymer onto gold nanocolloids. *Langmuir*, **17**, 957–60.

92 Stuart, D.A., Haes, A.J., Yonzon, C.R., Hicks, E.M. and Van Duyne, R.P. (2005) Biological applications of localised surface plasmonic phenomenae. *IEE Proceedings Nanobiotechnology*, **152**, 13–32.

93 Storhoff, J.J., Elghanian, R., Mucic, R.C., Mirkin, C.A. and Letsinger, R.L. (1998) One-pot colorimetric differentiation of polynucleotides with single base imperfections using gold nanoparticle probes. *Journal of the American Chemical Society*, **120**, 1959–64.

94 Reynolds, R.A., III, Mirkin, C.A. and Letsinger, R.L. (2000) Homogeneous, nanoparticle-based quantitative colorimetric detection of oligonucleotides. *Journal of the American Chemical Society*, **122**, 3795–6.

95 Elghanian, R., Storhoff, J.J., Mucic, R.C., Letsinger, R.L. and Mirkin, C.A. (1997) Selective colorimetric detection of polynucleotides based on the distance-dependent optical properties of gold nanoparticles. *Science*, **277**, 1078–80.

96 Sastry, M., Lala, N., Patil, V., Chavan, S.P. and Chittiboyina, A.G. (1998) Optical absorption study of the biotin-avidin interaction on colloidal silver and gold particles. *Langmuir*, **14**, 4138–42.

97 Aslan, K., Luhrs, C.C. and Perez-Luna, V.H. (2004) Controlled and reversible aggregation of biotinylated gold nanoparticles with streptavidin. *Journal of Physical Chemistry B*, **108**, 15631–9.

98 Thanh, N.T.K. and Rosenzweig, Z. (2002) Development of an aggregation-based immunoassay for anti-protein a using gold nanoparticles. *Analytical Chemistry*, **74**, 1624–8.

99 Raschke, G., Kowarik, S., Franzl, T., Sonnichsen, C., Klar, T.A. and Feldmann, J. (2003) Biomolecular recognition based on single gold nanoparticle light scattering. *Nano Letters*, **3**, 935–8.

100 Haynes, C.L., Van Duyne, R.P. (2003) Plasmon-sampled surface-enhanced Raman excitation spectroscopy. *Journal of Physical Chemistry B*, **107**, 7426–33.

101 Haynes, C.L., McFarland, A.D., Zhao, L., Van Duyne, R.P., Schatz, G.C., Gunnarsson, L., Prikulis, J., Kasemo, B. and Käll, M. (2003) Nanoparticle optics: the importance of radiative dipole coupling in two-dimensional nanoparticle arrays. *Journal of Physical Chemistry B*, **107**, 7337–42.

102 Maier, S.A., Kik, P.G., Atwater, H.A., Meltzer, S., Harel, E., Koel, B.E. and Requicha, A.A.G. (2003) Local detection of electromagnetic energy transport below the diffraction limit in metal nanoparticle plasmon waveguides. *Nature Materials*, **2**, 229–32.

103 Mucic, R.C., Storhoff, J.J., Mirkin, C.A. and Letsinger, R.L. (1998) DNA-directed synthesis of binary nanoparticle network materials. *Journal of the American Chemical Society*, **120**, 12674–5.

104 Haes, A.J. and Van Duyne, R.P. (2002) A nanoscale optical biosensor: sensitivity and selectivity of an approach based on the localized surface plasmon resonance spectroscopy of triangular silver nanoparticles. *Journal of the American Chemical Society*, **124**, 10596–604.

105 Hirsch, L.R., Jackson, J.B., Lee, A., Halas, N.J. and West, J.L. (2003) A whole blood immunoassay using gold nanoshells. *Analytical Chemistry*, **75**, 2377–81.

106 Haes, A.J., Stuart, D.A., Nie, S. and Van Duyne, R.P. (2004) Using solution-phase nanoparticles, surface-confined nanoparticle arrays and single nanoparticles as biological sensing platforms. *Journal of fluorescence*, **14**, 355–67.

107 Sönnichsen, C., Franzl, T., Wilk, T., von Plessen, G., Feldmann, J., Wilson, O. and Mulvaney, P. (2002) Drastic reduction of plasmon damping in gold nanorods. *Physical Review Letters*, **88** 077402.1–4.

108 Sönnichsen, C., Franzl, T., Wilk, T., von Plessen, G. and Feldmann, J. (2002) Plasmon resonances in large noble-metal clusters. *New Journal of Physics*, **4**, 93.1–8.

109 Mock, J.J., Bardic, D.R., Smith, D.R., Schultz, D.A. and Schultz, S. (2002)

Shape effects in plasmon resonance of individual colloidal silver nanoparticles. *Journal of Chemical Physics*, **116**, 6755–9.

110 Sherry, L.J., Chang, S.H., Schatz, G.C., Van Duyne, R.P., Wiley, B.J. and Xia, Y. (2005) Localized surface plasmon resonance spectroscopy of single silver nanocubes. *Nano Letters*, **5**, 2034–8.

111 Orendorff, C.J., Sau, T.K. and Murphy, C.J. (2006) Shape-dependent plasmon-resonant gold nanoparticles. *Small*, **2**, 636–9.

112 Nehl, C.L., Liao, H. and Hafner, J.H. (2006) Optical properties of star-shaped gold nanoparticles. *Nano Letters*, **6**, 683–8.

113 Prikulis, J., Svedberg, F. and Käll, M. (2004) Optical spectroscopy of single trapped metal nanoparticles in solution. *Nano Letters*, **4**, 115–18.

114 Su, K.H., Wei, Q.H., Zhang, H., Mock, J.J., Smith, D.R. and Schultz, S. (2003) Interparticle coupling effects on plasmon resonances of nanogold particles. *Nano Letters*, **3**, 1087–90.

10
Membrane-Active Natural and Synthetic Peptides and Peptidomimetics
Regine Willumeit

10.1
Introduction

Membrane-active natural and synthetic peptides and peptidomimetics are becoming the focus of interest because they might represent a promising approach to overcome increasing antibiotic resistance [1–4]. In parallel, since the late 1960s less research has been performed to find new antibacterial drugs and today, in 550 drugs currently under development, only six are new antibiotics [5, 6].

Pathogens can acquire resistance by various mechanisms [7, 8] that are based on modifications of the receptors specific for the drugs [9, 10]. However, there is one class of antibiotics that does destroy the membrane integrity without necessarily needing a specific protein or DNA receptor: antimicrobial peptides (AMPs) or membrane-active molecules that interact with the lipid bilayer of the cells. They change the biophysical properties of the bilayer, can dissolve lipids in a detergent-like manner or form transmembrane pores. In all cases cell lysis is eventually observed and the bacteria killed. These are the most interesting antibiotics because in can be expected that resistance against these molecules is not or hardly being developed [11, 12].

Several hundred membrane-active AMPs (A. Tossi, http://www.bbcm.univ.trieste.it/) [13–17] have been discovered over recent decades in plants [18, 19], insects [20] and mammals [21]. In addition, many non-natural membrane-active molecules have been designed for various application [22].

10.1.1
Structure of Cell Membranes

To understand the mechanisms of interaction of membrane-active antibacterial molecules it is essential to understand the structure, composition and function of the cell membrane, and the biophysical properties of the lipids involved.

The formation of a lipid bilayer in water is a self-assembling process that usually results in a lipid asymmetry between the outer and inner leaflet. In addition,

Cellular and Biomolecular Recognition: Synthetic and Non-Biological Molecules. Edited by Raz Jelinek
Copyright © 2009 WILEY-VCH Verlag GmbH & Co. KGaA, Weinheim
ISBN: 978-3-527-32265-7

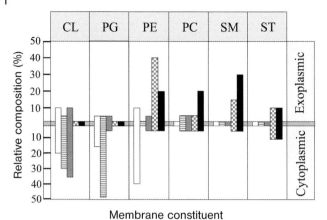

Figure 10.1 Comparison of the membrane composition for various species. CL = cardiolipin; PG = phosphatidylglycerol; PE = phosphatidylethanolamine; PC = phosphatidylcholine; SM = sphingomyelin; ST = sterols (cholesterol, ergesterol). Key: open, *E. coli*; horizontal hatching, *S. aureus*; shaded, *Bacillus subtilis*; checkered, *Candida albicans*;, solid, human erythrocyte. (With permission from [23]).

a membrane potential is built up. For mammalian cells the potential can range from −90 to −110 mV; it is larger for bacteria, where −130 to −150 mV can be found [23].

The lipid composition of cell membranes of prokaryotes and eukaryotes varies significantly; however, various differences can also be found between Gram-negative and Gram-positive bacteria. While Gram-negative bacteria such as *Escherichia coli* have an inner and an asymmetric outer membrane with different composition, Gram-positive bacteria such as *Staphylococcus aureus* only have one membrane, which is also asymmetric. Usually the outer membrane leaflet of the bacterial membrane contains glyco- and phospholipids, while the inner membrane leaflet and cytosolic membrane usually contain mainly phospholipids. Eukaryotic membranes are composed of several other lipids and have also include cholesterol. However, ergosterol, which is similar to mammalian cholesterol, can also be found in some prokaryotic membranes. See Figure 10.1.

10.1.2
Biophysical Properties of Phospholipids

Depending on their head group, acyl chain composition, solvent, temperature, pressure or concentration, phospholipids can exhibit a complicated phase behavior ranging from "typical" lipid bilayers to hexagonal or cubic phases that can have very interesting properties. For model membranes most scientists use 1,2-dipalmitoyl-*sn*-glycero-3-phosphocholine as a representative of a eukaryotic membrane and 1,2-dipalmitoyl-*sn*-glycero-3-phosphoethanolamine or 1-palmitoyl-2-oleoyl-*sn*-glycero-3-phosphoethanolamine as a bacterial model. These lipids – taken as prominent examples – show a quite different phase behavior.

While phosphatidylcholine lipids only form lamellar structures, phosphatidylethanolamine lipids can form inverse hexagonal structures. This so-called polymorphism of phosphatidylethanolamine lipids is a prerequisite for cell division or fusion and as such is essential for life. In addition, lipids modulate the membrane curvature (mainly based on the shape of lipid molecules), create phase-separated regions (mainly depending on head group and acyl chain composition) or influence the structure and function of membrane proteins [24, 25]. All these facts indicate that the biophysical, chemical and mechanical properties of lipids will also play a significant role when it comes to the interaction of peptide antibiotics or other membrane-active molecule with cell membranes.

10.2
Mode of Action of Membrane-Active Peptides

Membrane-active molecules cover a broad range of substances: from natural and synthetic peptides to small organic molecules or even larger polymer structures. The all have in common that they interact with the lipid of the cell membrane. This does not mean that no other interaction is possible. Indeed, it is well established that many peptide antibiotics also interact with proteins or nucleic acids inside the cells. However, in this overview we restrict ourselves to those molecules that target lipids and show activity by physically changing membrane properties rather than by receptor-specific interaction.

To summarize common features among membrane-active molecules is easy: there are none. However, we can list those properties that are mostly found for those molecules: they are amphipatic cationic molecules that interact rapid (within minutes) and lethal [26] resulting in cell lysis. There is agreement that the selectivity and activity are governed by the physicochemical properties of both the membrane-active molecule and the membrane itself [27]. When it comes to secondary structure there is a wide debate whether a specific three-dimensional structure is a requirement for membrane activity. The same is true for electrostatic interaction.

A variety of models exist that describe the possible mode of action. The most common models are the carpet, the barrel stave and the toroidal pore or wormhole model. They consider the α-helical peptides that assemble on the cell membrane. See Figure 10.2.

10.2.1
Carpet Model

The main aspect of the carpet model is that the peptides predominately orient at the surface of the membrane [28] and do not have intense contact with the hydrophobic core of the membrane. If a threshold concentration is reached, the membrane can be penetrated [29]. The carpet interaction usually is accompanied by disruption of acyl chain order [30], which in turn leads to a change in membrane fluidity [31, 32] and membrane thinning. This is also reported for various membrane-active molecules (i.e., anesthetics, organic solvents).

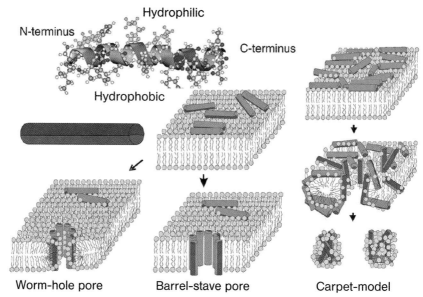

Figure 10.2 Summary of the most important models describing the mode of action of α-helical amphipatic peptides. Magainin is chosen as an example to show the amphipathic structure of the peptide (modified with permission from [104]). The α-helical conformation of magainin is separated into a hydrophilic and hydrophobic side (represented by the two shades of gray in the cylinder).

10.2.2
Barrel Stave Model

In the barrel stave model a transmembrane pore is formed by a bundle of transmembrane amphipathic α-helices. The hydrophobic side of the peptides will interact with the hydrophobic core of the membrane while the hydrophilic residues of the peptides will point inward, thus forming the pore [29]. This structural arrangement can only take place when assemblies of peptides are formed because it is energetically unfavorable for a single amphipatic peptide to have contact of its polar surface side with the membrane core region composed of hydrophobic acyl chains.

10.2.3
Toroidal or Wormhole Model

This model is very similar to the barrel stave model, but differs in the way the peptides finally interact with lipids. The lipids bend back on themselves like the inside of a torus. Some authors see these pores as transient holes, being an initial step before membrane collapse [29]. Another significant difference to the barrel stave model is the fact that, upon disintegration of the pore, the peptides finally translocate to the inner leaflet [33].

10.2.4
Two-State Model

The two-state model is a combination of the carpet and barrel stave/toroidal models [34] and a very general concept that only requires an amphipatic structure of the peptides. It explains the action of helical and β-sheet AMPs after they bind to the plasma membranes of cells with two distinct physical states of binding. (i) Below a threshold value of the peptide:lipid ratio the peptides adsorb physically in the head group region of the lipids in a functionally inactive state that, however, already shows the amphipathic nature of the peptides. This initial process is electrostatically driven and the function of the protein presumably follows the binding. (ii) If the number of peptides is increased and the threshold value reached, the peptides form lethal pores. It is the lipid composition of the cell membrane that determines the threshold value of the peptide:lipid ratio [35, 36].

10.2.5
"Detergent-Like" Model

This is a very general description of how peptides can interact with membranes [37]. Depending on their charge and hydrophobic volume, the peptides intercalate into the bilayer and behave like detergents. The interaction is quite complicated and most likely different from peptide monomers, especially if aggregates are formed. Comparable to detergent interaction, at low peptide:lipid ratio even membrane stabilization can occur. At higher peptide:lipid ratio temporary openings are formed, depending on temporal and local fluctuations of the peptide/detergent density, and membrane disintegration can be observed only for very high concentrations. The detergent-like model includes the above-mentioned wormhole and carpet models as "special cases" that can be considered to be supramolecular structures in a complex phase diagram.

10.2.6
Molecular Mechanism of Membrane Disruption

In some cases the antibacterial activity of membrane-active peptides is due to membrane depolarization [38, 39] or pore formation by molecular electroporation [40] (Figure 10.3). Upon peptide interaction the electrostatic potential of a significant area of the protein–membrane interface is found to be above the minimum threshold for electroporation. While for normal pore formation short pulses of 1 V across the membrane are used, 0.2 V can be sufficient for periods of time longer than 0.1 ms. These values can be reached for highly charged membrane-active peptides.

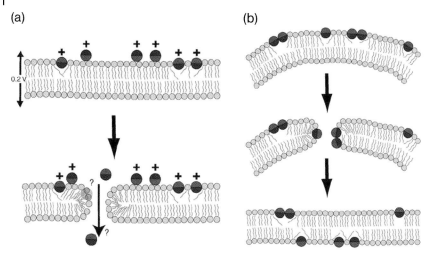

Figure 10.3 The molecular electroporation and the "sinking raft" model. (a) The cationic peptides associate with the bacterial membrane and generate an electrical potential difference across the membrane. Pores are formed for potential differences equal to or greater than 0.2 V. (b) The attached peptides cause a mass imbalance and thus an increase of local curvature. Upon self-association the peptides sink into the membrane and create transient pores. In the final state the peptides can be found on both sides of the membrane. (Modified with permission from [49]).

All models require a change in membrane curvature. By combination of the shape–structure concept of lipid polymorphism plus the position of peptides with respect to lipids it is possible to describe the feasible alteration of membrane structure by membrane-active molecules [41].

Connected with curvature is the bilayer elasticity resulting in a decrease of bilayer thickness by membrane-active molecules. When a membrane-active molecule inserts between the head groups of the phospholipids the cross-sectional area of the head groups is enlarged. This must be followed by changing the cross-sectional area of the acyl chains. As a result, thinning of the bilayer that is roughly proportional to the concentration of the peptides is observed [42–44]. The bound peptides cause an internal stress, or internal membrane tension, that can be sufficiently strong to create pores [36, 45]. See Figure 10.4.

10.3
Natural Peptides

It was shown that some amino acids (i.e., proline [46–48], arginine [49], tryptophan [49], phenylalanine [50] and glycine [51]) exhibit a specific function in the membrane rupture process.

However, it is generally accepted that in principle the peptide structure and their hydrophobic face properties as well as their cationic nature determine the

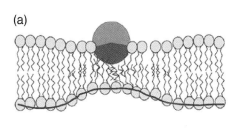 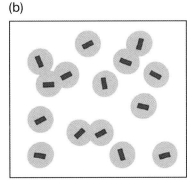

Figure 10.4 (a) Model of acyl chain arrangement that results in a nonuniform height profile of the bilayer. Grazing incidence X-ray diffraction elucidated that each peptide affects the bilayer in an area with an approximate radius of 6.2 nm. (Modified with permission from [104]). This means that the peptide significantly disturbs the acyl chain order of the lipids in an area of roughly 120 nm^2. (b) Top view on the membrane plane. The peptides are represented by dark gray rectangles, their regime of influence is highlighted in light gray.

membrane interaction. Helix formation permits an optimal spatial arrangement of aliphatic side-chains for membrane insertion. The strong hydrophobic interaction of these side-chains with the lipid layer, in turn, stabilizes the helical conformation, reducing main-chain hydrophobicity and thus allowing a deeper insertion into the bilayer [52]. This principle is usually mimicked by synthetic peptides or peptidomimetics.

10.3.1
α-Helical Peptides

Natural AMPs are mainly cationic and α-helical. Prominent examples for α-helical peptides are magainins [33, 53–58], melittin [59–61], alamethicin [56] or cecropins [56, 62–66].

Despite the fact that they are all α-helical and positively charged, their mode of action varies significantly. Their activity is mostly influenced by sequence, size, degree of helicity, charge, hydrophobicity and amphipathicity. However, while it is generally accepted that these factors play an important role for selectivity and activity, it not unambiguous how important they are for the antibacterial properties. Several authors have published experiments where amino acid residues were exchanged in favor of a higher or lower helicity and the resulting antibacterial and hemolytic activity was reported (for an overview, see [67]. However, the results are quite ambiguous and do not show a general rule that can predict the hemolytic or antibacterial activity based on the helicity.

10.3.2
β-Sheet Peptides

The group of β-sheet peptides is significantly smaller than the α-helical group and it is more difficult to find an exact explanation of their mode of action. The most prominent representatives are defensins [68]. They contain a high proportion of positively charged and hydrophobic amino acids, and exhibit six conserved cysteine residues. A specific structural feature is the stabilization of the peptide by three intramolecular disulphide (S–S) bonds. For some β-sheet peptides (polyphemusin and PV5) it was shown that they distinguish between the zwitterionic lipids phosphatidylcholine and phosphatidylethanolamine: They do not interact with phosphatidylcholine lipids, but alter the inverse hexagonal phase transition temperature by inducing a negative curvature strain to the membrane [69, 70].

10.3.3
Cyclic Peptides

The most prominent representative of naturally occurring cyclic AMP is gramicidin S, which strongly promotes the formation of bicontinuous inverted cubic phases by increasing the negative curvature stress in the total membrane lipids of *Acholeplasma laidlawii B* and *E. coli* [71]. In addition, it causes a thinning of the bilayer and a reduction in the lattice spacing of the inverted cubic phase.

10.4
Synthetic Peptides

There exist thousands of synthetic – mainly α-helical – peptides that are derived by modifications of the sequence of natural peptides (e.g., [72, 73]). The aim is to optimize the selectivity and activity, and to stabilize the peptides against enzymatic degradation by changing the character of the side-chain (charge, hydrophobicity, surface area), resulting in new peptide properties such as hydrophobic moment, hydrophobicity, and size of the polar/hydrophobic domain and charge distribution [74–76]. It is clear that the peptide hydrophobicity plays a significant, but not the most important role. Only the combination with electrostatic, van der Waals and hydrophobic interactions adds up to a complete picture of membrane perturbation. The fine tuning of this parameters and their interplay is of crucial importance [67, 77, 78].

Many synthetic peptides are diastereomeric lytic peptides containing D- and L-forms of leucine and lysine. Their average length is 12–17 amino acids [79]. The data suggest that the peptides disintegrate the cell membrane in a detergent-like manner. However, in contrast to native AMPs, the diastereomers bind and permeate similarly zwitterionic and phosphatidylserine-containing model membranes [80]. Using the sequence of natural peptides, but exchanging the D- and L-forms of the amino acid residues, results in a significant change of properties and a loss of α-helical structure [81, 82].

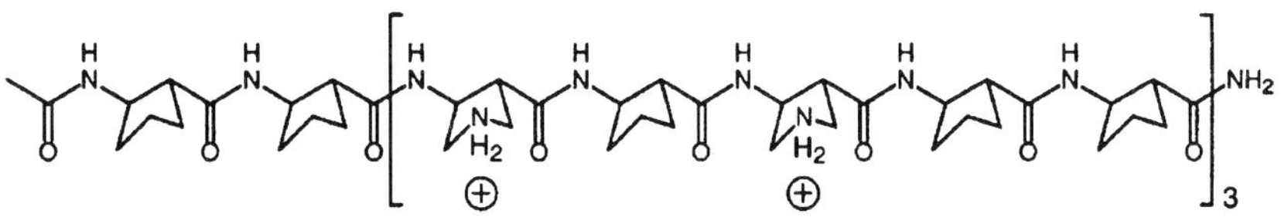

Figure 10.5 The structure of β-17. (With permission from [83]).

To a smaller extent synthetic β-peptides such as β-17 have been developed (Figure 10.5). This peptide decreases the hexagonal phase transition temperature and promotes negative membrane curvature [83].

α/β-Peptides are synthetic oligomers with discrete folding propensities that combine – usually alternating – α- and β-amino acid residues [84–87]. They form an amphiphilic α-helix that can induce phase segregation of anionic and zwitterionic lipids or convert lamellar to nonlamellar structures.

Most of the cyclic peptides are of synthetic origin. They are derived from linear peptides whose ends were connected [88–90]. Since they are usually too small to be transmembrane and their structural properties (e.g., charge, amphipathicity, etc.) seem to require an interaction parallel to the membrane surface, it was not clear how these molecules destabilized the membranes or if they form pores. In some cases it seems that these molecules stack together to form cylinder-like structures that resemble an α-helix and form transmembrane pores as in the case of α-helical peptides [91, 92].

10.5
Peptidomimetics

Peptidomimetics are non-natural membrane-active molecules such as polymers or small molecules that mimic some properties of antibacterial peptides. By combination of charges and the possibility to interact with the hydrophobic core of the membrane, a strong antibacterial activity is found irrespective the size of the non-natural molecules [22]. Contrary to observations stating that the distinct amino acid position in the sequence of the N- and C-terminal chain regions can influence the peptide activity and selectivity [93], the general idea here is that it is sufficient to produce an amphipatic and positively charged molecule without considering a special secondary structure [94] and lacking any resemblance to natural templates [95]. See Figure 10.6.

Two design lines are followed: the combination of biologically active groups (usually amides) with polymers and the synthesis of completely abiogenic molecules. Among the variety of peptidomimetics, arylamide oligomers and analogs, phenylene ethynylenes, polynorbornenes, and polymethacrylates show good antibacterial activity. In some cases it was observed that these molecules – like the corresponding peptide antibiotics – can influence the formation of hexagonal phases or form pores with the target lipid [1, 96–98]. X-ray diffraction and vesicle leakage experiments elucidated that for an interaction, the lipid composition

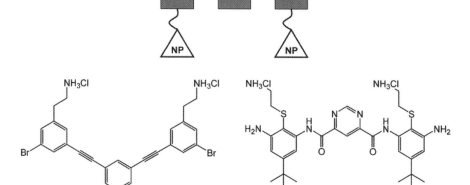

Figure 10.6 Principle features of the peptidomimetic structure of small synthetic antimicrobial molecules (P = polar, NP = nonpolar). The idea is to transfer the fundamental property of amphipathicity and charge the simple abiotic molecules, making them potent antimicrobial agents. (With permission from [94]).

and the overall concentration of a given lipid is more important for selectivity than overall net charge [94]. For water-soluble methacrylate polymers with pendant quaternary ammonium groups it was shown that the antibacterial activity against Gram-positive and Gram-negative strains was dependent on the alkyl chain length for the ammonium groups [99]. Long chains were more active against Gram-positive bacteria, while the effect decreased for Gram-negative bacteria. See Figure 10.7.

Another approach is to construct polymers that have a bacteriocidal end group. One example are polymers based on poly(2-methyl-1,3-oxazoline)s with the antimicrobial N,N-dimethyldodecylammonium end group [100]. Interestingly, the effects of these polymers are tunable by different satellite groups (i.e., methyl, decyl, hexadecyl groups). How they disturb the membrane is not fully understood; however, there is evidence that they can penetrate the cell membrane and induce lysis [101–103]. See Figure 10.8.

10.6
Conclusions

Membrane-active antimicrobial molecules cover a wide range of structures and their design follows a variety of approaches. They are promising candidates for various applications ranging from medical use as antibiotic drugs to surface

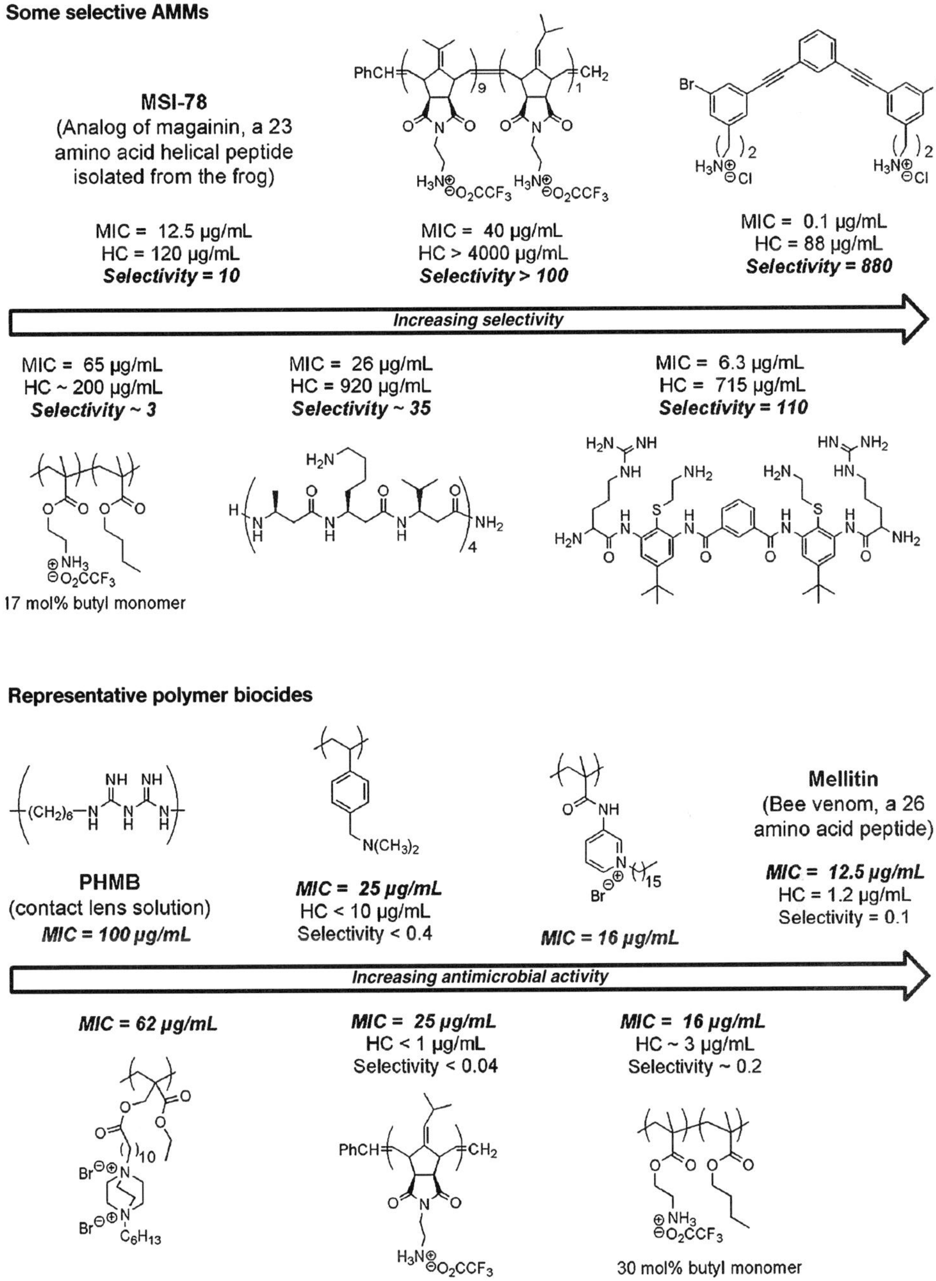

Some selective AMMs

MSI-78
(Analog of magainin, a 23 amino acid helical peptide isolated from the frog)

MIC = 12.5 µg/mL
HC = 120 µg/mL
Selectivity = 10

MIC = 40 µg/mL
HC > 4000 µg/mL
Selectivity > 100

MIC = 0.1 µg/mL
HC = 88 µg/mL
Selectivity = 880

→ Increasing selectivity →

MIC = 65 µg/mL
HC ~ 200 µg/mL
Selectivity ~ 3

MIC = 26 µg/mL
HC = 920 µg/mL
Selectivity ~ 35

MIC = 6.3 µg/mL
HC = 715 µg/mL
Selectivity = 110

17 mol% butyl monomer

Representative polymer biocides

PHMB
(contact lens solution)
MIC = 100 µg/mL

MIC = 25 µg/mL
HC < 10 µg/mL
Selectivity < 0.4

MIC = 16 µg/mL

Mellitin
(Bee venom, a 26 amino acid peptide)

MIC = 12.5 µg/mL
HC = 1.2 µg/mL
Selectivity = 0.1

→ Increasing antimicrobial activity →

MIC = 62 µg/mL
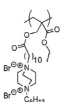

MIC = 25 µg/mL
HC < 1 µg/mL
Selectivity < 0.04

MIC = 16 µg/mL
HC ~ 3 µg/mL
Selectivity ~ 0.2

30 mol% butyl monomer

Figure 10.7 Summary of membrane-active natural [antimicrobial macromolecules (AMMs)] and abiotic molecules, and the minimal inhibitory concentration (MIC). HC = hemolytic concentration; PHMB = polyhexamethylene biguanide. (Modified with permission from [1]).

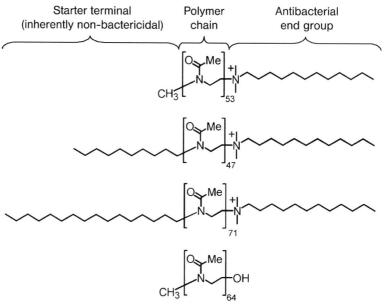

Figure 10.8 Scheme of the modifications of the poly(2-methyl-1,3-oxazoline) structure [100]. **[Permission requested.]**

coatings in health care or garment improvement. Mimicking natural molecules by using some of their fundamental properties is already successful and considering the biophysical characteristics of membranes such as curvature sensitivity and the capability of nonplanar bilayer formation, even more aspects could be taken into account to synthesize more powerful abiotic molecules that might help to overcome the threat of antibacterial resistance development.

References

1 Gabriel, G.J., Som, A., Madkour, A.E., Eren, T. and Tew, G.N. (2007) Infectious disease: connecting innate immunity to biocidal polymers. *Materials Science and Engineering: R: Reports*, **57**, 28–64.
2 Hancock, R.E. and Patrzykat, A. (2002) Clinical development of cationic antimicrobial peptides: from natural to novel antibiotics. *Current Drug Targets. Infectious Disorders*, **2**, 79–83.
3 Leeb, M. (2004) Antibiotics: a shot in the arm. *Nature*, **431**, 892–3.
4 Roghmann, M.C. and McGrail, L. (2006) Novel ways of preventing antibiotic-resistant infections: what might the future hold? *American Journal of Infection Control*, **34**, 469–75.
5 Overbye, K.M. and Barrett, J.F. (2005) Antibiotics: where did we go wrong? *Drug Discovery Today*, **10**, 45–52.
6 Sundriyal, S., Sharma, R.K., Jain, R. and Bharatam, P.V. (2008) Minimum requirements of hydrophobic and hydrophilic features in cationic peptide antibiotics (CPAs): pharmacophore generation and validation with cationic steroid antibiotics (CSAs). *Journal of Molecular Modeling*, **14**, 265–78.
7 Levy, S.B. (2005) Antibiotic resistance – the problem intensifies.

Advanced Drug Delivery Reviews, **57**, 1446–50.
8 Neu, H.C. (1995) Emergence and mechanisms of bacterial resistance in surgical infections. *American Journal of Surgery*, **169**, 13S–20S.
9 Alekshun, M.N. and Levy, S.B. (2007) Molecular mechanisms of antibacterial multidrug resistance. *Cell*, **128**, 1037–50.
10 Tenover, F.C. (2006) Mechanisms of antimicrobial resistance in bacteria. *American Journal of Medicine*, **119**, S3–10; discussion S62–70.
11 Peschel, A. (2002) How do bacteria resist human antimicrobial peptides? *Trends in Microbiology*, **10**, 179–86.
12 Peschel, A. and Sahl, H.G. (2006) The co-evolution of host cationic antimicrobial peptides and microbial resistance. *Nature Reviews Microbiology*, **4**, 529–36.
13 Boman, H.G. (2003) Antibacterial peptides: basic facts and emerging concepts. *Journal of Internal Medicine*, **254**, 197–215.
14 Hale, J.D. and Hancock, R.E. (2007) Alternative mechanisms of action of cationic antimicrobial peptides on bacteria. *Expert Review of Anti-Infective Therapy*, **5**, 951–9.
15 Meyer, J.E. and Harder, J. (2007) Antimicrobial peptides in oral cancer. *Current Pharmaceutical Design*, **13**, 3119–30.
16 Shai, Y. (1999) Mechanism of the binding, insertion and destabilization of phospholipid bilayer membranes by α-helical antimicrobial and cell non-selective membrane-lytic peptides. *Biochimica et Biophysica Acta*, **1462**, 55–70.
17 Tossi, A., Sandri, L. and Giangaspero, A. (2000) Amphipathic, α-helical antimicrobial peptides. *Biopolymers*, **55**, 4–30.
18 Thevissen, K., Kristensen, H.H., Thomma, B.P., Cammue, B.P. and Francois, I.E. (2007) Therapeutic potential of antifungal plant and insect defensins. *Drug Discovery Today*, **12**, 966–71.
19 Thomma, B.P., Cammue, B.P. and Thevissen, K. (2002) Plant defensins. *Planta*, **216**, 193–202.

20 Kuhn-Nentwig, L., Muller, J., Schaller, J., Walz, A., Dathe, M. and Nentwig, W. (2002) Cupiennin 1, a new family of highly basic antimicrobial peptides in the venom of the spider Cupiennius salei (Ctenidae). *Journal of Biological Chemistry*, **277**, 11208–16.
21 Eisenhauer, P.B., Harwig, S.S., Szklarek, D., Ganz, T., Selsted, M.E. and Lehrer, R.I. (1989). Purification and antimicrobial properties of three defensins from rat neutrophils. *Infection and Immunity*, **57**, 2021–7.
22 Tashiro, T. (2001) Antibacterial and bacterium adsorbing macromolecules. *Macromolecular Materials and Engineering*, **286**, 63–87.
23 Yeaman, M.R. and Yount, N.Y. (2003) Mechanisms of antimicrobial peptide action and resistance. *Pharmacological Reviews*, **55**, 27–55.
24 Drin, G., Casella, J.F., Gautier, R., Boehmer, T., Schwartz, T.U. and Antonny, B. (2007) A general amphipathic α-helical motif for sensing membrane curvature. *Nature Structural and Molecular Biology*, **14**, 138–46.
25 Teissier, E. and Pecheur, E.I. (2007) Lipids as modulators of membrane fusion mediated by viral fusion proteins. *European Biophysics Journal*, **36**, 887–99.
26 Zasloff, M. (2007) Antimicrobial peptides, innate immunity, and the normally sterile urinary tract. *Journal of the American Society of Nephrology*, **18**, 2810–16.
27 Toke, O. (2005) Antimicrobial peptides: new candidates in the fight against bacterial infections. *Biopolymers*, **80**, 717–35.
28 Schröder-Borm, H., Willumeit, R., Brandenburg, K. and Jörg, A. (2003) Molecular basis for membrane selectivity of NK-2, a potent peptide antibiotic derived from NK-lysin. *Biochimica Biophysica Acta*, **1612**, 164–71.
29 Oren, Z. and Shai, Y. (1998) Mode of action of linear amphipathic α-helical antimicrobial peptides. *Biopolymers*, **47**, 451–63.
30 Manson, A.J., Bechinger, B. and Kichler, A. (2007) Rational design of vector and

antibiotic peptides using solid-state NMR. *Mini-Reviews in Medical Chemistry*, **7**, 491–7.

31 Willumeit, R., Kumpugdee, M., Funari, S.S., Lohner, K., Pozo Navas, B., Brandenburg, K., Linser, S. and Andrä, J. (2005) Structural rearrangement of model membranes by the peptide antibiotic NK-2. *Biochimica Biophysica Acta*, **1669**, 125–34.

32 Zepik, H.H., Walde, P., Kostoryz, E.L., Code, J. and Yourtee, D.M. (2008) Lipid vesicles as membrane models for toxicological assessment of xenobiotics. *Critical Reviews in Toxicology*, **38**, 1–11.

33 Matsuzaki, K. (1998) Magainins as paradigm for the mode of action of pore-forming polypeptides. *Biochimica et Biophysica Acta*, **1376**, 391–400.

34 Huang, H.W. (2000) Action of antimicrobial peptides: two-state model. *Biochemistry*, **39**, 8347–52.

35 Boland, M.P. and Separovic, F. (2006) Membrane interactions of antimicrobial peptides from Australian tree frogs. *Biochimica et Biophysica Acta*, **1758**, 1178–83.

36 Huang, H.W. (2006) Molecular mechanism of antimicrobial peptides: the origin of cooperativity. *Biochimica et Biophysica Acta*, **1758**, 1292–302.

37 Bechinger, B. and Lohner, K. (2006) Detergent-like actions of linear amphipathic cationic antimicrobial peptides. *Biochimica et Biophysica Acta*, **1758**, 1529–39.

38 Shai, Y., Makovitzky, A. and Avrahami, D. (2006) Host defense peptides and lipopeptides: modes of action and potential candidates for the treatment of bacterial and fungal infections. *Current Protein and Peptide Science*, **7**, 479–86.

39 Silverman, J.A., Perlmutter, N.G. and Shapiro, H.M. (2003) Correlation of daptomycin bactericidal activity and membrane depolarization in Staphylococcus aureus. *Antimicrobial Agents and Chemotherapy*, **47**, 2538–44.

40 Miteva, M., Andersson, M., Karshikoff, A. and Otting, G. (1999) Molecular electroporation: a unifying concept for the description of membrane pore formation by antibacterial peptides, exemplified with NK-lysin. *FEBS Letters*, **462**, 155–8.

41 Batenburg, A.M. and de Kruijff, B. (1988) Modulation of membrane surface curvature by peptide–lipid interactions. *Bioscience Reports*, **8**, 299–307.

42 Huang, H.W. (1995) Elasticity of lipid bilayer interacting with amphiphilic helical peptides. *Journal de Physique II France*, **5**, 1427–31.

43 Lee, M.T., Chen, F.Y. and Huang, H.W. (2004) Energetics of pore formation induced by membrane-active peptides. *Biochemistry*, **43**, 3590–9.

44 Ludtke, S., He, K. and Huang, H. (1995) Membrane thinning caused by magainin 2. *Biochemistry*, **34**, 16764–9.

45 Huang, H.W., Chen, F.Y. and Lee, M.T. (2004) Molecular mechanism of peptide-induced pores in membranes. *Physical Review Letters*, **92**, 198304.

46 Chou, P.Y. and Fasman, G.D. (1978) Empirical predictions of protein conformation. *Annual Review of Biochemistry*, **47**, 251–76.

47 Tomasinsig, L., Skerlavaj, B., Papo, N., Giabbai, B., Shai, Y. and Zanetti, M. (2006) Mechanistic and functional studies of the interaction of a proline-rich antimicrobial peptide with mammalian cells. *Journal of Biological Chemistry*, **281**, 383–91.

48 Zhang, L., Benz, R. and Hancock, R.E. (1999) Influence of proline residues on the antibacterial and synergistic activities of α-helical peptides. *Biochemistry*, **38**, 8102–11.

49 Chan, D.I., Prenner, E.J. and Vogel, H.J. (2006) Tryptophan- and arginine-rich antimicrobial peptides: structures and mechanisms of action. *Biochimica et Biophysica Acta*, **1758**, 1184–202.

50 van Kan, E.J., Demel, R.A., van der Bent, A. and de Kruijff, B. (2003) The role of the abundant phenylalanines in the mode of action of the antimicrobial peptide clavanin. *Biochimica et Biophysica Acta*, **1615**, 84–92.

51 van Kan, E.J., van der Bent, A., Demel, R.A. and de Kruijff, B. (2001) Membrane activity of the peptide antibiotic clavanin

and the importance of its glycine residues. *Biochemistry*, **40**, 6398–405.

52 Zelezetsky, I., Pacor, S., Pag, U., Papo, N., Shai, Y., Sahl, H.G. and Tossi, A. (2005) Controlled alteration of the shape and conformational stability of α-helical cell-lytic peptides: effect on mode of action and cell specificity. *Biochemical Journal*, **390**, 177–88.

53 Duclohier, H., Molle, G. and Spach, G. (1989) Antimicrobial peptide magainin I from Xenopus skin forms anion-permeable channels in planar lipid bilayers. *Biophysical Journal*, **56**, 1017–21.

54 Gallucci, E., Meleleo, D., Micelli, S. and Picciarelli, V. (2003) Magainin 2 channel formation in planar lipid membranes: the role of lipid polar groups and ergosterol. *European Biophysics Journal*, **32**, 22–32.

55 Matsuzaki, K., Sugishita, K., Harada, M., Fujii, N. and Miyajima, K. (1997) Interactions of an antimicrobial peptide, magainin 2, with outer and inner membranes of Gram-negative bacteria. *Biochimica et Biophysica Acta*, **1327**, 119–30.

56 Matsuzaki, K., Sugishita, K., Ishibe, N., Ueha, M., Nakata, S., Miyajima, K. and Epand, R.M. (1998) Relationship of membrane curvature to the formation of pores by magainin 2. *Biochemistry*, **37**, 11856–63.

57 Powers, J.P. and Hancock, R.E. (2003) The relationship between peptide structure and antibacterial activity. *Peptides*, **24**, 1681–91.

58 Zasloff, M. (1987) Magainins, a class of antimicrobial peptides from Xenopus skin: isolation, characterization of two active forms, and partial cDNA sequence of a precursor. *Proceedings of the National Academy of Sciences of the United States of America*, **84**, 5449–53.

59 Chen, X. and Chen, Z. (2006) SFG studies on interactions between antimicrobial peptides and supported lipid bilayers. *Biochimica et Biophysica Acta*, **1758**, 1257–73.

60 Terwilliger, T.C., Weissman, L. and Eisenberg, D. (1982) The structure of melittin in the form I crystals and its implication for melittin's lytic and surface activities. *Biophysical Journal*, **37**, 353–61.

61 Vogel, H. and Jahnig, F. (1986) The structure of melittin in membranes. *Biophysical Journal*, **50**, 573–82.

62 Dubovskii, P.V., Volynsky, P.E., Polyansky, A.A., Karpunin, D.V., Chupin, V.V., Efremov, R.G. and Arseniev, A.S. (2008) Three-dimensional structure/hydrophobicity of latarcins specifies their mode of membrane activity. *Biochemistry*, **47**, 3525–33.

63 Gazit, E., Boman, A., Boman, H.G. and Shai, Y. (1995) Interaction of the mammalian antibacterial peptide cecropin P1 with phospholipid vesicles. *Biochemistry*, **34**, 11479–88.

64 Gazit, E., Miller, I.R., Biggin, P.C., Sansom, M.S. and Shai, Y. (1996) Structure and orientation of the mammalian antibacterial peptide cecropin P1 within phospholipid membranes. *Journal of Molecular Biology*, **258**, 860–70.

65 Hultmark, D., Steiner, H., Rasmuson, T. and Boman, H.G. (1980) Insect immunity. Purification and properties of three inducible bactericidal proteins from hemolymph of immunized pupae of *Hyalophora cecropia*. *European Journal of Biochemistry*, **106**, 7–16.

66 Lee, J.Y., Boman, A., Sun, C.X., Andersson, M., Jornvall, H., Mutt, V. and Boman, H.G. (1989) Antibacterial peptides from pig intestine: isolation of a mammalian cecropin. *Proceedings of the National Academy of Sciences of the United States of America*, **86**, 9159–62.

67 Dathe, M. and Wieprecht, T. (1999) Structural features of helical anti-microbial peptides: their potential to modulate activity on model membranes and biological cells. *Biochimica et Biophysica Acta*, **1462**, 71–87.

68 Clarke, D.J. and Campopiano, D.J. (2006) Structural and functional studies of defensin-inspired peptides. *Biochemical Society Transactions*, **34**, 251–6.

69 Doherty, T., Waring, A.J. and Hong, M. (2006) Peptide–lipid interactions of the β-hairpin antimicrobial peptide tachyplesin and its linear derivatives from solid-state NMR. *Biochimica et Biophysica Acta*, **1758**, 1285–91.

70 Powers, J.P., Tan, A., Ramamoorthy, A. and Hancock, R.E. (2005) Solution structure and interaction of the antimicrobial polyphemusins with lipid membranes. *Biochemistry*, **44**, 15504–13.

71 Staudegger, E., Prenner, E.J., Kriechbaum, M., Degovics, G., Lewis, R.N., McElhaney, R.N. and Lohner, K. (2000) X-ray studies on the interaction of the antimicrobial peptide gramicidin S with microbial lipid extracts: evidence for cubic phase formation. *Biochimica et Biophysica Acta*, **1468**, 213–30.

72 Epand, R.F., Lehrer, R.I., Waring, A., Wang, W., Maget-Dana, R., Lelievre, D. and Epand, R.M. (2003) Direct comparison of membrane interactions of model peptides composed of only Leu and Lys residues. *Biopolymers*, **71**, 2–16.

73 Stark, M., Liu, L.P. and Deber, C.M. (2002) Cationic hydrophobic peptides with antimicrobial activity. *Antimicrobial Agents and Chemotherapy*, **46**, 3585–90.

74 Braunstein, A., Papo, N. and Shai, Y. (2004) In vitro activity and potency of an intravenously injected antimicrobial peptide and its dl amino acid analog in mice infected with bacteria. *Antimicrobial Agents and Chemotherapy*, **48**, 3127–9.

75 Dathe, M., Meyer, J., Beyermann, M., Maul, B., Hoischen, C. and Bienert, M. (2002) General aspects of peptide selectivity towards lipid bilayers and cell membranes studied by variation of the structural parameters of amphipathic helical model peptides. *Biochimica et Biophysica Acta*, **1558**, 171–86.

76 Park, K.H., Park, Y., Park, I.S., Hahm, K.S. and Shin, S.Y. (2008). Bacterial selectivity and plausible mode of antibacterial action of designed Pro-rich short model antimicrobial peptides. *Journal of Peptide Science*, **14**, 876–82.

77 Dathe, M., Nikolenko, H., Meyer, J., Beyermann, M. and Bienert, M. (2001) Optimization of the antimicrobial activity of magainin peptides by modification of charge. *FEBS Letters*, **501**, 146–50.

78 Giangaspero, A., Sandri, L. and Tossi, A. (2001) Amphipathic α helical anti-microbial peptides. *European Journal of Biochemistry*, **268**, 5589–600.

79 Papo, N. and Shai, Y. (2004) Effect of drastic sequence alteration and d-amino acid incorporation on the membrane binding behavior of lytic peptides. *Biochemistry*, **43**, 6393–403.

80 Papo, N., Shahar, M., Eisenbach, L. and Shai, Y. (2003) A novel lytic peptide composed of dl-amino acids selectively kills cancer cells in culture and in mice. *Journal of Biological Chemistry*, **278**, 21018–23.

81 Oren, Z. and Shai, Y. (1997) Selective lysis of bacteria but not mammalian cells by diastereomers of melittin: structure–function study. *Biochemistry*, **36**, 1826–35.

82 Pag, U., Oedenkoven, M., Papo, N., Oren, Z., Shai, Y. and Sahl, H.G. (2004) In vitro activity and mode of action of diastereomeric antimicrobial peptides against bacterial clinical isolates. *Journal of Antimicrobial Chemotherapy*, **53**, 230–9.

83 Epand, R.F., Umezawa, N., Porter, E.A., Gellman, S.H. and Epand, R.M. (2003) Interactions of the antimicrobial β-peptide β-17 with phospholipid vesicles differ from membrane interactions of magainins. *European Journal of Biochemistry*, **270**, 1240–8.

84 Epand, R.F., Schmitt, M.A., Gellman, S.H. and Epand, R.M. (2006) Role of membrane lipids in the mechanism of bacterial species selective toxicity by two α/β-antimicrobial peptides. *Biochimica et Biophysica Acta*, **1758**, 1343–50.

85 Epand, R.F., Schmitt, M.A., Gellman, S.H., Sen, A., Auger, M., Hughes, D.W. and Epand, R.M. (2005) Bacterial species selective toxicity of two isomeric α/β-peptides: role of membrane lipids. *Molecular Membrane Biology*, **22**, 457–69.

86 Schmitt, M.A., Weisblum, B. and Gellman, S.H. (2004) Unexpected relationships between structure and function in α,β-peptides: antimicrobial foldamers with heterogeneous backbones. *Journal of the American Chemical Society*, **126**, 6848–9.

87 Schmitt, M.A., Weisblum, B. and Gellman, S.H. (2006) Interplay among folding, sequence, and lipophilicity in the antibacterial and hemolytic activities of α/β-peptides. *Journal of the American Chemical Society*, **129**, 417–28.

88 Appelt, C., Wessolowski, A., Dathe, M. and Schmieder, P. (2007) Structures of cyclic, antimicrobial peptides in a membrane-mimicking environment define requirements for activity. *Journal of Peptide Science*, **14**, 524-7 524–7.

89 Dathe, M., Nikolenko, H., Klose, J. and Bienert, M. (2004) Cyclization increases the antimicrobial activity and selectivity of arginine- and tryptophan-containing hexapeptides. *Biochemistry*, **43**, 9140–50.

90 Wessolowski, A., Bienert, M. and Dathe, M. (2004) Antimicrobial activity of arginine- and tryptophan-rich hexapeptides: the effects of aromatic clusters, d-amino acid substitution and cyclization. *Journal of Peptide Research*, **64**, 159–69.

91 Fernandez-Lopez, S., Kim, H.S., Choi, E.C., Delgado, M., Granja, J.R., Khasanov, A., Kraehenbuehl, K., Long, G., Weinberger, D.A., Wilcoxen, K.M., Ghadiri, M.R. (2001) Antibacterial agents based on the cyclic d,l-α-peptide architecture. *Nature*, **412**, 452–5.

92 Ganz, T. (2001) Chemistry: rings of destruction. *Nature*, **412**, 392–3.

93 Kuhn-Nentwig, L., Dathe, M., Walz, A., Schaller, J. and Nentwig, W. (2002) Cupiennin 1d∗: the cytolytic activity depends on the hydrophobic N-terminus and is modulated by the polar C-terminus. *FEBS Letters*, **527**, 193–8.

94 Arnt, L., Rennie, J.R., Linser, S., Willumeit, R. and Tew, G.N. (2006) Membrane activity of biomimetic facially amphiphilic antibiotics. *Journal of Physical Chemistry B*, **110**, 3527–32.

95 Gabriel, G.J. and Tew, G.N. (2008) Conformationally rigid proteomimetics: a case study in designing antimicrobial aryl oligomers. *Organic and Biomolecular Chemistry*, **6**, 417–23.

96 Beckloff, N., Laube, D., Castro, T., Furgang, D., Park, S., Perlin, D., Clements, D., Tang, H., Scott, R.W., Tew, G.N. and Diamond, G. (2007) Activity of an antimicrobial peptide mimetic against planktonic and biofilm cultures of oral pathogens. *Antimicrobial Agents and Chemotherapy*, **51**, 4125–32.

97 Som, A. and Tew, G.N. (2008) Influence of lipid composition on membrane activity of antimicrobial phenylene ethynylene oligomers. *Journal of Physical Chemistry B*, **112**, 3495–502.

98 Yang, L., Gordon, V.D., Mishra, A., Som, A., Purdy, K.R., Davis, M.A., Tew, G.N. and Wong, G.C. (2007) Synthetic antimicrobial oligomers induce a composition-dependent topological transition in membranes. *Journal of the American Chemical Society*, **129**, 12141–7.

99 Dizman, B., Elasri, M.O. and Mathias, L.J. (2006) Synthesis and antibacterial activities of water-soluble methacrylate polymers containing quaternary ammonium compounds. *Journal of Polymer Science: Part A: Polymer Chemistry*, **44**, 5965–73.

100 Waschinski, C.J., Barnert, S., Theobald, A., Schubert, R., Kleinschmidt, F., Hoffmann, A., Saalwachter, K. and Tiller, J.C. (2008) Insights in the antibacterial action of poly(methyloxazoline)s with a biocidal end group and varying satellite groups. *Biomacromolecules*, **9**, 1764–71.

101 Lin, J., Tiller, J.C., Lee, S.B., Lewis, K. and Klibanov, A.M. (2002) Insights into bactericidal action of surface-attached poly(vinyl-N-hexylpyridinium) chains. *Biotechnology Letters*, **24**, 801–5.

102 Tiller, J.C., Liao, C.J., Lewis, K. and Klibanov, A.M. (2001) Designing surfaces that kill bacteria on contact.

Proceedings of the National Academy of Sciences of the United States of America, **98**, 5981–5.

103 Waschinski, C.J. and Tiller, J.C. (2005) Poly(oxazoline)s with telechelic antimicrobial functions. *Biomacromolecules*, **6**, 235–43.

104 Salditt, T., Li, C. and Spaar, A. (2006) Structure of antimicrobial peptides and lipid membranes probed by interface-sensitive X-ray scattering. *Biochimica et Biophysica Acta*, **1758**, 1483–98.

11
Luminescent Quantum Dot Fluorescence Resonance Energy Transfer-Based Probes in Cellular and Biological Assays

Lifang Shi, Nitsa Rosenzweig, and Zeev Rosenzweig

11.1
Introduction

This chapter describes the development of luminescent quantum dot (QD)-based bioanalytical probes for the analysis of target analytes in biological samples. The newly developed QDs respond to changes in biological systems by changing their luminescence properties. More specifically, the chapter focuses on studies carried out in our laboratory towards the development of QD fluorescence (or Förster) resonance energy transfer (FRET)-based protease sensors for real-time monitoring of proteolytic activity as a means to discriminate between normal and cancerous cells. These studies are a part of a large effort by the research community to develop sensors with high sensitivity and high specificity for cancer research [1]. To date, fluorescence methods based on the use of molecular fluorophores have been the most common methods of detecting biomolecules. Many of these molecular fluorophores suffer from serious chemical and photophysical limitations, such as broad emission spectra, low photobleaching thresholds and poor chemical stability. The field of optical imaging was recently advanced by the development of nanoparticle-based agents like luminescent QDs that exhibit higher photostability and more desirable photophysical properties. The development of water-soluble and biocompatible QDs was first reported in 1998 [2, 3]. Since then, luminescent QDs have emerged as new and promising components in fluorescence sensors for cancer detection [4]. The QD FRET-based probes that have been developed in our laboratory are prepared by linking molecular fluorophores to the surface of CdSe/ZnS luminescent QDs. Their analyte response is based on FRET interactions between the QDs and molecular acceptors, and on the attenuation of these FRET interactions in the presence of target analytes. The following sections briefly review recent developments in the field of QDs and the phenomenon of FRET to enable better understanding of the working principle and analytical utility of QD FRET-based nanosensors.

Cellular and Biomolecular Recognition: Synthetic and Non-Biological Molecules. Edited by Raz Jelinek
Copyright © 2009 WILEY-VCH Verlag GmbH & Co. KGaA, Weinheim
ISBN: 978-3-527-32265-7

11.2
Luminescent QDs

QDs are semiconductor nanoparticles of 1–10 nm in diameter. These luminescent nanocrystals are composed of atoms from the II–VI (CdS, CdSe, CdTe, ZnO, ZnSe), III–V (InP, InAs, GaN, GaP, GaAs) and IV–VI (PbS, PbSe, PbTe) groups of the periodic table. QDs are spherical, crystalline particles of a given material consisting of hundreds to thousands of atoms. Their diameter is smaller than their electron–hole pair (exciton) Bohr radius. When the size of a semiconductor particle is small enough to approach the size of the material exciton Bohr radius, the electron energy levels are no longer treated as a continuum and instead are treated as discrete – a condition defined as quantum confinement. Quantum confinement leads to increased stress on the exciton, which results in increased energy of the emitted photon. The smaller the QDs, the higher the energy required to form the exciton. The behavior of the excited electron can be described by a simple "particle-in-a-box" model [5]. The quantum confinement increases the probability of overlap between the electron and hole, which increases the rate of radiative recombination. This results in QDs with unique optical and electronic properties [6–10].

CdX QDs (X = S, Se, Te) have attracted great interest due to their emission in the ultraviolet/visible/near-infrared range of the electromagnetic spectrum. These QDs have several optical characteristics that distinguish them from conventional organic fluorophores. These include size-dependent luminescence [2, 11], in which the emitted light wavelength is determined by the energy band gap between the valence and conduction band of the QDs. As the size of the QDs decreases, the energy band gap increases. Since the energy band gap of the QDs is size dependent, the emission color of the QDs is also size dependent [12, 13]. Luminescent QDs have a much wider absorption spectrum than molecular fluorophores. This enables excitation of QDs using a wide range of wavelengths. It also enables the excitation of QDs of different size with a single wavelength, which make them suitable for multiplexing or simultaneous detection of QDs of different emission colors [8]. Additionally, the molar extinction coefficients of QDs are larger than that of organic dyes [14, 15]. QDs have symmetric and narrow emission spectra without a red tail. This reduces cross-talk between emission signals of QDs of different emission colors [16]. The long fluorescence lifetime of QDs is also an advantage since it enables their use in time-gated detection to separate their signal from that of shorter lifetime species (e.g., the autofluorescence of cells) [17, 18]. Finally, a major advantage of QDs is their high photostability and chemical stability compared to molecular fluorophores, which enables their use in imaging applications that require long exposure times [19–25].

The size and shape of QDs are controlled by altering the duration, temperature and ligand molecules used in their synthesis. To date, QDs, such as CdSe, CdS and CdTe, have been synthesized in various media including aqueous solution [26, 27], reverse micelles [28], polymer films [29, 30], sol-gel systems [31] and trioctylphosphine oxide (TOPO)/trioctylphosphine (TOP) [11, 32–34]. High-quality QDs have been achieved by pyrolysis of organometallic precursors in TOP/TOPO media, which was first reported by Murray in 1993 [11]. The synthesis is carried out by injecting dimethylcadmium [Cd(CH$_3$)$_2$] and sulfur, selenium or tellurium dissolved in TOP solution to hot TOPO media. However, Cd(CH$_3$)$_2$ is very toxic, pyrophoric, unstable and expensive. The synthesis procedure was later refined by Peng *et al.* who replaced the toxic cadmium precursor Cd(CH$_3$)$_2$ with CdO, Cd(Ac)$_2$ and CdCO$_3$, which led to a more user-friendly green synthesis [32–34]. Confining the electrons to the bulk of luminescent QDs is imperative to their bright luminescence. The excited electron or hole could be trapped by surface defects like vacancies, local lattice mismatches, dangling bonds or adsorbates at the surfaces. These lead to nonradiative recombination and to low emission quantum yield [5]. Additionally, the uncapped QDs are so reactive that they readily undergo photochemical degradation. To decrease the effect of surface defects and to protect surface atoms from oxidation and other chemical reactions, an additional thin layer made of a higher energy band gap semiconductor material (e.g., ZnS) is grown on the surface of the QDs [35–39]. This process, often described in the literature as surface passivation, increases the emission quantum yield, improves chemical stability and photostablility, and reduces the toxicity by preventing leakage of cadmium or selenium to the surrounding environment. Due to the availability of precursors and the simplicity of crystallization, CdSe/ZnS core/shell QDs have been widely used in biological applications.

High-quality QDs that are synthesized in organic solvents are not water soluble, not biocompatible and do not have the functional groups required for bioconjugation. To facilitate their application in aqueous biological systems, the hydrophobic TOPO molecules that serve as capping ligands of luminescent QDs must be replaced with bifunctional hydrophilic capping ligands or coated with amphiphilic protective layer to impart water solubility and potential bioconjugation sites. Various methods of QD surface functionalization to facilitate their water solubility, stability in aqueous systems and biocompatibility were developed in recent years. These solubilization strategies can be divided into three categories.

(1) Ligand exchange is a process involving the replacement of hydrophobic ligands with bifunctional ligands in which one end binds to the QD surface and an opposing end imparts water solubility via hydrophilic groups. Thiols (SH) are often used to bind the capping ligand to QD ZnS surface. The TOPO ligands are often exchanged with thiol-functionalized compounds like mercaptoacetic acid [3], dihydrolipoic acid [40], dithiothreitol [41] and dendrons [42]. In our laboratory, we found that the amino acid cysteine is also an effective capping ligand to create hydrophilic QDs [43]. Weiss *et al.* reported that cysteine-containing peptides can also be used as effective capping ligands to facilitate water solubility of QDs [44]. Since the bond between thiol and ZnS is not particularly strong, the ligands often fall off the surface, which leads aggregation of the QDs [45]. In addition, the ligand-exchange process often disturbs the chemical and physical state of the surface atoms of QDs and reduces their emission quantum yield [46].

(2) Silica encapsulation involves the growth of a silica layer on the surface of QDs. Functional organosilane molecules are incorporated into the shell to provide surface functionalities for bioconjugation [2, 47–50]. The silica-coated QDs are

extremely stable because the silica layer is highly cross-linked. However, the method is very laborious and the silica layer may be hydrolyzed [51].

(3) Another approach is to coat QDs with an amphiphilic polymer or phospholipids, which interleave with the hydrophobic TOPO ligands through hydrophobic attraction and provide a hydrophilic exterior to ensure aqueous solubility [25, 52–54]. This process maintains the native ligands (TOPO) on the surface of the QDs. This retains the high emission quantum yield of the QDs and protects the QDs surface from deterioration in biological solutions. However, the final size of the polymer coated QDs is large, which could limit many biological applications [51].

In spite of these limitations, water-soluble QDs have been widely used in protein assays [55–59], and DNA and RNA hybridization assays [60–64]. They have also been used as labels in *in vitro* imaging of cells and tissues [25, 65–68], and in *in vivo* imaging applications in whole animals [69–73]. As mentioned previously the focus of our work is the development of QD FRET-based probes for biological applications. A brief description of FRET leading to our results in this area is given in the following sections.

11.3
FRET

FRET involves nonradiative energy transfer from an excited donor to an acceptor via a through-space dipole–dipole interaction [74–76]. Molecular FRET donor–acceptor pairs satisfy the following conditions: (i) spectral overlap between the absorption spectrum of the acceptor and the fluorescence emission spectrum of the donor, and (ii) the donor and acceptor molecules must be in close proximity (typically 10–100 Å). The rate of energy transfer depends on the extent of spectral overlap between the emission spectrum of the donor and the absorption spectrum of the acceptor, the relative orientation of donor/acceptor transition dipoles, and the distance between the donor and acceptor. The rate of energy transfer is given by:

$$k_t = \tau_D^{-1} \left(\frac{R_0}{R} \right)^6 \tag{1}$$

where τ_D is the measured lifetime of the donor in the absence of the acceptor, R_0 is termed the critical radius of the transfer or the Förster distance, which is the distance at which the energy transfer efficiency is 50%. The Förster distance, R_0, depends on the spectral characteristics of the donor–acceptor pair and is expressed as:

$$R_0 = \left(\frac{3000}{4\pi N |A|_{1/2}} \right)^{1/3} \tag{2}$$

where N is Avogadro's number and $|A|_{1/2}$ is the concentration of the acceptor at which the energy transfer efficiency E is 50%. For a donor and acceptor pair that is covalently bound, E is expressed as:

$$E = \frac{R_0^6}{R_0^6 + R^6} \tag{3}$$

The FRET efficiency can be experimentally measured by monitoring changes in the donor or/and acceptor fluorescence intensities, or changes in the fluorescent lifetimes of fluorophores. As a result of FRET interactions, the fluorescence intensity and lifetime of donor decrease, whereas the acceptor fluorescence is sensitized and its lifetime is longer.

Molecular fluorophores have been widely used as donors and acceptors in FRET-based assays and sensors. However, they have several limitations as FRET agents [51,77,78]. These include narrow and overlapping absorption spectra, which make it difficult to avoid direct excitation of the acceptor, and broad emission spectrum of the donor with long red tailing, which often overlaps with the emission spectrum of the acceptor and results in spectral cross-talk. In addition, molecular fluorophores have low photobleaching thresholds, which prevent real-time monitoring of FRET signals over long durations under conditions of continuous exposure.

QDs have been investigated as FRET donors as alternatives to traditional molecular fluorophores because of their high photostability and their unique spectral properties [52, 77, 79]. In 1996, Kagan et al. first reported energy transfer between QDs [80, 81]. In 2001, several research groups reported FRET between QDs and molecular fluorophores and quenchers [82–84]. For example, Willard et al. developed QDs as FRET donors in a protein–protein binding assay [82]. In 2003, Medinta et al. reported the development of QD FRET-based biosensors for maltose, which was realized by coating CdSe/ZnS QDs capped with dihydrolipoic acid with maltose-binding protein (MBP) molecules [85]. The FRET assay was based on competitive interactions between maltose and molecular quenchers on the MBP binding site. Maltose molecules displaced the molecular quenchers from the MBP binding sites, which resulted in a maltose concentration-dependent increase in the emission of the MBP-coated QDs. Similar QD FRET-based sensors were developed for 2,4,6-trinitrotoluene [86], toxins [87], β-lactamase [88], collagenase [89], DNA [64], RNA [90] and proteins [91]. In all of these probes, the QDs were used as donors while the organic fluorophores served as fluorescent acceptors. The FRET mechanism allows the QDs to respond to environmental changes while avoiding direct chemical interaction with the QDs that could negatively affect their photophysical properties and decrease their brightness.

11.4
QD FRET-Based Protease Probes

Recent studies in our laboratory focused on the development of QD FRET-based protease probes [92, 93]. The working principle of these probes is illustrated in Scheme 11.1. Luminescent QDs that are coated with unlabeled RGDC

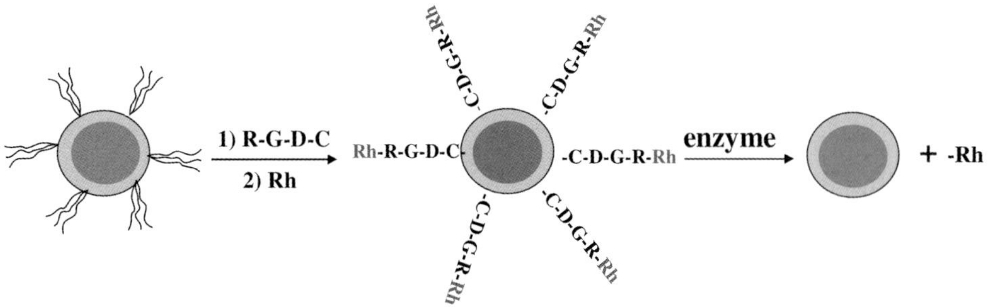

Scheme 11.1 Schematic representation of a QD FRET-based protease sensor.

(Arg–Gly–Asp–Cys) peptide molecules emit green light. When capped with rhodamine-labeled RGDC molecules the emission color turns orange due to FRET interactions between the QDs and rhodamine molecules. The RGDC peptide molecules are cleaved by proteolytic enzymes to release the rhodamine molecules from the surface. This in turn restores the green emission of the QDs. To prepare the QD FRET-based probes, TOPO-capped CdSe/ZnS QDs were first synthesized following a method developed by Peng with slight modifications [32, 94]. A ligand-exchange reaction was then used to replace the TOPO ligands with RGDC peptide molecules, some labeled with rhodamine and some unlabeled. The reaction was carried out in a mixture of pyridine and dimethyl sulfoxide following a method developed by Pinaud et al. [44]. Unbound peptide molecules were removed by spin dialysis. The ratio between rhodamine-labeled RGDC and unlabeled RGDC peptide molecules was varied to maximize the FRET signal between the QDs and attached rhodamine molecules. Figure 11.1 shows the emission spectra of rhodamine-labeled peptide coated at increasing rhodamine-labeled RGDC concentration. When excited at 445 nm, the emission peak of the QDs at 545 nm decreased with increasing rhodamine concentration; the emission peak of rhodamine at 590 nm also increased progressively with increasing rhodamine-labeled RGDC concentration. These observations indicated the occurrence of FRET between the QDs and the rhodamine molecules. Digital fluorescence microscopy images were used to provide another visual evidence of FRET between the QDs and rhodamine molecules. The emission color of unlabeled peptide coated QDs was green (Figure 11.1b). The emission color turned yellow/orange when the unlabeled RGDC peptide molecules were replaced with rhodamine-labeled RGDC molecules (Figure 11.1c). This emission color change from green to orange also indicated the occurrence of FRET between QDs and bound rhodamine molecule.

The QD FRET-based probes were first used to determine the activity of the proteolytic enzyme trypsin, which cleaves peptides and proteins at the carboxyl end of lysine (K) and arginine (R). Figure 11.2a shows the temporal dependence of the ratio F_d/F_a at increasing trypsin concentration ranging from 0 to 500 μg/ml. The ratio F_d/F_a is a direct measure of the FRET efficiency between the QDs and molecular acceptors. High F_d/F_a values indicate low FRET efficiency. The F_d/F_a ratio was normalized to $(F_d/F_a)_0$, which is the value of F_d/F_a prior to the addition

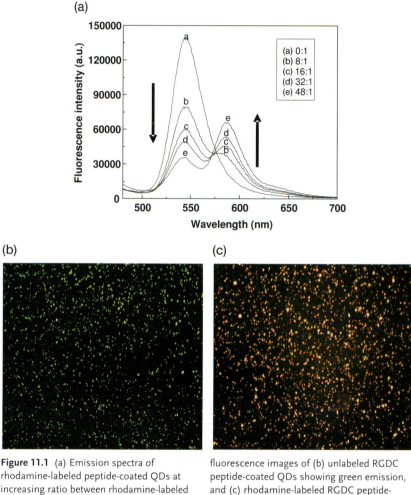

Figure 11.1 (a) Emission spectra of rhodamine-labeled peptide-coated QDs at increasing ratio between rhodamine-labeled RGDC peptide and unlabeled RGDC peptide molecules (see inset for details). Digital fluorescence images of (b) unlabeled RGDC peptide-coated QDs showing green emission, and (c) rhodamine-labeled RGDC peptide-coated QDs showing yellow/orange emission due to FRET between the QDs and rhodamine.

of trypsin to the QD solutions. It can be seen that the ratio F_d/F_a increased faster at higher trypsin concentrations. For example, at 250 µg/ml trypsin the enzymatic reaction was completed in less than 15 min. The short assay time is a significant advantage over previously reported QD FRET-based probes in which longer reaction times were reported [89]. The QD FRET-based probes were also used successfully to determine the activity of collagenase, another proteolytic enzyme from the family of extracellular matrix metalloproteinases (MMPs). Following the demonstration of the ability of QD FRET-based probes to monitor the activity of collagenase in solution we employed the same probes to measure the activity of MMPs in cell cultures. The QD FRET-based probes were embedded in the

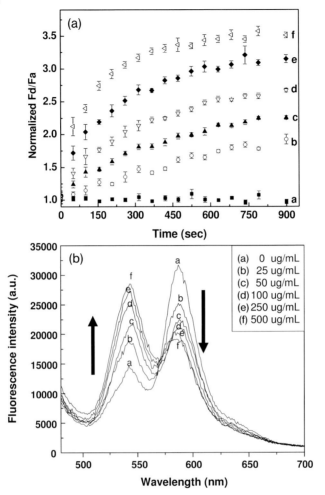

Figure 11.2 (a) Temporal dependence of the rhodamine-labeled peptide-coated QDs at increasing trypsin concentration (as indicated in inset to (b)). The ratio F_d/F_a was normalized to $(F_d/F_a)_0$, which is the ratio F_d/F_a prior to adding trypsin to the QDs probes solutions. (b) Emission spectra of the QD FRET-based probes at increasing trypsin concentration ($\lambda_{ex} = 445$ nm).

extracellular matrix of normal and cancerous breast cells. Figure 11.3 shows images of the QD FRET-based probes in normal (Figure 11.3a and b) and cancerous (Figure 11.3c and d) breast cells when taken at $t = 0$ and $t = 15$ min following the addition of the probes. It can be seen that the emission color of the QDs (orange) did not change when incubated with normal breast cells. On the other hand, a clear change of emission color from orange to green was observed when the QD FRET-based probes were incubated with breast cancer cells, which is

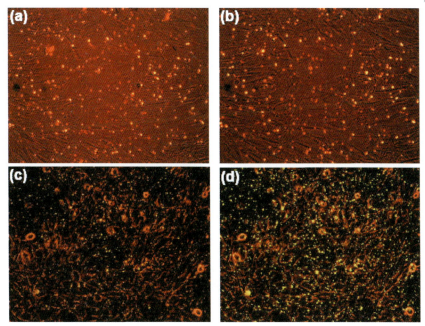

Figure 11.3 Digital fluorescence microscopy images of QD FRET-based probes in breast cell cultures. (a) Normal HTB-125 cells at $t = 0$; (b) normal HTB-125 cells at $t = 15$ min; (c) cancer HTB-126 cells at $t = 0$; (d) cancer HTB-126 cells at $t = 15$ min.

attributed to the overexpression of MMPs in breast cancer cells. Similarly to soluble collagenase, the MMPs cleave the peptide molecules and release the rhodamine molecules from the QDs. This results in rapid FRET signal changes and in emission color change of the QDs. These newly developed cellular assays provide valuable tools in the quest of the research community to develop new and improved methods for cancer detection and monitoring.

11.5
Summary and Conclusions

This chapter describes the development of QD FRET-based sensors with a particular focus on protease activity probes. The QD FRET-based sensors are based on FRET interactions between QDs, which serve as donors, and molecular fluorophores that are attached to the QD surface and serve as fluorescent acceptors. This unique geometry has enabled the use of QDs for the first time in nanosensing applications. It must be noted, however, that FRET interactions between QDs and fluorescent acceptor molecules are not fully understood, and need to be studied in greater detail. Unlike in FRET interactions between molecular donors and acceptors, the distance between QDs and molecular acceptors is not well defined. Our studies show that the FRET efficiency is high even when a short tetrapeptide links the QDs and the acceptor molecules. It is possible that the accumulative interaction between a single QDs and multiple acceptor molecules compensates for the low FRET efficiency between QDs and individual acceptor molecules when these are bound through a short linker. It should also be noted that the Forster theory commonly used to describe energy transfer between molecular donor and acceptor molecules was never tested in heterogeneous systems consisting of a luminescent nanoparticles as donors and fluorescent molecules as acceptors. The heterogeneity in QDs size can affect the precision of single-molecule FRET measurements.

In broader terms it is clear that QDs have considerable advantages when compared to molecular fluorophores. However, QDs have their limitations in biological applications. Surface modification of QDs to realize aqueous solubility and the presence of functional groups suitable for bioconjugation often decrease the emission quantum yield of QDs. Also, QDs have limited pH stability and tend to aggregate in biological media. While it is possible to attach a number of biomolecules to a single QD, it is difficult to quantify the actual number of biomolecules on the QD surface. This requires the development of a new generation of instrumentation with the capability to interrogate the surface of nanoparticles to quantify molecular surface overages at the single particle level. Perhaps the biggest concern associated with large-scale use of luminescent QDs is their cytotoxicity. It is difficult to envision widespread use of cadmium-based QDs given their toxicity properties. Nevertheless, given the superb photophysical properties of luminescent QDs it is likely that research will continue to minimize if not eliminate their toxicity, and increase their stability in biological systems and biocompatibility while maintaining their unique photophysical properties intact.

Acknowledgments

Studies described in this chapter were supported by NSF award CHE-0717526 and DoD/DARPA award HR0011-07-01-0032.

References

1 Murcia, M.J. and Naumann, C.A. (2005) Biofunctionalization of fluorescent nanoparticles, in *Biofunctionalization of Nanomaterials* (ed. C.S.S.R Kumar), Wiley-VCH, Weinheim, pp. 1–38.

2 Bruchez, M., Moronne, M., Gin, P., Weiss, S. and Alivisatos, A.P. (1998) Semiconductor nanocrystals as fluorescent biological labels. *Science*, 281, 2013–16.

3 Chan, W.C.W. and Nie, S.M. (1998) Quantum dot bioconjugates for ultrasensitive nonisotopic detection. *Science*, 281, 2016–18.

4 Willard, D.M., Mutschler, T., Yu, M., Jung, J. and Van Orden, A. (2006) Directing energy flow through quantum dots: towards nanoscale sensing. *Analytical and Bioanalytical Chemistry*, 384, 564–71.

5 Murphy, C.J. (2002) Optical sensing with quantum dots. *Analytical Chemistry*, 74 (19), 520A–526A.

6 Alivisatos, A.P. (1996) Semiconductor clusters, nanocrystals, and quantum dots. *Science*, 271 (5251), 933–7.

7 Efros, A.L. and Rosen, M.M. (2000) The electronic structure of semiconductor nanocrystals. *Annual Reviews of Materials Science*, 30, 475–521.

8 Chan, W.C.W., Maxwell, D.J., Gao, X., Bailey, R.E., Han, M. and Nie, S. (2002) Luminescent quantum dots for multiplexed biological detection and imaging. *Current Opinion in Biotechnology*, 13, 40–6.

9 Gao, X.H., Yang, L., Petros, J.A., Marshall, F.F., Simons, J.W. and Nie, S.M. (2005) *In vivo* molecular and cellular imaging with quantum dots. *Current Opinion in Biotechnology*, 16, 63–72.

10 Michalet, F.X., Pinaud, F., Bentolila, L.A., Tsay, J.M., Doose, S., Li, J.J., Sundaresan, G., Wu, A.M., Gambhir, S. and Weiss, S.S. (2005) Quantum dots for live cells, *in vivo* imaging, and diagnostics. *Science*, 307, 538–44.

11 Murray, C.B., Norris, D.J. and Bawendi, M.G. (1993) Synthesis and characterization of nearly monodisperse CdE (E = sulfur, selenium, tellurium) semiconductor nanocrystallites. *Journal of the American Chemical Society*, 115, 8706–15.

12 Alivisatos, A.P. (1996) Perspectives on the physical chemistry of semiconductor nanocrystals. *Journal of Physical Chemistry*, 100, 13226–39.

13 Eailey, R.E., Smith, A.M. and Nie, S.M. (2004) Quantum dots in biology and medicine. *Physica E*, 25, 1–12.

14 Striolo, A., Ward, J., Prausnitz, J.M., Parak, W.J., Zanchet, D., Gerion, D., Milliron, D. and Alivisatos, A.P. (2002) Molecular weight, osmotic second virial coefficient, and extinction coefficient of colloidal CdSe nanocrystals. *Journal of Physical Chemistry B*, 106, 5500–5.

15 Leatherdale, C.A., Woo, W.K., Mikulec, F.V. and Bawendi, M.G. (2002) On the absorption cross section of CdSe nanocrystal quantum dots. *Journal of Physical Chemistry B*, 106, 7619–22.

16 Parak, W.J., Pellogrino, T. and Plank, C. (2005) Labeling of cells with quantum dots. *Nanotechnology*, 16, R9–25.

17 Dahan, M., Laurence, T., Pinaud, F., Chemla, D.S. and Alivisatos, A.P. (2001) Time-gated biological imaging using colloidal quantum dots. *Optics Letters*, 26, 825–7.

18 Hohng, S. and Ha, T. (2004) Near-complete suppression of quantum dot blinking in ambient conditions. *Journal of the American Chemical Society*, 126, 1324–5.

19 Michalet, X., Pinaud, F., Lacoste, T.D., Dahan, M. and Bruchez, M. (2001) Properties of fluorescent semiconductor nanocrystals and their application to biological labeling. *Single Molecules*, 2, 261–76.

20 Wu, X., Liu, H., Liu, J., Haley, K.N. and Treadway, J.A. (2003) Immuno-fluorescent labeling of cancer marker Her2 and other cellular targets with semiconductor quantum dots. *Nature Biotechnology*, 21, 41–6.

21 Xiao, Y. and Barker, P.E. (2004) Semiconductor nanocrystal probes for human metaphase chromosomes. *Nucleic Acids Research*, 32, e28.

22 Ness, J.M., Akhtar, R.S., Latham, C.B. and Roth, K.A. (2003) Combined tyramide signal amplification and quantum dots for sensitive and photostable immunofluorescence. *Journal of Histochemistry and Cytochemistry*, 51, 981–7.

23 Jaiswal, J.K., Mattoussi, H., Mauro, J.M. and Simon, S.M. (2003) Long-term multiple color imaging of lice cells using quantum dot bioconjugates. *Nature Biotechnology*, 21, 47–51.

24 Hanaki, K., Momo, A., Oku, T., Komoto, A. and Maenosono, S. (2003) Semiconductor quantum dot/albumin complex is a long-life and highly photostable endosome marker. *Biochemical and Biophysical Research Communications*, 302, 496–501.

25 Dubertret, B., Skourides, P., Norris, D.J., Noireaux, V., Brivanlou, A.H. and Libchaber, A. (2002) *In vivo* imaging of quantum dots encapsulated in phospholipid micelles. *Science*, 298, 1759–62.

26 Gaponik, N., Talapin, D.V., Rogach, A.L., Eychmuller, A. and Weller, H. (2002) Efficient phase transfer of luminescent thiol-capped nanocrystals: from water to nonpolar organic solvents. *Nano Letters*, **2** (8), 803–6.

27 Ni, T., Nagesha, D.K., Robles, J., Materer, N.F., Mussig, S. and Kotov, N.A. (2002) CdS nanoparticles modified to chalcogen sites: new supramolecular complexes, butterfly bridging, and related optical effects. *Journal of the American Chemical Society*, **124** (15), 3980–92.

28 Hirai, T., Watanabe, T. and Komasawa, I. (1999) Preparation of semiconductor nanoparticle–polyurea composites using reverse micellar systems via an *in situ* diisocyanate polymerization. *Journal of Physical Chemistry B*, **103** (46), 10120–6.

29 Farmer, S.C. and Patten, T.E. (2001) Photoluminescent polymer/quantum dot composite nanoparticles. *Chemistry of Materials*, **13** (11), 3920–6.

30 Mansur, H.S., Vasconcelos, W.L., Grieser, F. and Caruso, F. (1999) Photoelectrochemical behavior of CdS "Q-state" semiconductor particles in 10,12-nonacosadiynoic acid polymer Langmuir–Blodgett films. *Journal of Materials Science*, **34** (21), 5285–91.

31 Zelner, M., Minti, H., Reisfeld, R., Cohen, H. and Tenne, R. (1997) Preparation and characterization of CdS films synthesized *in situ* in zirconia sol-gel matrix. *Chemistry of Materials*, **9** (11), 2541–3.

32 Peng, Z.A. and Peng, X. (2001) Formation of high-quality CdTe, CdSe, and CdS nanocrystals using CdO as precursor. *Journal of the American Chemical Society*, **123** (1), 183–4.

33 Peng, Z.A. and Peng, X. (2001) Mechanisms of the shape evolution of CdSe nanocrystals. *Journal of the American Chemical Society*, **123** (7), 1389–95.

34 Yu, W.W., Qu, L., Guo, W. and Peng, X. (2003) Experimental determination of the extinction coefficient of CdTe, CdSe, and CdS nanocrystals. *Chemistry of Materials*, **15** (14), 2854–60.

35 Hines, M.A. and Guyot-Sionnest, P. (1996) Synthesis and characterization of strongly luminescing ZnS-capped CdSe nanocrystals. *Journal of Physical Chemistry*, **100**, 468–71.

36 Dabbousi, B.O., Rodriguez-Viejo, J., Mikulec, F.V., Heine, J.R., Mattoussi, H., Ober, R., Jensen, K.F. and Bawendi, M.G. (1997) (CdSe)ZnS core–shell quantum dots: synthesis and characterization of a size series of highly luminescent nanocrystallites. *Journal of Physical Chemistry B*, **101**, 9463–75.

37 Li, J.J., Wang, Y.A., Guo, W., Keay, J.C., Mishima, T.D., Johnson, M.B. and Peng, X. (2003) Large-scale synthesis of nearly monodisperse CdSe/CdS core/shell nanocrystals using air-stable reagents via successive ion layer adsorption and reaction. *Journal of the American Chemical Society*, **125**, 12567–75.

38 Mekis, I., Talapin, D.V., Kornowski, A., Haase, M. and Weller, H. (2003) One-pot synthesis of highly luminescent CdSe/CdS core–shell nanocrystals via organometallic and "greener" chemical approaches. *Journal of Physical Chemistry B*, **107**, 7454–64.

39 Malik, M.A., O'Brien, P. and Revaprasadu, N. (2002) A simple route to the synthesis of core/shell nanoparticles of chalcogenides. *Chemistry of Materials*, **14**, 2004–10.

40 Mattoussi, H., Mauro, J.M., Goldman, E.R., Anderson, G.P., Sundar, V.C., Mikulec, F.V. and Bawendi, M.G. (2000) Self-assembly of CdSe–ZnS quantum dot bioconjugates using an engineered recombinant protein. *Journal of the American Chemical Society*, **122**, 12142–50.

41 Pathak, S., Choi, S.K., Arnheim, N. and Thompson, M.E. (2001) Hydroxylated quantum dots as luminescent probes for *in situ* hybridization. *Journal of the American Chemical Society*, **123**, 4103–4.

42 Guo, W., Li, J.J., Wang, Y.A. and Peng, X. (2003) Conjugation chemistry and bioapplications of semiconductor box nanocrystals prepared via dendrimer bridging. *Chemistry of Materials*, **15**, 3125–33.

43 Chen, Y. and Rosenzweig, Z. (2002) Luminescent CdS quantum dots as selective ion probes. *Analytical Chemistry*, **74** (19), 5132–8.
44 Pinaud, F., King, D., Moore, H.P. and Weiss, S. (2004) Bioactivation and cell targeting of semiconductor CdSe/ZnS nanocrystals with phytochelatin-related peptides. *Journal of the American Chemical Society*, **126**, 6115–23.
45 Aldana, J., Wang, Y.A. and Peng, S. (2001) Photochemical instability of CdSe nanocrystals coated by hydrophilic thiols. *Journal of the American Chemical Society*, **123**, 8844–50.
46 Yu, W.W., Chang, E., Drezek, R. and Colvin, V.L. (2006) Water-soluble quantum dots for biomedical applications. *Biochemical and Biophysical Research Communications*, **348**, 781–6.
47 Mulvaney, P., Liz-Marzan, L.M., Giersig, M. and Ung, T. (2000) Silica encapsulation of quantum dots and metal cluster. *Journal of Materials Chemistry*, **10**, 1259–70.
48 Gerion, D., Pinaud, F., Williams, S.C., Parak, W.J., Zanchet, D., Weiss, S. and Alivisatos, A.P. (2001) Synthesis and properties of biocompatible water-soluble silica-coated CdSe/ZnS semiconductor quantum dots. *Journal of Physical Chemistry B*, **105**, 8861–71.
49 Nann, T. and Mulvaney, P. (2004) Single quantum dots in spherical silica particles. *Angewandte Chemie (International Edition in English)*, **43**, 5393–6.
50 Rogach, A.L., Nagesha, D., Ostrander, J.W., Giersig, M. and Kotov, N.A. (2000) "Raisin bun"-type composite spheres of silica and semiconductor nanocrystals. *Chemistry of Materials*, **12**, 2676–85.
51 Alivisatos, A.P., Gu, W.W. and Larabell, C. (2005) Quantum dots as cellular probes. *Annual Review of Biomedical Engineering*, **7**, 55–76.
52 Gao, X., Cui, Y., Levenson, R.M., Chung, L.W. and Nie, S. (2004) *In vivo* cancer targeting and imaging with semiconductor quantum dots. *Nature Biotechnology*, **22**, 969–76.
53 Pellegrino, T., Manna, L., Kudera, S., Liedl, T., Koktysh, D., Rogach, A.L., Keller, S., Raedler, J., Natile, G. and Parak, W.J. (2004) Hydrophobic nanocrystals coated with an amphiphilic polymer shell: a general route to water soluble nanocrystals. *Nano Letters*, **4**, 703–7.
54 Fan, H., Leve, E.W., Scullin, C., Gabaldon, J., Tallant, D., Bunge, S., Boyle, T., Wilson, M.C. and Brinker, C.J. (2005) Surfactant-assisted synthesis of water-soluble and biocompatible semiconductor quantum dot micelles. *Nano Letters*, **5**, 645–8.
55 Goldman, E.R., Balighian, E.D., Mattoussi, H., Kuno, M.K., Mauro, J.M., Tran, P.T. and Anderson, G.P. (2002) Avidin: a natural bridge for quantum dot–antibody conjugates. *Journal of the American Chemical Society*, **124** (22), 6378–82.
56 Lingerfelt, B.M., Mattoussi, H., Goldman, E.R., Mauro, J.M. and Anderson, G.P. (2003) Preparation of quantum dot–biotin conjugates and their use in immunochromatography assays. *Analytical Chemistry*, **75** (16), 4043–9.
57 Aoyagi, S. and Kudo, M. (2005) Development of fluorescence change-based, reagent-less optic immunosensor. *Biosensors and Bioelectronics*, **20** (8), 1680–4.
58 Ravindran, S., Kim, S., Martin, R., Lord, E.M. and Ozkan, C.S. (2005) Quantum dots as bio-labels for the localization of a small plant adhesion protein. *Nanotechnology*, **16** (1), 1–4.
59 Zhang, Y., So, M.-K., Loening, A.M., Yao, H., Gambhir, S.S. and Rao, J. (2006) HaloTag protein-mediated site-specific conjugation of bio-luminescent proteins to quantum dots. *Angewandte Chemie (International Edition in English)*, **45** (30), 4936–40.
60 Tholouli, E., Hoyland, J.A., Di Vizio, D., O'Connell, F., MacDermott, S.A., Twomey, D., Levenson, R., Yin, J.A., Liu, G., Todd, R., Loda, M. and Byers, R. (2006) Imaging of multiple mRNA targets using quantum dot based in situ hybridization and spectral deconvolution in clinical biopsies. *Biochemical and Biophysical Research Communications*, **348** (2), 628–36.

61 Srinivasan, C., Lee, J., Papadimitrako-poulos, F., Silbart, L.K., Zhao, M. and Burgess, D.J. (2006) Labeling and intracellular tracking of functionally active plasmid DNA with semiconductor quantum dots. *Molecular Therapy*, **14** (2), 192–201.

62 Crut, A., Geron-Landre, B., Bonnet, I., Bonneau, S., Desbiolles, P. and Escude, C. (2005) Detection of single DNA molecules by multicolor quantum-dot end-labeling. *Nucleic Acids Research*, **33** (11), e98/1–9.

63 Gill, R., Willner, I., Shweky, I. and Banin, U. (2005) Fluorescence resonance energy transfer in CdSe/ZnS–DNA conjugates: probing hybridization and DNA cleavage. *Journal of Physical Chemistry B*, **109**, 23715–19.

64 Zhang, C.Y., Yeh, H.C., Kuroki, M.T. and Wang, T.H. (2005) Single-quantum-dot-based DNA nanosensor. *Nature Materials*, **4**, 826–31.

65 Wu, X., Liu, H., Liu, J., Haley, K.N., Treadway, J.A., Larson, J.P., Ge, N., Peale, F. and Bruchez, M.P. (2003) Immunofluorescent labeling of cancer marker Her2 and other cellular targets with semiconductor quantum dots. *Nature Biotechnology*, **21** (1), 41–6.

66 Fu, A., Gu, W., Larabell, C. and Alivisatos, A.P. (2005) Semiconductor nanocrystals for biological imaging. *Current Opinion in Neurobiology*, **15** (5), 568–75.

67 Smith, A.M., Dave, S., Nie, S., True, L. and Gao, X. (2006) Multicolor quantum dots for molecular diagnostics of cancer. *Expert Review of Molecular Diagnostics*, **6** (2), 231–44.

68 Weng, J., Song, X., Li, L., Qian, H., Chen, K., Xu, X., Cao, C. and Ren, J. (2006) Highly luminescent CdTe quantum dots prepared in aqueous phase as an alternative fluorescent probe for cell imaging. *Talanta*, **70** (2), 397–402.

69 Akerman, M.E., Chan, W.C.W., Laakkonen, P., Bhatia, S.N. and Ruoslahti, E. (2002) Nanocrystal targeting in vivo. *Proceedings of the National Academy of Sciences of the United States of America*, **99** (2), 12617–21.

70 Larson, D.R., Zipfel, W.R., Williams, R.M., Clark, S.W., Bruchez, M., Wise, P., Webb, F.W. and W.W. (2003) Water-soluble quantum dots for multiphoton fluorescence imaging in vivo. *Science*, **300** (5624), 1434–7.

71 Hoshino, A., Hanaki, K., Suzuki, K. and Yamamoto, K. (2004) Applications of T-lymphoma labeled with fluorescent quantum dots to cell tracing markers in mouse body. *Biochemical and Biophysical Research Communications*, **314** (1), 46–53.

72 So, M.-K., Xu, C., Loening, A.M., Gambhir, S.S. and Rao, J. (2006) Self-illuminating quantum dot conjugates for in vivo imaging. *Nature Biotechnology*, **24** (3), 339–43.

73 Cai, W., Shin, D.-W., Chen, K., Gheysens, O., Cao, Q., Wang, S., Gambhir, S.S. and Chen, X. (2006) Peptide-labeled near-infrared quantum dots for imaging tumor vasculature in living subjects. *Nano Letters*, **6** (4), 669–76.

74 Lakowicz, J.R. (1999) *Principles of Fluorescence Spectroscopy*, Plenum Press, New York.

75 Vekshin, N. (1997) *Energy Transfer in Macromolecules*, SPIE Press, Belligham, WA.

76 Van Der Meer, B.W., Coker, G. and Chen, S.-Y.S. (1994) *Resonance Energy Transfer: Theory and Data*, VCH, New York.

77 Clapp, A.R., Medintz, I.L. and Mattoussi, H. (2006) Förster resonance energy transfer investigations using quantum-dot fluorophores. *ChemPhysChem*, **7**, 47–57.

78 Clapp, A.R., Medintz, I.L., Mauro, J.M., Fisher, B.R., Bawendi, M.G. and Mattoussi, H. (2004) Fluorescence resonance energy transfer between quantum dot donors and dye-labeled protein acceptors. *Journal of the American Chemical Society*, **126** (1), 301–10.

79 Willard, D.M. and Van Orden, A. (2003) Quantum dots: resonant energy-transfer sensor. *Nature Materials*, **2** (9), 575–6.

80 Kagan, C.R., Murra, C.B. and Bawendi, M.G. (1996) Long-range resonance

transfer of electronic excitations in close-packed CdSe quantum-dot solids. *Physical Review. B, Condensed Matter*, **54**, 8633–43.
81 Kagan, C.R., Murray, C.B., Nirmal, M. and Bawendi, M.G. (1996) Electronic energy transfer in CdSe quantum dot solids. *Physical Review Letters*, **76**, 1517–20.
82 Willard, D.M., Carillo, L.L., Jung, J. and Orden, A.V. (2001) CdSe–ZnS quantum dots as resonance energy transfer donors in a model protein-protein binding assay. *Nano Letters*, **1**, 469–74.
83 Finlayson, C.E., Ginger, D.S. and Greenham, N.C. (2001) Enhanced Forster energy transfer in organic/inorganic bilayer optical microcavities. *Chemical Physics Letters*, **338**, 83–7.
84 Mamedova, N.N., Kotov, N.A., Rogach, A.L. and Studer, J. (2001) Albumin–CdTe nanoparticle bioconjugates: preparation, structure, and interunit energy transfer with antenna effect. *Nano Letters*, **1**, 281–6.
85 Medintz, I.L., Clapp, A.R., Matoussi, H., Goldman, E.R., Fisher, B. and Mauro, J.M. (2003) Self-assembled nanoscale biosensors based on quantum dot FRET donors. *Nature Materials*, **2**, 630–8.
86 Goldman, E.R., Medinta, I.L. and Whitley, J.L. (2005) A hybrid quantum dot–antibody fragment fluorescence resonance energy transfer-based TNT sensor. *Journal of the American Chemical Society*, **127**, 6744–51.
87 Goldman, E.R., Clapp, A.R., Anderson, G.P., Uyeda, H.T., Mauro, J.M., Medintz, I.L. and Mattoussi, H. (2004) Multiplexed toxin analysis using four colors of quantum dot fluororeagents. *Analytical Chemistry*, **76** (3), 684–8.
88 Xu, C., Xing, B. and Rao, J. (2006) A self-assembled quantum dot probe for detecting beta-lactamase activity. *Biochemical and Biophysical Research Communications*, **344**, 931–5.
89 Chang, E., Miller, J.S., Sun, J., Yu, W.W., Colvin, V.L., Drezek, R. and West, J.L. (2005) Protease-activated quantum dot probes. *Biochemical and Biophysical Research Communications*, **334**, 1317–21.
90 Bakalova, R., Zhelev, Z., Ohba, D. and Baba, Y. (2005) Quantum dot-conjugated hybridization probes for preliminary screening of siRNA sequences. *Journal of the American Chemical Society*, **127**, 11328–35.
91 Medintz, I.L., Konnert, J.H. and Clapp, A.R. (2004) A fluorescence resonance energy transfer-derived structure of a quantum dot–protein bioconjugate nanoassembly. *Proceedings of the National Academy of Sciences of the United States of America*, **101**, 9612–17.
92 Shi, L., De Pauli, V., Rosenzweig, N. and Rosenzweig, Z. (2006) Synthesis and application of quantum dots FRET-based protease sensors. *Journal of the American Chemical Society*, **128**, 10378–9.
93 Shi, L., Rosenzweig, N. and Rosenzweig, Z. (2007) Luminescent quantum dots fluorescence resonance energy transfer-based probes for enzymatic activity and enzyme inhibitors. *Analytical Chemistry*, **79** (1), 208–14.
94 Wang, D., He, J., Rosenzweig, N. and Rosenzweig, Z. (2004) Superparamagnetic Fe_2O_3 beads–CdSe/ZnS quantum dots core–shell nanocomposite particles for cell separation. *Nano Letters*, **4** (3), 409–13.

12
New Proteins for New Sensing Methodologies: The Case of the Protein-Binding Family

Vincenzo Aurilia, Maria Staiano, Mosè Rossi, and Sabato D'Auria

12.1
Introduction

The solute transport systems in prokaryotes are based on two types depending on the energetic requirement of the transport system: (i) the secondary transporters where proton or sodium motive force drives the transport and (ii) the ATP-binding cassette (ABC) primary transporters that use the hydrolysis of ATP as fuel. High-affinity binding protein-dependent ABC transporters were originally discovered in Gram-negative bacteria. These transport proteins play an important physiological role in the transport of different molecules through biological membranes. Bacterial ABC transporters predominantly import essential nutrients that are delivered to them by specific binding proteins. A typical ABC is composed of five domains or subunits, two of which are hydrophobic and are predicted to span the membrane multiple times in an α-helical conformation, and two of which bind nucleotide and are exposed to the cytoplasm. The fifth component is the periplasmic soluble binding protein, a high-affinity receptor, that interacts with the substrate to be transported. The soluble ligand-bound binding protein interacts with the transporters proteins, stimulates the ATPase activity and initiates the transport process. These periplasmic binding proteins have two globular domains attached by a flexible hinge and in the ligand-bound structures the ligand is buried deep within the cleft between the two domains. Solute-binding proteins for a variety of ligands have been identified, including carbohydrates, amino acids, anions, metal ions, dipeptides and oligopeptides. Conformational changes involving the hinge are thought to be necessary for ligands to get in and out of the protein-binding site. Differences in the structures of the ligand-bound and ligand-free proteins are essential for their proper recognition by the membrane components [1] (Figure 12.1).

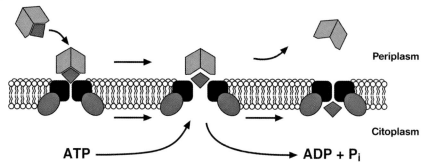

Figure 12.1 Domain structure of bacterial transport systems. Genes coding for the components of solute protein-dependent transporters are almost invariably organized into operons to achieve a coordinated regulation of their expression [2].

12.2
Galactose/Glucose-Binding Protein from *Escherichia coli*

The D-galactose/D-glucose-binding protein (GGBP) from *E. coli* is a monomeric periplasmic protein that serves as a high-affinity receptor for the active transport and chemotaxis towards both sugars. It can binds several types of monosaccharides (e.g., D-glucose, D-galactose, L-arabinose and L-xylose) with different affinity constants. In fact, a K_d value of 0.21 mM for D-glucose dissociation from GGBP has been reported by Zukin *et al.* [3]; for the other sugars, the affinity constants are 100- to 1000-fold weaker than glucose [4]. Genetic analyses reveal that the galactose operon contains three open reading frames (ORFs): *mglB* that codes for the galactose-binding protein, *mglA* encoding a protein that possesses a putative nucleotide-binding site and, finally, *mglC* that codes for a hydrophobic protein believed to generate the transmembrane pore [5]. GGBP is a monomeric protein of molecular weight of about 32 kDa, composed of two globular domains attached by a flexible hinge. The sugar is bound within the cleft. The X-ray structure [6, 7] of GGBP in the absence and in the presence of D-glucose shows a close view of the binding site as well as the amino acid residues directly involved in glucose binding. Specificity and affinity are conferred primarily by polar planar side-chain amino acid residues that form an intricate network of cooperative and bidentate hydrogen bonds with the sugar, and secondarily by aromatic amino acid residues that sandwich the pyranose ring of the sugar [8]. Upon interaction with the ligands, GGBP undergoes a conformational change to facilitate the interaction of ligand-bound GGBP to the membrane-anchored chemoreceptor. Details of structural properties as well as information on the conformational stability of proteins are needed for developing biotechnological applications. Thus, it becomes important to understand how physical phenomena, such as high temperature and pressure, can impact on the structural and spectroscopic properties of GGBP.

The effect of the temperature on the secondary structure and stability of GGBP from E. coli in the absence and in the presence of the ligands was investigated by differential scanning calorimetry (DSC), circular dichroism (CD), Fourier transform infrared (FTIR) spectroscopy and fluorescence spectroscopy [9]. The data revealed that the glucose binding induces a moderate change in the secondary structure of the protein and increases the thermal stability of GGBP. In addition, the results also suggested that some of the GGBP structural stretches, involved in α-helices and β-strand conformations, were particularly sensitive to the increase of temperature. In particular, DSC data showed that temperature induced two unfolding transitions of the protein structure and a change in the character of the folding/unfolding process in the absence or in the presence of D-glucose. Intrinsic tryptophan fluorescence could be ascribed without doubt to the C-terminal domain of GGBP, allowing for a detailed identification of thermal transitions of the protein. When D-glucose binds to the ligand-binding site of GGBP, the relative positions of domains change, involving the participation of some amino acids involved in a network of hydrogen bonds [10]. These interactions result in a large T_m increase of around 13 °C that can be observed by calorimetric (DSC) and optical (CD and tryptophan fluorescence) methods. Four of the five GGBP tryptophan residues are located in the C-terminal domain of the protein. The fifth tryptophan residue, at position 284, is located in a C-terminal loop headed toward the N-terminal domain [10]. Measurements of changes in the tryptophan fluorescence as a function of temperature should therefore reflect, predominantly, conformational changes in the C-terminal domain of the protein. The fluorescence data indicated that the unfolding of the C-terminal domain of GGBP occurs with a lower T_m value than the temperature needed for the unfolding of the protein N-terminal domain. The refined crystallographic structures of GGBP revealed the presence of a calcium site located about 30 Å from the sugar-binding site. In order to better understand the stability of GGBP, especially connected to the C-terminal domain, the role of calcium on the conformational dynamics and the thermal stability of GGBP in the absence and in the presence of glucose was investigated by molecular simulation experiments, fluorescence spectroscopy, CD and FTIR. The entire set of experiments demonstrated that the absence of calcium promotes a decrease of the protein thermal stability. However, it was also noticed that the presence of glucose stabilizes the calcium-free GGBP protein structure and restores the cooperativity of the temperature-induced transition. Since the calcium-binding site is located close to the C-terminal domain, the observed stability changes of the GGBP structure caused by the removal of calcium and/or glucose binding could be related mainly to the GGBP C-terminal domain [11, 12].

In order to share out the entire protein in different portions and gain more information about the structure and the function of the protein, a genetic variant of GGBP was obtained with a single cysteine in the proximity of the glucose-binding site. This allowed to achieve information not only on the C-terminal domain of the protein, but also on the protein-binding site. In fact, the single cysteine residue present in GGBP was labeled with an extrinsic fluorescence probe and investigated by fluorescence spectroscopy, CD, FITR and molecular dynamics

(MD). In this context, it was documented that the binding of glucose to GGBP results in no stabilizing effect on the N-terminus of GGBP and in a moderate stabilization of the protein matrix close to the sugar-binding site of GGBP. On the contrary, the binding of glucose has a strong stabilization effect on the C-terminal domain of the GGBP [13]. In an attempt to deeply characterize the structural features of GGBP and its reaction to physical perturbation, the effect of pressure on the stability and conformational dynamics of GGBP in the absence and in the presence of glucose was also studied, as monitored by steady-state and time-resolved fluorescence experiments. In addition, MD simulations studies were applied to provide a molecular portrait of the effect of the pressure increase on the secondary and tertiary structure of the protein. The results showed that GGBP undergoes a major alteration of its structural properties at a moderate increase of external pressure and the binding of glucose to the protein is able to enhance the resistance of GGBP to pressure stress. In particular, with increasing external pressure the fluorescence intensity of GGBP significantly decreases [14, 15]. The protein conformational changes are usually accompanied with variations of the tryptophan quantum yield [16]; the observed difference in the pressure-induced GGBP fluorescence intensity quenching suggests a different stability of GGBP structure in the absence and in the presence of the ligand. At atmospheric pressure both GGBP and the GGBP/Glc complex exhibited broad emissions with the same spectral shape and emission peak. Although the GGBP/Glc complex exhibits negligible spectra dependent on pressure, the spectrum of unliganded GGBP slightly shifts to red with increasing pressure. This indicates a pressure-induced change of protein conformation causing some of the tryptophan residues to become more exposed to the solvent. Consistently, with the steady-state data, the effect of pressure on the fluorescence decay is more pronounced in the absence of glucose compared to the GGBP/Glc complex, causing a marked decrease of the mean fluorescence lifetime. In agreement with the fluorescence results, differences in the rate of loss of secondary structure between GGBP and GGBP/Glc are evident. Furthermore, phosphorescence spectroscopy was also used to investigate the effects of glucose and calcium on the dynamics and stability of GGBP [17]. The binding of glucose modifies the phosphorescence lifetime values as well as the spectrum of GGBP, shifting the blue band 0.54 nm to the blue and the red band 1 nm to the red. Finally, the removal of the calcium from GGBP structure causes variations in lifetime values and spectral shifts similar to those induced by glucose binding to the native protein. Based on this conformational change, a sensing system for glucose was developed. Several research laboratories are studying the biotechnological applications of GGBP as a glucose sensor and, recently, in our laboratory we have demonstrated that the glucose-binding protein can be used as a probe to develop a glucose sensor [18, 19]. E. coli GGBP seems to be a good candidate as a probe for the development of a nonconsuming glucose biosensor. Glucose binding was performed by steady-state and life-time fluorescence spectroscopy and fluorescence (or Förster) resonance energy transfer (FRET) measurements. In particular, glucose titration was performed by using an extrinsic fluorophore (acrylodan) labeled to Cys182 of the GGBP mutant, so that it was

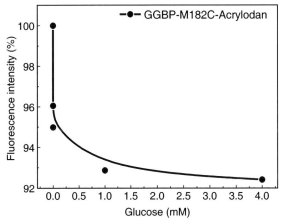

Figure 12.2 Titration curve of GGBP-M182C labeled with acrylodan with increasing amount of glucose.

possible to monitor environmental changes upon glucose binding. Fluorescence data showed that the intensity of the acrylodan emission was sensitive to the addition of glucose, suggesting that the binding of glucose to the GGBP-M182C/acrylodan displaces the acrylodan into a more polar environment as a result of conformational changes of the protein (Figure 12.2).

FRET analysis was performed by using acrylodan and rhodamine for labeling the Cys182 and the N-terminus of the GGBP-M182C, respectively, as donor–acceptor pair fluorophores. Upon glucose binding, an increase was observed in the fluorescence emission of rhodamine of about 10%. In addition, the real-time measurements of glucose concentration by using GGBP-M182C/acrylodan shows that, upon glucose binding, the lifetime slightly changes, which correlates well with the intensity changes observed for steady-state measurements [20].

12.3
Glutamine-Binding Protein from *Escherichia coli*

Glutamine-binding protein (GlnBP) from *E. coli* is a monomeric protein composed of 224 amino acid residues (26 kDa) responsible for the first step in the active transport of L-glutamine across the cytoplasmic membrane. GlnBP belongs to a large family of ligand-binding proteins that share the same architecture [21] and is localized in the periplasmic space of *E. coli* [22, 23]. GlnBP from *E. coli* consists of two similar globular domains – the large domain (residues 1–84 and 186–224) and the small domain (residues 90–180), which are linked by two peptide hinges. Each domain contains a central core of β-sheets flanked by α-helices. The deep cleft formed between the two domains contains the ligand-binding site. X-ray crystallographic data provide evidence that, compared to the GlnBP/Gln complex, the ligand-free GlnBP exhibits a large-scale movement of the two hinges upon ligand binding, which occurs in the so-called flap region [24]. Among the naturally occurring amino acids, only glutamine is bound by GlnBP, with a K_d of 3×10^{-7} M. An elaborate analysis of the structural properties of the recombinant GlnBP in the absence and in the presence of glutamine can contribute to a better understanding of the transport-related functions of the protein, and of other structurally similar periplasmic transport proteins, as well as to the design and develop new biotechnological applications for this class of protein. In order to obtain detailed information about the structure, the thermal stability and thermal unfolding of GlnBP in the absence and in the presence of glutamine was investigated by means of FTIR spectroscopy [25], fluorescence spectroscopy [26, 27], CD and computational analysis [28]. The spectroscopic data showed that the interaction of GlnBP with glutamine resulted in a marked change of the structural and conformational dynamics features of the protein. In particular, the fluorescence and CD data showed that the presence of glutamine resulted in a dramatic increase of the protein thermal stability of about 10 °C. In addition, the fluorescence time-resolved data pointed out that both in the absence and in the presence of glutamine, the protein structure was highly rigid with a small amplitude of segmental motion up to 65 °C and a low accessibility of the protein tryptophan residues to acrylamide. The spectroscopic data were in good agreement with the data obtained by X-ray experiments on the crystals of GlnBP and GlnBP/Gln. The binding of the ligand causes rearrangement of the two domains of the protein in a closed-cleft conformation. This allows the formation of strong interactions between the two parts of the binding site. The formation of these links increases the stability of secondary structures involved in ligand binding because they are embedded in a more rigid and compact conformation with a lower extent of fluctuation. As for other periplasmic binding proteins, an interesting property of GlnBP from *E. coli* is that, upon interaction with glutamine, the protein undergoes a conformational change to facilitate the interaction of bound GlnBP to the membrane-anchored chemoreceptor. Based on this conformational change, a sensing system for glutamine can be developed. In fact, the molecular binding between GlnBP from *E. coli* and L-glutamine was optically transduced by means of a biosensor based on porous silicon technology [29]. GlnBP was used as a molecular probe for glutamine for developing a sensor that operates by the measurement of the interferometric fringes in the reflectivity spectrum of a porous silicon Fabry–Perot layer. The binding event was revealed as a shift in wavelength of the fringes [29].

The binding of GlnBP to gliadin peptides, which are considered toxic for celiac patients, was investigated by mass spectrometry experiments and optical techniques, demonstrating that GlnBP binds the following amino acid sequence XXQPQPQQQQQQQQQQQQL, present only into the toxic prolamines. The binding of GlnBP to gliadin suggested to design a new optical biosensor based on nanostructured porous silicon (PSi) for the detection of trace amounts of gliadin in food. GlnBP was covalently linked to the surface of the PSi wafer by a proper passivation process and the GlnBP/gliadin interaction was revealed as a shift in wavelength of the fringes in the reflectivity spectrum of the PSi layer. The GlnBP, covalently bonded to the PSi chip, selectively recognized the toxic peptide. Finally, the sensor response to the protein concentration was measured in the range 2.0–40.0 g/l and the sensitivity of the sensor was determined. Moreover, in Staiano et al. 2006 [30], a competitive resonance energy transfer assay for an easy and rapid detection of traces of gluten in raw and cooked foods was developed.

12.4
Trehalose/Maltose-Binding Protein from the Hyperthermophilic Archaeon *Thermococcus litoralis*

The D-trehalose/D-maltose-binding protein (TMBP) is a component of the trehalose and maltose uptake system, and it is the first ABC transporter studied in archaea, specifically in the hyperthermophilic archaeon *Thermococcus litoralis*. In this bacterium it exhibits a high affinity for maltose and trehalose at 85 °C – the optimum growth temperature of this organism [31]. The gene coding for TMBP was cloned and expressed in *E. coli*. The recombinant protein carries at the N-terminus the natural signal sequence of the *E. coli* maltose-binding protein that allows the export of the protein into the periplasmic space. The recombinant TMBP was purified to homogeneity. It is a monomer with a molecular weight of 48 kDa. It binds both maltose and trehalose with a K_d of 0.160 µM, and it is highly thermostable [32]. The X-ray structure [33] of TMBP shares common structural motifs with several other sugar-binding proteins. These proteins consist of two globular domains connected by a hinge region composed of two or three short polypeptide segments. TMBP was also investigated by fluorescence spectroscopy, FTIR spectroscopy and MD simulation experiments for studying the effect of trehalose, maltose and also glucose binding on the structural properties and the physical parameters of the protein. The temperature dependence of fluorescence spectra of TMBP in the absence and in the presence of maltose exhibits a broad emission with a spectral shape essentially independent of the presence of ligand. In particular, the protein fluorescence intensity in the presence of maltose increases about 12%, whereas it decreases about 11% in the presence of trehalose. The emission peaks of the unliganded TMBP and TMBP/Mal complex are centered near 342 nm. The fluorescence spectrum of TMBP/Tre is slightly blue-shifted with the emission maximum centered near 339 nm. Consistent with the lower tryptophan quantum yield, this suggests a different conformation of TMBP/Tre compared with the TMBP/Mal complex. In the temperature range of 25–90 °C, the spectral positions of all the protein spectra are essentially temperature independent. This indicates that TMBP possesses a remarkable temperature stability. Frequency domain emission decays of TMBP in the absence and in the presence of the two ligands were performed. Modulation and phase frequency responses reveal that the fluorescence lifetimes depend on the type of the sugar bound. A lower frequency of the intersection indicates a longer fluorescence lifetime. It can be seen that the mean lifetime increases as τ (TMBP/Tre) < τ (TMBP) < τ (TMBP/Mal). This behavior is consistent with the steady-state data that show the protein fluorescence quenching upon the trehalose binding (Figure 12.3).

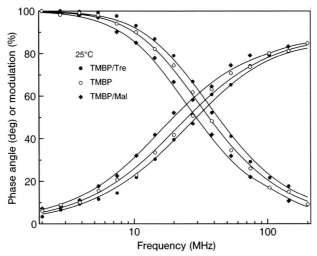

Figure 12.3 Phase and modulation frequency responses of TMBP (○), TMBP/Mal (♦) and TMBP/Tre (●) at 25 °C.

An opposite protein fluorescence behavior was recorded upon maltose binding. This is an additional indication that TMBP can exist in three different conformational states: TMBP alone, TMBP/Mal and TMBP/Tre. In the protein/sugar complex, tryptophan accessibility did not change at all temperatures (25–90 °C). This indicates the absence of a large tryptophan fraction with significantly different accessibilities to the quencher as consequence of a dense protein packing. Instead, the quenching efficiency of fluorescence gradually increases in the sugar-free TMBP with increasing temperature. As a consequence the residue packing of some domains in the sugar-free TMBP significantly loosens with increasing temperature and some of the protein tryptophan residues become more accessible to acrylamide. To gain more information on temperature-induced structural changes of TMBP in the absence and in the presence of maltose or trehalose, FTIR absorption measurements were performed. A comparison of resolution-enhanced deconvoluted spectra of unliganded TMBP and the TMBP/Mal and TMBP/Tre complexes at room temperature shows that both sugars induce small, but significant changes in the protein FTIR spectra. In particular, the main β-sheet band shifts slightly to a higher wavenumber, suggesting a lower accessibility of the solvent (^2H_2O) to this structural element. Comparison of the spectra collected at 20 and 99.5 °C indicates that the temperature induces only partial denaturation of the protein because the α-helix and β-sheet bands dominate the spectra at all temperatures. The spectra collected at 85 °C show that the sugar binding induces conformational change causing the buried α-helices to become more solvent-exposed. Molecular dynamics simulation experiments on unliganded TMBP and TMBP/Tre, TMBP/Mal complexes were performed. In all systems, the secondary structures are largely intact after exposure to high temperatures; there are only insignificant differences in percentages of the secondary structure elements between all calculated structures. This supports the conclusions from the FTIR spectroscopy. Moreover, tryptophan accessibility to the solvent and interactions of the sugars with TMBP were analyzed. The tryptophan residues of TMBP are more accessible to the solvent then the TMBP/Mal and TMBP/Tre structures, and it should not be surprising that in the absence of the sugar their accessibility to solvent significantly increases. This can also explain the obtained quenching data at 90 °C that indicated the presence of both accessible and inaccessible classes of tryptophan.

The analyses of the interactions of the sugars with TMBP by means of the HBPLUS program showed that the binding mode of the two disaccharides is different. In addition, fluorescence spectroscopy was used to verify whether the protein can bind D-glucose in order to use this protein for the development of a glucose biosensor. It was observed that the addition of glucose results in a modest, although a highly significant and reproducible increase of TMBP fluorescence intensity and this strongly indicates that glucose binds to TMBP [34]. In conclusion, the recombinant TMBP constitutes a promising basis for the design of a novel nonconsuming substrate fluorescent biosensor for monitoring the level of glucose in fluids that are free of other sugars (e.g., in blood). A sensor based on TMBP/Tre, TMBP/Mal or TMBP/Glc interactions would have excellent sensibility, stability and shelf-life.

12.5
Lipocalins and Odorant-Binding Protein

The odorant-binding proteins (OBPs) are the most abundant class of proteins found in the olfactory apparatus. Olfactory perception, in all invertebrate and vertebrate organisms, is based on the activation by odorant molecules of G-coupled receptors, designed as olfactory receptors and located at the cilia of olfactory neuronal endings. The olfactory receptor cells are bipolar neurons with dendrites that protrude into fluid medium. At the opposite pole of the cell, the axons project directly into the central nervous system. The OBPs are defined by their properties of reversibly binding volatile chemicals, called odorants. This definition is not correct for many reason [35]. Odorants are not an homogenous class of compounds, but they are a large class of molecules with the same characteristics, such as low molecular weight (300–400 Da), hydrophobicity and high volatility. Owing to their capacity to bind a nonspecific class of compound, OBPs appear to play a specific and important role in olfaction, in particular in the first step of olfaction recognition. More than 20 years of studies has not been enough to clarify completely their role. There are several hypotheses suggesting their function, such as their involvement in facilitating the movement of hydrophobic odorant across the aqueous mucus layer to accesses to the olfactory receptors or their action in the termination of the olfactory signal by "removing" odorants from the receptors once they have been stimulated [36, 37]. See Scheme 12.1.

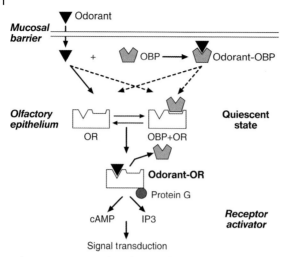

Scheme 12.1 Proposed mechanism for OBP and olfactory receptor (OR) interaction in the presence and absence of odorant. cAMP = cyclic adenosine monophosphate; IP$_3$ = inositol 1,4,5-trisphosphate.

The OBPs are a subclass of lipocalins – a protein family composed of many members. They are a large group of proteins that exhibit great structural and functional variations among species. Most lipocalin proteins have highly conserved structures; some of them maintain three characteristic sequence motifs while others only one or two. They are extracellular small proteins; typically they bind small hydrophobic molecules, but they bind also to specific cell surface receptors or they create covalent and noncovalent complexes with other soluble macromolecules.

It is clear that members of the lipocalin family fulfill a wide variety of different functions. The common structure of the lipocalin protein fold is a highly symmetrical all-β structure dominated by a single eight-stranded antiparallel β-sheet closed back on itself to form a continuously hydrogen-bonded β-barrel. The β-barrel encloses a ligand-binding site composed of both an internal cavity and an external loop scaffold that gives rise to a variety of different binding modes, each capable of accommodating ligands of different size, shape and chemical character. Together with three other distinct protein families – the fatty acid-binding proteins, avidin and metalloproteinase inhibitors, the lipocalin family belongs to a large structural superfamily: the calycins – a superfamily that includes a set of proteins with closely related three-dimensional structures that show no significant similarity at the sequence level. Members of the calycin protein family share a β-barrel motif, like lipocalins, that shows the ability to bind ligands with a strongly hydrophobic character. For these reasons the OBPs have been attributed to this family. Although sequence similarity among different OBPs is very low, different OBP subtypes have been reported to simultaneously occur in the same animal species – three in pig, four in mouse, three in rat, three in rabbit and at least eight in porcupine. The first vertebrate OBP was isolated from bovine nasal mucus and characterized as pyrazine-binding protein using 2-isobutyl-3-methoxypyrazine as ligand [38–40]. Subsequently, OBPs were identified in a variety of species, including pig, rabbit, mouse and rat [35, 41–43].

12.5.1
Structural Characterization of Pig OBP

In our laboratory we investigated the biotechnological application on OBPs, in particular we studied the structural and function characteristics of pig OBP (pOBP) and its possible uses as a probe to develop a biosensor for explosive detection. Previous fluorescence measurements and FTIR studies [44] showed that pOBP is surprisingly resistant at high temperatures (up to 80 °C), although it derives from a mesophilic organism. This intrinsic stability is very intriguing, since pOBP could be used as a biological probe for a fluorescence biosensor [45]. We used fluorescence spectroscopy, CD and MD simulations to obtain a "molecular portrait" of the protein, and to dissect the phenomena related to protein stability and thermal perturbations at a molecular level. The crystallographic of pOBP (Figure 12.4) revels that the β-barrel structure is characterized by the presence of a high number of hydrophobic and aromatic residues (Val, Leu, Ile, Met, Phe, Trp) that form a large cluster in the core of the protein and, in particular, they create the internal cavity of the β-barrel itself.

Analysis with NACCES reveals that a few hydrophobic residues are exposed to solvent and generally they are isolated on the protein surface. The vast majority of hydrophobic residues are almost all buried in the protein matrix and therefore these hydrophobic amino acids could create a strong network of hydrophobic interactions, which in general contribute to the high thermostability of pOBP

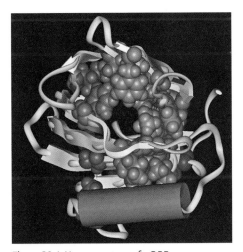

Figure 12.4 X-ray structure of pOBP.

[46, 47]. Furthermore, the pOBP structure is also characterized by the presence of large amounts of charged amino acid residues (30 acid residues and 14 basic residues) that are responsible for a salt bridge that it is important for structure and stability of the protein.

The experimental fluorescence data with the results of analysis of the location of the single tryptophan residue shows three phenomena interesting for the theory of protein intrinsic fluorescence. (1) pOBP has a red-shifted fluorescence spectrum, although the density of the Trp16 microenvironment is high and there is only one polar group in its microenvironment (Lys120). (2) The intensity of tryptophan fluorescence increases on protein unfolding, as though in its microenvironment there is only one polar group (Lys120), which is not accepted as an effective quenching group. (3) The changes of fluorescence intensity and fluorescence lifetime are not correlated. The native structure of pOBP is highly resistant to heating. The denaturing curve recorded by monitoring the ratio of fluorescence intensities at two emission wavelengths (parameter A) shows a transition midpoint at 70 °C. It was found that the thermal denaturation of pOBP is irreversible as a consequence of the protein aggregation. Interestingly, the far-ultraviolet (UV) CD spectrum practically remains unchanged in the range of temperature between 20 and 95 °C [48].

We also studied the effect of chemical compounds on the stability of pOBP. The denaturation process was studied by intrinsic fluorescence analysis and far- and near-UV CD measurements. The results showed a reversible one-step denaturation process induced by GdnHCl. The midpoint of the transition (i.e., the point where the free energies of protein in the native and unfolded states are equal) corresponds to 2.3 M GdnHCl. The difference in free energy between native and unfolded states of pOBP is -5.95 kcal/mol in the absence of GdnHCl, indicating that the protein molecule is very stable to the denaturing action of GdnHCl. A 15% increase in fluorescence intensity accompanied by a 25% decrease of fluorescence decay lifetime, recorded in the range of 0.0–1.4 M GdnHCl, was explained by the destruction of the complex between Trp16 and the positively charged atom NZ of Lys120, localized over the center of the Trp16 indole ring, with concurrent formation of a complex between Trp16 and bound water molecules also located in its close vicinity [49].

12.5.2
Functional Characterization and Biotechnological Application of pOBP

In order to study the effective possibility to use this OBP as a biological element in the fluorescence biosensor, we assayed the ligand binding of two OBPs (from pig and bovine) with explosive components like diphenylamine (DFA), dimethylphthalate (DMF), resorcinol (RES) and dinitrotoluene (DNT) [45]. The affinities for DMF, RES, DNT and DFA were determined measuring the progressive chasing of 1-amminoantrhacene bound to OBPs in response to increasing concentration of each compound. The values of the dissociation constants for all the OBP forms are listed in Table 12.1.

Table 12.1 Dissociation constants (K_d; µM) of the binding complexes between OBP and explosive components.

	AMA	DMF	Resorcitol	DFA	DNT
Pig OBP	1.2	32	2100	0.11	18
Bovine OBP	0.6	800	5000	0.12	10560

The experimental results showed that all the OBP forms can bind to all the molecules tested, and this suggest that the protein scaffold of OBP can be considered as a promising platform for production of biological recognition elements to be employed in biosensors for the detection and identification of explosive substances. In particular, the high affinity that DNT, DMF and DFA show for pOBP suggests that the binding site of this form should be primarily considered for the development of mutants for the detection of these hazardous compounds.

Finally, we purified a new OBP from the nasal mucosa of *Canis familiaris* OBP (*CfOBP*) [50] and we performed the first experiments to use *CfOBP* as a probe for the development of a refractive index-based biosensor for biohazard assessment. Interestingly, *CfOBP* was able to bind odorant molecules that are usually recognized and bound by OBP proteins purified from several sources. In fact, ligand-binding experiments showed that *CfOBP* binds the odorant molecule 3,7-dimethyloctan-1-ol as demonstrated by fluorescence experiments with *CfOBP* previously labeled with the fluorescence probe 8-anilinonaphthalene sulfate (ANS). The ANS–*CfOBP* complex exhibits two emission maxima centered at 340 and 500 nm upon excitation at 290 nm. The emission maximum at 340 nm due to tryptophan emission suggests that the indolic residues are partially buried in the protein matrix. The emission maximum at 500 nm indicates a quite strong interaction between ANS and ANS–*CfOBP*. Upon odorant molecule addition, we observed an increase of the tryptophan emission and a moderate decrease of the ANS emission, indicating an energy transfer process between the indolic residues of the protein and the ANS molecules. This result suggests a possible structural change of *CfOBP* caused by the binding of the odorant molecule, prompting us to use *CfOBP* as a probe for the development of a refractive index biosensor. The possibility to exploit OBPs to realize an integrated biosensor based on refractive index measurements has been devised. Preliminary results show a peculiar behavior of *CfOBP* when exposed to pyrazine solution and pyrazine vapors in comparison with the bovine serum albumin response. This work is now in progress in obtaining a large amount of OBPs from different vertebrates in order to perform a more complete set of measurements.

12.6
Conclusions

There is a wealth of knowledge on biomolecules that specifically bind numerous substances of biochemical interest. Hence, the possibility to use biomolecules belonging to the "protein-binding family" as probes for reversible optical sensors will greatly expand the range of biochemically relevant analytes that can be measured using protein-based sensors.

The results described in this chapter represent a first attempt to use signaling proteins as probes for glucose, glutamine, gliadin and explosive compounds sensors. These proteins can be engineered for covalent labeling by insertion of cysteine residues at appropriate locations in the sequence or for improving their specificity, affinity or stability. For example, the described glucose-induced spectral changes may be larger with other polarity-sensitive probes or by the use of different donor–acceptor pairs in the FRET measurements. In summary, transporter proteins and sensing proteins appear to be a valuable source of biomolecule probes for the development of advanced optical biosensors.

Acknowledgments

This work was conducted in the frame of the CNR Commessa "Diagnostica avanzata ed Alimentazione" (M.S., S.D., V.A.). This work was also supported by the ASI project MoMa 1/014/06/0 (M.S., S.D., M.R.), the NATO grant CBP.EAP.CLG 982437 (S.D.) and the NATO grant CBP.NR.NRCLG.983088 (M.S.).

References

1 Driessen, A.J.M., Rosen, B.P. and Konings, W.N. (2000) Diversity of transport mechanisms: common structural principles. *Trends in Biochemical Sciences*, **25**, 397–401.
2 Plantinga, T.H., van der Does, C. and Driessen, A.J.M. (2004) Transporter's evolution and carbohydrate metabolic clusters. *Trends in Microbiology*, **12**, 4–7.
3 Zukin, R.S., Strange, P.G., Heavey, L.R. and Koshland, D.E., Jr (1977) Properties of the galactose binding protein of *Salmonella typhimurium* and *Escherichia coli*. *Biochemistry*, **16**, 381–6.
4 D'Auria, S., Alfieri, F., Staiano, M., Pelella, F., Rossi, M., Scire, A., Tanfani, F., Bertoli, E., Gryczynski, Z. and Lakowicz, J.R. (2004) Structural and thermal stability characterization of *Escherichia coli* D-galactose/D-glucose-binding protein. *Biotechnology Progress*, **20**, 330–7.
5 Hogg, R.W., Voelker, C. and Von Carlowitz, I. (1991) Nucleotide sequence and analysis of the mgl operon of Escherichia coli K12. *Molecular and General Genetics*, **229**, 453–9.
6 Vyas, N.K., Vyas, M.N. and Quiocho, F.A. (1991) Comparison of the periplasmic receptors for L-arabinose, D-glucose/D-galactose, and D-ribose. Structural and functional similarity. *Journal of Biological Chemistry*, **266**, 5226–37.
7 Borrok, M.J., Kiessling, L.L. and Forest, K.T. (2007) Conformational changes of D-glucose/D-galactose binding protein illuminated by apo and ultrahigh

resolution ligand-bound structures. *Protein Science*, **16** (6), 1032–41.

8 Vyas, N.K., Vyas, M.N. and Quiocho, F.A. (1988) Sugar and signal-transducer binding sites of the Escherichia coli galactose chemoreceptor protein. *Science*, **242**, 1290–5.

9 Piszczek, G., D'Auria, S., Staiano, M., Rossi, M. and Ginsburg, A. (2004) Conformational stability and domain coupling in D-glucose/D-galactose-binding protein from *Escherichia coli*. *Biochemical Journal*, **381**, 97–103.

10 Magnusson, U., Chaudhuri, B.N., Ko, J., Park, C., Jones, T.A. and Mowbray, S.L. (2002) Hinge-bending motion of d-allose-binding protein from *Escherichia coli*: three open conformations. *Journal of Biological Chemistry*, **277**, 14077–84.

11 Herman, P., Vecer, J., Barvik, I., Jr, Scognamiglio, V., Staiano, M., de Champdore, M., Varriale, A., Rossi, M. and D'Auria, S. (2005) The role of calcium in the conformational dynamics and thermal stability of the D-galactose/D-glucose-binding protein from *Escherichia coli*. *Proteins*, **61**, 184–95.

12 D'Auria, S., Ausili, A., Marabotti, A., Varriale, A., Scognamiglio, V., Staiano, M., Bertoli, E., Rossi, M. and Tanfani, F. (2006) Binding of glucose to the D-galactose/D-glucose-binding protein from *Escherichia coli* restores the native protein secondary structure and thermostability that are lost upon calcium depletion. *Journal of Biochemistry*, **139**, 213–21.

13 Scognamiglio, V., Scirè, A., Aurilia, V., Staiano, M., Crescenzo, R., Palmucci, C., Bertoli, E., Rossi, M., Tanfani, F. and D'Auria, S. (2007) A strategic fluorescence labeling of D-galactose/D-glucose-binding protein from *E. coli* for a better understanding of the protein structural stability and dynamics. *Journal of Proteome Research*, **6**, 4119–26.

14 Marabotti, A., Ausili, A., Staiano, M., Scire, A., Tanfani, F., Parracino, A., Varriale, A., Rossi, M. and D'Auria, S. (2006) Pressure affects the structure and the dynamics of the D-galactose/D-glucose-binding protein from *Escherichia coli* by perturbing the C-terminal domain of the protein. *Biochemistry*, **45**, 11885–94.

15 Marabotti, A., Herman, P., Staiano, M., Varriale, A., de Champdore, M., Rossi, M., Gryczynski, Z. and D'Auria, S. (2006) Pressure effect on the stability and the conformational dynamics of the D-galactose/D-glucose-binding protein from *Escherichia coli*. *Proteins*, **62**, 193–201.

16 Lakowicz, J.R. (1999) *Principles of Fluorescence Spectroscopy*, 2nd edn, Kluwer, New York.

17 D'Auria, S., Varriale, A., Gonnelli, M., Saviano, M., Staiano, M., Rossi, M. and Giovanni, B. (2007) Strambini tryptophan phosphorescence studies of the D-galactose/D-glucose-binding protein from *Escherichia coli* provide a molecular portrait with structural and dynamics features of the protein. *Journal of Proteome Research*, **6**, 1306–12.

18 Salins, L.L., Ware, R.A., Ensor, C.M. and Daunert, S. (2001) A novel reagentless sensing system for measuring glucose based on the galactose/glucose-binding protein. *Analytical Biochemistry*, **294**, 19–26.

19 Staiano, M., Sapio, M.R., Scognamiglio, V., Marabotti, A., Facchiano, A.M., Bazzicalupo, P.M., Rossi, M. and D'Auria, S. (2004) A putative thermo-stable sugar-binding protein from the archaeon *Pyrococcus horikoshii* as a probe for the development of a fluorescence biosensor for diabetic patients. *Biotechnology Progress*, **5**, 1572–7.

20 Scognamiglio, V., Aurilia, V., Cennamo, N., Ringhieri, P., Iozzino, L., Tartaglia, M., Staiano, M., Ruggiero, G., Orlando, P., Labella, T., Zeni, L., Vitale, A. and D'Auria, S. (2007) The D-galactose/D-glucose-binding protein from *Escherichia coli* as probe for a non-consuming glucose implantable fluorescence biosensor. *Sensors Journal*, **7**, 2484–791.

21 Hsiao, C.D., Sun, Y.J., Rose, J. and Wang, B.C. (1996) The crystal structure of glutamine-binding protein from *Escherichia coli*. *Journal of Molecular Biology*, **262**, 225–42.

22 Higgins, CF. (1992) ABC transporters: from microorganisms to man. *Annual Review of Cell Biology*, **8**, 67–113.

23 Adams, M. and Oxender, D. (1989) Bacterial periplasmic binding protein tertiary structures. *Journal of Biological Chemistry*, **264**, 15739–42.

24 Sun, Y.J., Rose, J., Wang, B.C. and Hsiao, C.D. (1998) The structure of glutamine-binding protein complexed with glutamine at 1.94 angstrom resolution: comparisons with other amino acid binding proteins. *Journal of Molecular Biology*, **278**, 219–29.

25 D'Auria, S., Scirè, A., Varriale, A., Scognamiglio, V., Staiano, M., Ausili, A., Marabotti, A., Rossi, M. and Tanfani, F. (2004) Binding of glutamine to glutamine-binding protein from Escherichia coli induces changes in protein structure and increases protein stability. *Proteins: Structure, Function, and Bioinformatics*, **58**, 80–7.

26 Staiano, M., Scognamiglio, V., Rossi, M., D'Auria, S., Stepanenko, O.V., Kuznetsova, I.M. and Turoverov, K.K. (2005) Unfolding and refolding of the glutamine-binding protein from *Escherichia coli* and its complex with glutamine induced by guanidine hydrochloride. *Biochemistry*, **44**, 5625–33.

27 Kuznetsova, I.M., Stepanenko, O.V., Turoverov, K.K., Staiano, M., Scognamiglio, V., Rossi, M. and D'Auria, S. (2005) Fluorescence properties of glutamine-binding protein from *Escherichia coli* and its complex with glutamine. *Journal of Proteome Research*, **4**, 417–23.

28 Herman, P., Vecer, J., Scognamiglio, V., Staiano, M., Rossi, M. and D'Auria, S. (2004) A recombinant glutamine-binding protein from *Escherichia coli*: effect of ligand-binding on protein conformational dynamics. *Biotechnology Progress*, **20**, 1847–54.

29 De Stefano, L., Rotiroti, L., Rendina, I., Moretti, L., Scognamiglio, V., Rossi, M. and D'Auria, S. (2006) Porous silicon-based optical microsensor for the detection of L-glutamine. *Biosensors and Bioelectronics*, **21**, 1664–7.

30 Staiano, M., Scognamiglio, V., Mamone, G., Rossi, M., Parracino, A., Rossi, M. and D'Auria, S. (2006) Glutamine-binding protein from *Escherichia coli* specifically binds a wheat gliadin peptide. 2. Resonance energy transfer studies suggest a new sensing approach for an easy detection of wheat gliadin. *Journal of Proteome Research*, **5**, 2083–6.

31 Xavier, K.B., Martins, L.O., Peist, R., Kossmann, M., Boos, W. and Santos, H. (1996) High-affinity maltose/trehalose transport system in the hyperthermophilic archaeon *Thermococcus litoralis*. *Journal of Bacteriology*, **178**, 4773–7.

32 Horlacher, R., Xavier, K.B., Santos, H., DiRuggiero, J., Kossmann, M. and Boos, W. (1998) Archaeal binding protein-dependent ABC transporter: molecular and biochemical analysis of the trehalose/maltose transport system of the hyperthermophilic archaeon *Thermococcus litoralis*. *Journal of Bacteriology*, **180**, 680–9.

33 Diez, J., Diederichs, K., Greller, G., Horlacher, R., Boos, W. and Welte, W. (2001) The crystal structure of a liganded trehalose/maltose-binding protein from the hyperthermophilic archaeon *Thermococcus litoralis* at 1.85 Å. *Journal of Molcular Biology*, **305**, 905–15.

34 Herman, P., Staiano, M., Marabotti, A., Varriale, A., Scirè, A., Tanfani, F., Vecer, J., Rossi, M. and D'Auria, S. (2006) D-Trehalose/D-maltose-binding protein from the hyperthermophilic archaeon *Thermococcus litoralis*: the binding of trehalose and maltose results in different protein conformational states. *Proteins*, **63**, 754–67.

35 Breer, H. (1994) Odor recognition and secondary messenger signaling olfactory receptor neurons. *Seminars in Cell Biology*, **5**, 25–32.

36 Tegoni, M., Pelosi, P., Vincent, F., Spinelli, S., Campanacci, V., Grolli, S., Ramoni, R. and Cambillau, C. (2000) Mammalian odorant binding proteins. *Biochim Biophys Acta*, **1482**, 229–40.

37 Pevsner, J. and Snyder, S.H. (1990) Odorant-binding protein: odorant transport function in the vertebrate nasal epithelium. *Chemical Senses*, **15**, 217–22.

38. Flower, D.R., North, A.C.T. and Sansom, C.E. (2000) The lipocalin protein family: structural and sequence overview. *Biochimica et Biophysica Acta*, **1482**, 9–24.
39. Avanzini, F., Bignetti, E., Bordi, C., Carfagna, G., Cavaggioni, A., Ferrari, G., Sorbi, R.T. and Tirindelli, R. (1987) Immunocytochemical localization of pyrazine-binding protein in bovine nasal mucosa. *Cell and Tissue Research*, **247**, 461–4.
40. Pelosi, P. and Tirindelli, R. (1989) Structure/activity studies and characterization of an odorant-binding protein, in *Receptor Events and Transduction in Taste and Olfaction, Chemical Senses*, Vol. 1 (eds J.G. Brand, J.H. Teeter, R.H. Cagan and M.R. Kare), Marcel Dekker, New York, pp. 207–26.
41. Pelosi, P., Baldaccini, N.E. and Pisanelli, A.M. (1982) Identification of a specific olfactory receptor for 2-isoabutyl-3-methoxypyrazine. *Biochemical Journal*, **201**, 245–8.
42. Garibotti, M., Navarrini, A., Pisanelli, A.M. and Pelosi, P. (1997) Three odorant-binding proteins from rabbit nasal mucosa. *Chemical Senses*, **22**, 383–90.
43. Dal Monte, M., Andreini, I., Revoltella, R. and Pelosi, P. (1991) Purification and characterization of two odorant-binding proteins from nasal tissue of rabbit and pig. *Comparative Biochemistry and Physiology. B, Comparative Biochemistry*, **99**, 445–51.
44. Paolini, S., Tanfani, F., Fini, C., Bertoli, E. and Pelosi, P. (1999) Porcine odorant-binding protein: structural stability and ligand affnities measured by Fourier-transform infrared spectroscopy and fluorescence. *Biochimica et Biophysica Acta*, **1431**, 179–88.
45. Ramoni, R., Bellucci, S., Grycznyski, I., Grycznyski, Z., Grolli, S., Staiano, M., De Bellis, G., Micciulla, F., Pastore, R., Tiberia, A., Conti, V., Merli, E., Varriale, A., Rossi, M. and D'Auria, S. (2007) The protein scaffold of the lipocalin odorant-binding protein is suitable for the design of new biosensors for the detection of explosive components. *Journal of Physics – Condensed Matter*, **19**, 395012.
46. Danson, M.J. and Hough, D.W. (1998) Structure, function and stability of enzymes from the Archaea. *Trends in Microbiology*, **6**, 307–14.
47. Vieille, C. and Zeikus, G.J. (2001) Hyperthermophilic enzymes: sources, uses, and molecular mechanisms for thermostability. *Microbiology and Molecular Biology Reviews*, **65**, 1–43.
48. Stepanenko, O.V., Marabotti, A., Kuznetsova, I.M., Turoverov, K.K., Fini, C., Varriale, A., Staiano, M., Rossi, M. and D'Auria, S. (2007) Hydrophobic interactions and ionic networks play an important role in thermal stability and denaturation mechanism of the porcine odorant binding protein. *Proteins*, **71**, 35–44.
49. Staiano, M., D'Auria, S., Varriale, A., Rossi, M., Marabotti, A., Fini, C., Stepanenko, O.V., Kuznetsova, I.M. and Turoverov, K.K. (2007) Stability and dynamics of the porcine odorant-binding protein. *Biochemistry*, **46**, 11120–7.
50. D'Auria, S., Staiano, M., Varriale, A., Scognamiglio, V., Rossi, M., Parracino, A., Campopiano, S., Cennamo, N. and Zeni, L. (2006) The odorant-binding protein from Canis familiaris: purification, characterization and new perspectives in biohazard assessment. *Protein and Peptide Letters*, **13**, 349–52.

13
Methods of Analysis for Imaging and Detecting Ions and Molecules
Sung Bae Kim, Hiroaki Tao, and Yoshio Umezawa

13.1
Fluorescent and Luminescent Proteins

Eukaryotic cells are organized into complex structures enclosed within membranes and the structures are filled with numerous proteins. The molecular actions within the membranes are elaborate and highly specialized in every process. To trace the molecular events allowing advanced knowledge of the life phenomenon, we should access a key protein ruling the events with a high time and space resolution. However, it is nearly impossible to noninvasively determine the dynamics of a key protein that generally exists in a small amount without any illumination in the complex context of eukaryotic cells. Alternatively, researchers label the key protein with a functional protein emitting fluorescence or luminescence. Green fluorescent proteins (GFP) variants and luciferases have been commonly utilized in a broad range of analytical studies as the functional proteins owing to their excellent character to generate a highly visible, efficiently emitting, internal chromophore or active site. In practice, many of the latest bioanalyses directly or indirectly take advantage of the functional proteins. In this section, we briefly review the fluorescent and luminescent proteins facilitating the latest bioanalysis.

13.1.1
GFP and Its Variants

Fluorescent proteins such as GFP are ideal for use in live cell assays because they are extremely bright and require no additional substrates or cofactors for their fluorescence. Owing to their distinctive spectra, they can be readily combined for the simultaneous detection of two or more events in a single cell or cell population.

GFP from the jellyfish *Aequorea victoria* is one of the most widely studied and exploited proteins in biochemistry and cell biology [1]. Since it was discovered by Shimomura *et al.* [2], GFP has become well established as a marker of gene expression and molecular events in intact cells and organisms. The crucial breakthroughs

Table 13.1 Properties of the representative fluorescent proteins.

Mutation[a]	Common name	Excitation maxima (nm) (absorbance extinction coefficient: $10^3 M^{-1} cm^{-1}$)	Emission maxima (nm) (quantum yield)	References
None or Q80R	wild-type GFP	395–397 (25–30); 470–475 (9.5–14)	504–508 (0.79)	[6]
F64L, S65T	enhanced green fluorescent protein (EGFP)	488 (55–57)	507–509 (0.60)	[6]
S65G, S72A, V68L, T203Y	enhanced yellow fluorescent protein (EYFP)	513 (83.4)	527 (0.61)	[7]
F64L, S65T, Y66W, N149I, M153T, V163A, H231L	enhanced cyan fluorescent protein (ECFP)	434 (major peak) and 453 (minor peak) (26)	477 (major peak) and 496–505 (minor peak) (0.4)	[7, 8]
Y66H	blue fluorescent protein (BFP)	384 (21)	448 (0.24)	[9]
F64L, S65T, Y66H, Y145F	enhanced blue fluorescent protein (EBFP)	380–383 (26.3–31)	440–447 (0.17–0.26)	[6, 9]

a) Substitutions from the primary sequence of GFP are given as the single-letter code for the amino acid being replaced.

Some variants of *Aequorea* fluorescent proteins contain additional mutations believed to be neutral (such as K26R, Q80R, N146H, H231L, etc.).

for analytical use were made by Prasher *et al.* with the cloning of the gene [3], and by Chalfie *et al.* and Inouye *et al.* showing that the gene is expressed in other organisms with fluorescence [4, 5].

In the early days, only fluorescent proteins emitting short wavelengths were purified, which practically exist abundantly in nature. The characters of wild-type fluorescent proteins were improved through mutagenesis to exert brighter emission and faster chromophore maturation (Table 13.1). In addition, their genes have been human codon-optimized to enhance their translation in mammalian cells. Now these proteins, ranging in color from cyan to far-red, are practically available as functional proteins. Figure 13.1a shows the palette of monomeric fluorescent proteins. Although many fluorescent protein variants were found to emit highly tissue-permeable red light, their quantum yields are generally poor compared to

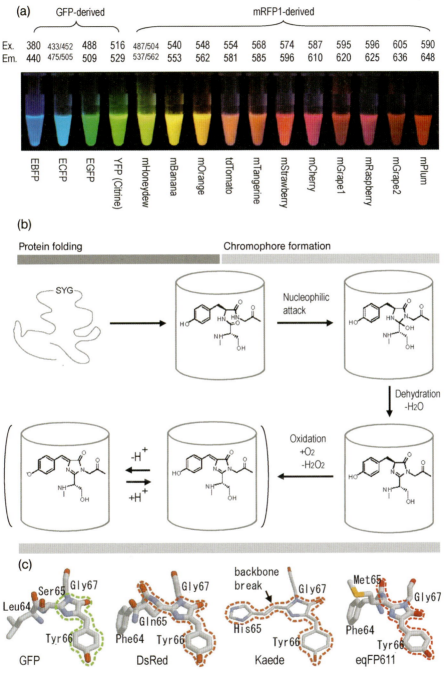

Figure 13.1 (a) The palette of monomeric fluorescent proteins (adapted from [10]). (b) Diagram representing folding and chromophore formation of *Aequorea* GFP (modified from [11]). (c) The structures show extended π-conjugation for visible-light absorption highlighted with dotted lines. (Modified from [12]).

those of fluorescent proteins illuminating green light. Another disadvantage of many red fluorescent proteins (RFPs) is the general requirement of quaternary structure for light emission. According to a recent study, the dimerization of GFP variants properties may be relieved by mutating the variants at A206, L221 and F223 [13].

Up to now, many GFP variants have been made by mutagenesis studies to exert (i) fast folding properties, (ii) enhanced fluorescence intensity, (iii) low sensitivity to pH and Cl^-, and (iv) temperature stability of GFP (Tables 13.1 and 13.2). In addition, genetic engineering of GFP into chimera proteins is opening new vistas in physiological indicators, biosensors and photochemical memories.

After the discovery and cloning of GFP from $Aequorea$, GFP was soon found to produce sufficient fluorescence in different organisms, serving as a fluorescent marker for cell biology studies. The chromophore of $Aequorea$ GFP is a p-hydroxybenzylideneimidazolinone formed from residues 65–67, which are Ser65 (or Thr65)–Tyr66–Gly67 (SYG or TYG) in the native protein [3, 35]. The chromophore is surrounded by 11 strands of β-barrels and an α-helix. After protein synthesis, many fluorescent proteins mature slowly through a multistep process that consists of folding, chromophore formation and chromophore modification. The chromophore is matured through the following self-catalytic steps as illustrated in Figure 13.1b: (i) the conformation changes of Ser65 (or Thr65)–Tyr66–Gly67, (ii) the cyclization and dehydration between Ser65 (or Thr65) and Gly67, and (iii) the oxidation of Tyr66 [1]. In addition to wild-type GFP and its variants, the discovery of novel GFP-like proteins from Anthozoans (coral animals) has significantly extended the range of colors available for cell biological applications. The GFP-like protein emitting red light follows an additional maturation process for changing its emission color from green to red. The extended π-conjugation in the chromophore is thought to be responsible for the red shift of emission light. Ando $et\ al.$ demonstrated a photoconvertible fluorescent protein, Kaede. The photoconversion is interpreted as the break or extension of π-conjugation in the chromophore of Kaede in response to specific light (Figure 13.1b and c).

Despite the fact that the wild-type GFP possesses excellent properties as a reporter, it has some shortcomings that limit its application: (i) formation of weak dimers through homophilic interactions, (ii) slow folding and maturation at 37 °C, (iii) poor extinction efficiency, (iv) pH sensitivity, (v) long half-life of the protein in mammalian cells, and (v) existence of two excitation maxima: a neutral chromophore excitable at 395 nm (major) and an anionic chromophore excitable at 475 nm (minor) [29–37]. To eliminate the negative features, cDNA of wild-type GFP has been genetically modified with point and/or random mutagenesis as shown in Table 13.1.

Some fluorescent proteins have been found from nonbioluminescent Anthozoa species. Among them, an RFP isolated from $Discosoma$ (DsRed) exhibited largely red-shifted excitation and emission peaks – 558 nm (excitation) and 583 nm (emission). The α-carbon and nitrogen of DsRed are dehydrogenated, reasoning the extension of π-conjugation and the subsequent red shift (Figure 13.1c) [38]. DsRed exerts elevated an extinction coefficient, high quantum yield, and resistance to pH

Table 13.2 Spectral properties of representative fluorescent proteins.

Name and origin (acronym)	Excitation maxima (nm) (molar extinction coefficient: $10^3 M^{-1} cm^{-1}$)	Emission maxima (nm) (quantum yield)	In vivo structure and molecular weight	Characteristics	References
Blue fluorescent proteins					
EBFP	380–383 (29)	440–445 (0.31)	monomer*	poor photostability	[14]
Sapphire	399 (29)	511 (0.64)	monomer*	a mutant of wild-type GFP	[9]
Cyan fluorescent proteins					
ECFP	433–439 (33)	475–476 (0.40)	monomer*	fair photostability	[15]
Cerulean	433 (43)	475 (0.62)	monomer*	excellent brightness, poor photostability efficient folding at 37 °C	[16]
Green fluorescent proteins					
GFP (wild-type)	395/475 (21)	509 (0.77)	monomer*	27 kDa, wild-type GFP from jellyfish A. victoria	[17]
S65T-GFP	489	511	monomer*	widely used for live cell imaging	[8]
EGFP	484–488 (56)	507 (0.60)	monomer*	excellent photostability	Clontech
Yellow fluorescent proteins					
EYFP	514 (83)	527 (0.61)	monomer*	employed in FRET together with ECFP	[15]
Venus	515 (92)	528 (0.57)	monomer*	improved YFP variant; less chemically sensitive	[18]

Table 13.2 (Continued)

Name and origin (acronym)	Excitation maxima (nm) (molar extinction coefficient: $10^3\,M^{-1}\,cm^{-1}$)	Emission maxima (nm) (quantum yield)	In vivo structure and molecular weight	Characteristics	References
Orange fluorescent proteins					
mOrange	548 (71)	562 (0.69)	monomer	excellent brightness; poor photostability; rapid maturation	[19]
Red fluorescent proteins					
DsRed (drFP583)	558 (75)	583 (0.79)	tetramer	poor cellular viability in mice	[20]
DsRed-Monomer	556 (35)	586 (0.10)	monomer	very soluble, rapid maturation time	Clontech
mRFP1	584 (44–50)	607 (0.25)	monomer*	first monomeric RFP; good viability in mice; poor photostability	[21, 22]
Red proteins with large Stokes shift					
dKeima-Red	440 (25)	616 (0.31)	dimer	–	[23]
Far-red proteins					
HcRed	592	645	tetramer	poor extinction efficiency and quantum yield	[24]
mPlum	590 (41)	649 (0.10)	monomer	excellent brightness and photostability	[25]

Table 13.2 (*Continued*)

Name and origin (acronym)	Excitation maxima (nm) (molar extinction coefficient: $10^3 \, M^{-1} \, cm^{-1}$)	Emission maxima (nm) (quantum yield)	*In vivo* structure and molecular weight	Characteristics	References
Photoactivatable					
PA-GFP	475, 504 (photo-activated; 17)	517 (0.79)	monomer	switched at 413 nm	[26]
PA-mRFP1	578 (10)	605 (0.08)	monomer	activated by wavelengths between 380 and 400 nm	[27]
Photoconvertible					
PS-CFP	402 (pre; 34)/490 (post; 27)	468 (pre: 0.16)/511 (post: 0.19)	monomer	switched at 405 nm; 1500 times contrast change	[28]
Kaede (green or red)	508 (pre; 99)/572 (post; 60)	518 (pre: 0.88)/580 (post: 0.33)	tetramer	less efficient photoconversion; switching at 405 nm	[29]
Photoswitchable (reversibly switchable)					
KFP1	580	600	tetramer	the A148G mutant of asFP595; green light activates; ultraviolet quenches	[30, 31]
asFP595	572	595	tetramer	activated by green and quenched by blue	[32]

Table 13.2 (Continued)

Name and origin (acronym)	Excitation maxima (nm) (molar extinction coefficient: $10^3 \, M^{-1} \, cm^{-1}$)	Emission maxima (nm) (quantum yield)	In vivo structure and molecular weight	Characteristics	References
Dronpa	503 (95)	518 (0.85)	monomer	intense 488 nm illumination switches emission off and 405 nm illumination switches on	[33]

"m" means a monomer. An asterisk "*" means that it makes a weak dimer. "d" means a dimer.

The present table was summarized from various sources, including pages from the websites of Nikon (http://www.microscopyu.com/articles/livecellimaging/fpimaging.html) and Olympus (http://www.olympusconfocal.com/applications/fpcolorpalette.html) in addition to [34].

variations and photobleaching. In spite of these merits, it tends to make tetramers and requires a long maturation time, typically days, for the protein to shift from the premature greenish to the mature red emission [38].

A DsRed variant named DsRed-Express (or -T1) showed a fast maturation rate [29]. Alternative mutation of DsRed to disrupt the oligomerization tendency and enhance its folding efficiency made a monomeric RFP (mRFP1) [21].

To overcome the relatively poor extinction coefficient, quantum yield and photostability of mRFP1, a number variants were constructed, including mHoneydew, mBanana, mOrange, mTangerine, mStrawberry, mCherry, mRaspberry, mPlum, mGrape and tdTomato (Figure 13.1a; "m" indicates the monomeric character of each fluorescent protein; "td" means a tandem dimer; the nomenclature indicates their emission colors [39]).

A photoconvertible fluorescent protein was first found by Tsien *et al.* during GFP mutation [40]. Since then, fluorescent proteins with changing fluorescence intensity (photoactivation) or color (photoconversion) in response to light of a specific wavelength have been discovered. These proteins are categorized into three groups: (i) irreversible change in intensity upon light stimulation, (ii) irreversible change in color upon light stimulation, and (iii) reversible change in intensity and color upon light stimulation.

The proteins in the first category include photoactivatable (PA)-GFP and PA-mRFP1, exerting an irreversible change of the fluorescence intensity in response to strong illumination of a color. A mutation of wild-type GFP at T203 with histidine (T203H) causes loss of its fluorescence at 488 nm excitation. The preillumination of PA-GFP with strong light at 413 nm enables the protein to emit 100-times the fluorescence by excitation with 488 nm [26].

The proteins in the second category comprise photoswitchable (photoconvertible) CFP (PS-CFP), Kaede, KikGR and Dendra. These fluorescent proteins undergo an irreversible change in fluorescence color. PS-CF as a mutant of GFP emits cyan color (468 nm) before photoconversion. By an intense illumination of blue light at 405 nm, the excitation and emission wavelengths are converted to 490 and 511 nm, respectively, whereas the original cyan color is weakened [28].

The proteins in the third category include asFP595, kindling fluorescent protein (KFP1) and Dronpa. asFP595 derived from the sea anemone *Anemonia sulcata* enhances or suppresses the red fluorescence emission in response to green or blue light, respectively [32]. Dronpa purified from a coral pectiniidae was engineered to become monomeric by mutations [33]. Dronpa can be photobleached at 490 nm to a dim protonated state and deprotonated to regain bright green fluorescence upon illumination at 400 nm.

These fluorescence conversions can be very useful for monitoring time- and space-resolved determinations of molecular dynamics in the complex context of living mammalian cells. For instance, molecular actions of photoconverted proteins may be determined by time at a specific compartment in mammalian cells.

Although we have only briefly reviewed fluorescent proteins, countless numbers of GFPs and RFPs have been discovered from various living organisms. Some of the fluorescent proteins provide excellent brightness and photostability as monomers. The large number of commercial fluorescent proteins may confuse the user. Tsien *et al.* previously suggested a guide to choosing fluorescent proteins. He mentioned five, general requirements for fluorescent proteins: (i) fluorescent proteins should express efficiently and without toxicity in the chosen system; (ii) fluorescent proteins should be photostable and bright; (iii) fluorescent proteins should not be oligomerized; (iv) fluorescent proteins should be insensitive to environmental effects; and (v) in multiple-labeling experiments, cross-talk in their excitation and emission channels could be minimized [41]. In addition to his claims, we consider that (vi) fluorescent proteins themselves should not be a stimulator modifying the signal transductions of the host cells or organisms.

The large population of fluorescent proteins gives researchers a variety of high-performance options. GFP and its variants have been utilized in various fusion protein probes for molecular imaging. The representative applications are summarized in the following groups: (i) monitoring expression, localization and motility of GFP itself or GFP-fused proteins [26, 42]; (ii) studying protein dynamics based on the bleaching of GFP variants [29, 28, 43]; (iii) evaluating protein–protein interactions using fluorescence (or Förster) resonance energy transfer (FRET) [44, 45] and split GFP variants [46]; (iv) tracing the rate of degradation using PA-GFP variants [47]; and (v) determining protein aging with a timer GFP exhibiting time-dependent changes in fluorescence [28, 48]. In the future, monomeric fluorescent proteins with greater brightness and photostability will be discovered from various organisms. In addition, the optical properties of fluorescent proteins will be improved by point and/or random mutations. The number of fluorescent proteins providing long-wavelength spectra will be also expanded, allowing for more sensitive and efficient imaging in thick tissues and animal models. However, how to use fluorescent proteins with excellent optical properties is another topic to be addressed. Although many fluorescent proteins facilitate the development of fusion protein probes, the large variety of fluorescent proteins does not guarantee the success of the applications. Only upon combination with a smart and theoretically robust methodology will the fluorescent proteins enhance the brightness in real research. In later sections, we will review the practical usage of fluorescent proteins.

13.1.2
Luciferases

Engineered functional proteins have contributed to advances in bioindustry and breakthroughs for the treatment of human disease. Manipulation of enzymes as functional proteins is largely supported by knowledge of the principles that govern enzyme catalysis and the enzyme structures solved by X-ray crystallography [49]. Recent revolutionary advances in enzyme manipulation technologies now allow researchers to carry out quantitative examination of molecular dynamics and cell signaling in living cells [50]. While GFP as an excellent functional protein has been widely used, enzymes catalyzing light emission are now proving their distinguished merits for quantitative, specific, signal-enhanced, noninvasive and real-time investigation of intracellular molecular events [51, 52].

Luciferases are a family of photoproteins that can be isolated from a large variety of insects, marine organisms and prokaryotes [53, 54]. A variety of organisms regulate their light production using different luciferases in a variety of light-emitting reactions. The emission spectrum ranges between 400 and 620 nm. Firefly luciferase (FLuc) and *Renilla* luciferase (RLuc) are widely used by researchers to investigate diverse biological events of cells.

Luciferases are enzymes that catalyze the oxidation of a luciferin in the presence of cofactors. Beetle luciferases including FLuc and click beetle luciferase (CBLuc) mediate oxidization of D-luciferin in the presence of ATP, Mg^{2+} and oxygen. On the other hand, many deep-sea organisms emit bioluminescence with luciferases that catalyze oxidation of coelenterazine. In addition, luminous bacteria emit continuous light generally at 490 nm. The bacterial luciferase is a flavin mixed-function monooxygenase. Reduced flavin mononucleotide (FMN) reacts with oxygen in the presence of bacterial luciferase to produce an intermediate peroxy flavin, which then reacts with a long-chain aldehyde to form the acid and the luciferase-bound hydroxy flavin in its excited state. Luminescence in dinoflagellates is emitted by oxidation of dinoflagellate luciferin, which is mediated by the luciferase and luciferin-binding protein [55]. The specific chemical reactions are summarized in Figure 13.2.

Beetle luciferases produce bioluminescence of different colors ranging from green to red using a common luciferin substrate in the presence of ATP and Mg^{2+}. They are categorized into pH-sensitive and pH-insensitive luciferases. pH-sensitive luciferases such as FLuc undergo a red shift with decreasing pH or increasing temperature or concentration of heavy metal ions, whereas pH-insensitive luciferases like click beetle- and railroad worm-derived luciferases are insensitive to those conditions [53, 57].

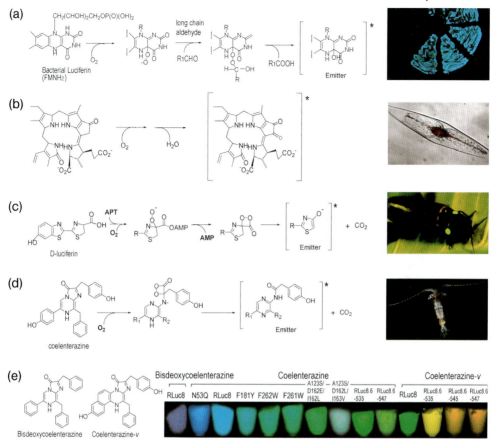

Figure 13.2 Chemical reactions of representative luciferins for light emission. (a) Catalytic oxidation of FMN by bacteria luciferases. The photo shows *Vibrio fischeri* [141]. (b) The chemical reaction occurs when the dinoflagellate (a plankton) luciferin is oxidized by the enzyme luciferase in the presence of ATP and O_2 The red glow in the picture is chlorophyll fluorescence emitted by *Pyrocystis fusiformis* (www.lifesci.ucsb.edu/~biolum). (c) Oxidation reaction of D-luciferin by beetle luciferases to emit luminescence (www.lifesci.ucsb.edu/~biolum). (d) Oxidation of coelenterazine by marine luciferases. The photo shows the copepod *Gaussia* (www.nanolight.com). (e) Palette of bioluminescence emitted by RLuc and coelenterazine variants. (From [56]).

CBLuc emits a stable light in various physiological circumstances inside cells or tissues, due to the insensitivity to pH, temperature and heavy metals. In addition, an engineered CBLuc for mammalian expression emits red light (λ_{max} = 615 nm), which is a highly red-shifted wavelength compared to those from other insect and marine organism luciferases. The red light, close to near-infrared, is highly tissue transparent, thus especially attractive for the molecular imaging of a signal transduction of interest in living animals [50, 57]. A railroad worm is a larva or larviform female adult of a beetle of the genus Phrixothrix in the family of Phengodidae. The luminescent organs of their body can glow green, while those on their head can glow red; this is due to two different kinds of luciferin and luciferase in their bodies. The luciferases of the Phrixothrix railroad worms are the only luciferases that naturally catalyze bioluminescence in the red region of the spectrum [58].

Marine luciferases (e.g., those derived from Gaussia princeps and Metridia longa) catalyze the oxidation of the small molecule coelenterazine to produce light. Unlike the beetle luciferase systems, these coelenterazine-utilizing luciferases do not require accessory high-energy molecules such as ATP and Mg^{2+} for their signal, simplifying their use in a number of new reporter applications.

Several features of Gaussia luciferase (GLuc) make it very attractive as a novel reporter system for studying gene expression. Humanized GLuc emits around 1000-fold brighter than native RLuc or FLuc, and is stable at elevated temperatures. In addition, it is one of the smallest luciferases discovered. The detailed properties of known luciferases are summarized in Table 13.3.

Much of the bioluminescence in the sea comes from single-celled algae such as the tropical dinoflagellate Pyrocystis fusiformis. Ninety percent of all dinoflagellates are marine plankton. They can emit short flashes of light (bioluminescence) when disturbed. The chemical reaction itself occurs when the compound luciferin (a substrate chemically similar to a chlorophyll precursor) is oxidized by the enzyme luciferase in the presence of ATP and oxygen [66].

The emission of bioluminescence occurs in many phylogenetically different groups. However, the evolutionary origins of bioluminescence still remain obscure. One of the most striking characteristics of bioluminescence is the very high diversity of mechanisms, structures and functions that bioluminescent organisms have achieved. This high diversity suggests multiple independent origins of bioluminescence over the course of evolution [67].

Some researchers have elucidated the functions and origin of luminous reactions in organisms. Hastings et al. suggested the functions of luminescence be divided in three different classes: defensive (to help deter predators), offensive to aid in predation and communication (e.g., for courtship or mating [55]). Rees et al. previously proposed that the luminescent substrates of the luminous reactions (luciferins) were the evolutionary core of most luminescent systems [68]. McElroy et al. also suggested that luciferases evolved to detoxify molecular oxygen in early anaerobic life forms at the time when photosynthetic processes had begun releasing oxygen into the primitive atmosphere [69]. In practice, phylogenetic classification of bioluminescent organisms according to their luciferin extremely simplifies the categories. The most accepted reactions of the luciferases are shown in Figure 13.2. It is also interesting that all of the known bioluminescence reactions consume molecular oxygen.

Luciferases are nearly ideal reporters for bioanalysis and molecular imaging. (i) The assays are potentially very simple (just mixing the cell extract with an assay buffer and placing the mixture in a luminometer). (ii) The assay time is usually less than 1 min. (iii) Luciferase assays are generally very sensitive. Luciferase assays are 10- to 1000-fold more sensitive than the standard chloramphenicol

Table 13.3 Optical properties of the well-known luciferases.

Luciferase	Origin	Substrate	Molecular weight (kDa)	λ_{max} (nm)[a]	Reference
firefly luciferase (FLuc)	*Photinus pyralis*	D-luciferin	61	562	[59]
Renilla luciferase (RLuc)	*Renilla reniformis*	coelenterazine	36	480	[60]
Gaussia luciferase (GLuc)	*Gaussia princeps*	coelenterazine	20	480	[61]
click beetle luciferase (CBLuc)	*Pyrophorus plagiophthalamus*	D-luciferin	65	540 (green) or 613 (red)	[62]
railroad worm luciferase (RWLuc)	*Phrixothrix vivianiand Phrixothrix hirtus*	D-luciferin	61 (green) and 60 (Red)	546 (green) or 623 (red)	[63]
Metridia luciferase (MLuc)	*Metridia longa*	coelenterazine	24	480	[58]
Vargula luciferase (VLuc)	*Vargula hilgendorfii*	coelenterazine	62	460	[64]
Oplophorus luciferase (OLuc)	*Oplophorus gracilorostris*	coelenterazine	complex of 35 and 19	454	[65]

a) Maximum wavelength of bioluminescence spectrum.

transacetylase (CAT) assay [70]. (iv) The luciferase assays can be broadly applicable to various organisms, from bacteria to living subjects. (v) The reagents used for the luciferase assay are not hazardous. By contrast, CAT and *neo* assays utilize radioisotopes and/or hazardous solvents. (vi) The luciferase assay is economical – it is generally 10–50 times cheaper than other conventional assays such as the CAT assay. (vii) Luciferases are generally monomeric and do not require dimerization or tetramerization for red light emission, different from many fluorescent proteins. (viii) Luciferases generally consist of two main domains and the domains are linked with a hydrophilic, flexible linker. This conformational character makes it easy to split luciferases into two portions with resulting functionally inactive fragments. These split luciferases are very useful for constructing new assay systems (e.g., measuring protein–protein interactions in living cells or animals). (ix) Some marine luciferases contain a secretion signal at the N-terminal end. It enables us to measure gene expression extracellularly without destroying the cells or tissues. This is particularly useful for time-course studies of transcription activity or receptor/channel response kinetics.

However, luciferases have some intrinsic drawbacks when applied in bioanalytical systems. (i) It is not easy to keep an intracellular concentration of luciferin at a constant level in the case of application of FLuc and RLuc. In the case of beetle luciferases, a consideration may be required to facilitate the passage of the substrate across the cell membrane. (ii) The intensity and stability of the bioluminescent response of beetle luciferases in living cells can be affected by concentrations of ATP, luciferin and luciferin–luciferase complex [71, 72]. Considering that catalyzation reactions of luciferins by luciferases are basically enzymatic reactions, any factors influencing enzyme reactions may affect the luminescence intensities by luciferases. Researchers should note the limitations upon the construction of new bioluminescent assay systems.

As described above, there is no doubt that luciferases are excellent reporter proteins shooing great potential to be utilized in various assay systems. Readers may design their own new assay systems utilizing the merits of luciferases. The recent bioanalytical advances with luciferases are summarized in Sections 13.3.3 to 13.3.6.

13.2
Functional Peptides

Upon designing a new fusion protein probe, an inevitable ingredient is a small functional peptide. Through incorporating appropriate functional peptides into the protein probes, researchers add some specific properties to the probes. Selecting appropriate functional peptides is directly connected to the success of the construction of fusion protein probes. In practice, many kinds of small functional peptides with diverse properties generate a great number of choices. Researchers can consider two strategies for the successful choice of the functional peptides: (i) to mimic the success stories of fusion protein probes that were previously well examined and (ii) cumulative experience. Trial-and-error may guide the users finally to the successful use of a functional peptide.

The desired properties of functional peptides can be summarized in eight categories. (i) With a localization signal, the probe can be genetically targeted to distinct locations within cells, such as organelles and membranes (e.g., a nuclear localization signal). (ii) With a flexible linker such as a GS linker, we can create an artificial hinge region between the domains in a host probe backbone. (iii) A secretion signal enables us to measure gene expression extracellularly without destroying the cells or tissues; conversely, some peptides can provoke endocytosis of the host protein. (iv) Addition of a small epitope sequence (e.g., a FLAG epitope) to fusion proteins enables us to easily recognize the expression of the proteins with the specific antibody. (v) With an affinity tag (e.g., a GST or His tag), we can easily purify the probes through an affinity column. (vi) With a small α-helical coactivator peptide such as a LXXLL motif, we can examine a protein–peptide binding that may be important for cellular signaling. (vii) A functional peptide substrate sequence recognized by endogenous enzymes enables us to trigger proteolysis; it may include DEVD and PEST peptides, which are, respectively, recognized by caspase-3 and calpain proteases. (viii) Some peptides can be phosphorylated by specific kinases. Thus, the phosphorylation produces an index of the kinase activity. A tyrosine phosphorylation domain within insulin receptor substrate-1 (IRS-1) can be exemplified as the substrate peptide which is recognized by the insulin receptor.

13.3
Representative Technologies for Molecular Imaging

An intrinsic merit of the aforementioned functional proteins is to generate fluorescence or luminescence as an optical signature, and produce a stable fusion protein with other lineages of proteins. Such distinctive advantages of functional proteins have been utilized for constructing fusion protein probes for determining the localization, kinetics and molecular dynamics of a target protein of interest.

Current technologies for molecular imaging of a target substance in cell lines have largely depended on synthetic organic indicators and fusion protein probes. The advantages of synthetic organic indicators were well proved for illuminating heavy metal ions, whereas those of fusion protein probes are for determining bioactive small molecules, or sensing protein interactions or conformation changes of a protein. The following sections review recent advances in molecular imaging probes. See Figure 13.3.

13.3.1
Classical Methods for Sensing Bioactive Small Molecules

Recognition of ions or molecules is now a topic of pivotal interests in organic and inorganic chemistry in addition to biological chemistry. One of the powerful methodologies for molecular recognition is to develop synthetic chemical sensors that emit an optical signature like fluorescence in response to target substances.

The first, and probably most successful, chemical sensor may be litmus paper for recognizing H^+ ions (first used by a Spanish alchemist around 1300 AD). Currently, most successful modern chemical sensors have been synthesized to determine metal ions. Metal ions are required for the proper function of all cells within every living organism, and disruptions in localizations and concentrations of metal ion pools are a major contributor to aging and disease [73, 74]. From this point of view, both synthetic chemistry and molecular biology continue to

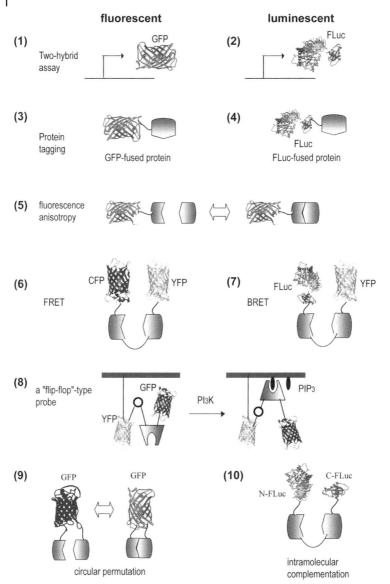

Figure 13.3 (1 and 2) Yeast or mammalian two-hybrid assay. Two-hybrid systems are well established as tools for determining intracellular protein–protein interactions by means of transcriptional activation of a reporter. In this system, a transcription factor is split into two domains: a DNA-binding and a transcription activation domain. The domains are, respectively, linked with proteins of interest. Close physical approximation between the proteins of interest triggers reassembly of the transcriptional machinery and the subsequent expression of a marker protein, typically emitting fluorescence or luminescence. (3 and 4) Protein tagging with a functional protein. A target protein can be tagged with a functional protein emitting fluorescence or luminescence. Molecular dynamics can be traced with the tagged target protein. (5) Fluorescence anisotropy. Fluorescence anisotropies differ according to the molecular weight of fluorescent proteins,

13.3 Representative Technologies for Molecular Imaging | 315

fluorescent

(11) N-GFP C-GFP

luminescent

(12) N-FLuc C-FLuc

(13) N-GFP C-GFP
N-intein C-intein
Protein splicing

(14) N-FLuc C-FLuc
N-intein C-intein
Protein splicing

(15) N-GFP
conformation change
C-GFP

(16) N-FLuc
conformation change
C-FLuc
intramolecular folding sensor

(17) CFP
DEVD
YFP
cleavage by caspatase
protease-induced acceptor dequenching

(18) C-FLuc
DEVD
N-FLuc
cleavage by caspatase

before and after protein interactions. This difference can be imaged differently in cell lines. The main strength of this approach is the ease of measuring fluorescence polarization parallel and perpendicular to the excitation with a high signal:noise ratio. As the data can be acquired rapidly and minimal image processing is needed, this approach is well suited for applications in high-content screening. (6) FRET. (7) BRET. (8) A "flip-flop"-type FRET. (9) CP. (10) An integrated molecule-format (IMF) bioluminescent probe assay. (11) Determining protein–protein interactions with split GFP. (12) Determining protein–protein interactions with split FLuc. (13) Intein-mediated split fluorescent PSA. (14) Intein-mediated split luciferase splicing assay. (15) Intramolecular folding sensor (fluorescent protein). (16) Intramolecular folding sensor (luciferase). (17) Fluorescent cleavage probe. (18) Cyclic luciferase assay. See relevant sections for further details.

promote synergistic advancements in both imaging probe design and instrumentation [74, 75]. While synthetic chemical probes are important in sensing metal ions, fusion protein probes are emerging as important in molecular recognition science.

The distinctive advantages of fusion protein probes are summarized as follows. (i) Fusion protein probes are expressed by host cells and therefore they do not require external addition or microinjection. Thus, they do not leak out of cells in long-term experiments. (ii) A stably transfected cell line uniformly expresses the probes and this reliably standardizes an experiment. (iii) Fusion protein probes can be directed to any particular organelle of living cells by employing a small localization signal. (iv) The scale of experiments can be expanded to transgenic organisms by introducing the cDNA into the organisms.

In the early days, the protein of interest was simply fused to GFP. The fusion proteins look extremely simple in terms of structure, but are an attractive tool for determining the molecular dynamics of target proteins of interest. Nuclear receptors (NRs) are one of the most abundant classes of transcriptional regulators in living subjects. The initial study on the dynamics of NRs was conducted with a GFP-fused NR [42]. The GFP-fused androgen receptor (AR) traffics into the nucleus in response to agonist. The ratio of receptors in cytoplasmic and nuclear compartments was related to both the affinity and concentration of ligand. Similarly, glucocorticoid receptor and estrogen receptor were tagged with GFP for tracing their cellular dynamics [76, 77].

Another simple method is to measure fluorescence polarization [78], which offers particular advantages for high-contrast discrimination of protein–protein interactions [79–81]. First of all, a protein is genetically fused to a fluorescent protein. The fusion protein may interact with other proteins in response to a signal. Before and after protein–protein interactions, fluorescence anisotropies differ according to the molecular weight of fluorescent proteins. This difference can be imaged with color variance in cell lines.

The advantages of these methods are summarized as follows. (i) These fusion protein probes eliminate complex assay procedures that are normally required when using immunocytochemistry techniques (e.g., the use of fixation, cell permeabilization and antibody incubation). (ii) The use of functional proteins permits kinetic studies of protein localization and trafficking due to the intracellular stability of the functional proteins. These analyses are effective for imaging the spatial and temporal dynamics of proteins of interest within single living cells. Automated fluorescence microscopy is effective for high-throughput analysis of the protein movement inside the cells [82].

Although the strategies described above provide the simplest and straightforward methodology for determining molecular dynamics, some limitations still exist. (i) The functional proteins such as GFP should not influence the nature and function of the host proteins. However, functional proteins are as large as 27 kDa (GFP variants) and 61 kDa (beetle luciferases), and thus inevitably are a steric burden for the host protein. (ii) Proteins smaller than 60 kDa can traffic freely between the nucleus and the cytosol. Localization of proteins is an important determinant for the protein functions. As mislocalization of various nuclear proteins is found in cancer cells, there is intense interest in identifying small molecules that direct the proteins to the correct compartments. Tagging the large functional proteins can destroy their native functions and/or provide a distorted readout. In spite of the disadvantages, protein tagging provides the simplest strategy for tracing molecular dynamics of a target protein. Tagging an extremely small functional protein may largely relieve the limitations.

As a more complex assay system, the two-hybrid assay was previously introduced [142]. In the assay, a transcription factor is dissected into two fragments, each of which is, respectively, linked with a protein of interest. Stimulation-based interactions between the fusion proteins triggers reassembly of the transcription factor and the subsequent transcription of a reporter protein. The mammalian two-hybrid assay is conventionally limited in terms of (i) monitoring temporal dynamics of a target protein because it takes hours to days for the reporters to accumulate sufficiently to be determined with the detectors. In addition, it requires (ii) nuclear translocation of hybrid fusion proteins or their localization within the nucleus, because their proximity to DNA and the transcriptional machinery is needed for expression of marker proteins. (iii) Typical two-hybrid assays express fluorescent proteins or luciferases as a reporter. Thus, the results of the assays are affected by the intrinsic limitations of those reporters. For instance, the fluorescent reporter proteins generally require from hours to days for intramolecular processing for the fluorophore maturation [83, 84]. In contrast to fluorescent proteins, luciferases are active immediately after translation. However, the relative stability of luciferase ($t_{1/2}$ = 3–6 h) limits its applicability to kinetics and real-time imaging. Activities of both fluorescent and luminescent proteins can be influenced by (i) levels of expression, (ii) protein stability, (iii) protein degradation and (iv) local environment in each intracellular compartment.

To address such drawbacks of the above methods, many new strategies have been developed. In the following sections, we review molecular imaging schemes of fusion protein probes.

13.3.2
Fluorescence (or Förster) Resonance Energy Transfer

Molecular imaging with fluorescence (or Förster) resonance energy transfer (FRET) may be one of the most powerful techniques for observing the spatial and temporal dynamics of molecular processes in living cells. Advances in molecular biology and the use of the versatile GFP have led to new and more powerful applications of FRET.

The principle of FRET is based on the ability of a higher energy donor fluorophore to transfer energy directly to a lower energy acceptor molecule [7].

The efficiency of energy transfer (E) is highly dependent on the distance (r) between the donor and acceptor chromophores, as described by the Förster equation:

$$E = R_o^6/(R_o^6 + r^6)$$

where the Förster radius R_o is the distance at which the efficiency of energy transfer is 50% of maximum, and can be calculated from the spectral properties and relative orientation of each fluorescent molecule [85].

For constructing an efficient FRET system, the emission spectrum of the donor and the excitation spectrum of the acceptor should overlap sufficiently to enable energy transfer. At the same time, they should be separated enough to exert minimal basal fluorescence. The efficiency of a RET reaction depends on five factors: (i) spectral overlap between the emission of donor and the absorbance of acceptor, (ii) relative orientation between the emission and absorption dipoles of the respective donor and acceptor pair, (iii) spatial distance between the donor and acceptor, (iv) quantum yield of the donor, and (v) extinction coefficient of the acceptor [83, 86].

FRET provides a very sensitive measure of small changes in intermolecular distances. In most cases, no FRET can be observed at distances greater than 100 Å, so the presence of FRET is a good indicator of close proximity, implying biologically meaningful protein–protein interactions. The typical measurement of RET is to calculation of the ratio of light emitted by the donor to that emitted by the acceptor and a change in this ratio signifies changes in the distance between the two proteins of interest.

There are several potential FRET pairs among the currently available fluorescent proteins. The first pair developed was blue fluorescent protein (BFP) coupled with GFP [8]. However, the poor photophysical properties of BFP made this pair impractical. The first effective pair comprised cyan fluorescent protein (CFP) as the donor and yellow fluorescent protein (YFP) as the acceptor [87]. Among two pairs of GFP mutants, this CFP/YFP pair is generally superior because of the greater extinction coefficient, quantum yield, photostability of CFP compared with BFP and greater Förster radius R_o. As there are numerous methods for imaging and measuring FRET-based probes, useful comparisons of FRET partners and methods have been carried out by many groups. These quantitative comparisons have shown that the most efficient pair is mCerulean to mVenus, although substitution of the optimized CFP SCFP3A for mCerulean, or optimized YFPs SYFP2 or YPET for mVenus, produces statistically similar results [85]. The potential pairs of donor and acceptor were well reviewed by Souslova et al. [88].

The classical, but most important ionic species to be determined would be calcium. A prominent strategy for illuminating the endogenous calcium level is to use a genetically encoded indicator based on FRET [87]. The energy transfer varies according to subtle alterations in interactions between the indicator components when bound to the analytes. This method was originally introduced by Tsien et al. by the development of "cameleons". The energy transfer in the indicator is enhanced between GFP variants upon intramolecular binding between calmodulin (CaM) and its binding peptide M13 from myosin light chain kinase by elevation of the Ca^{2+} level [87].

This pioneering study has been progressively modified in several ways, including (i) a decrease in pH sensitivity of the photon acceptor in the yellow chameleons [89], (ii) a shift to longer wavelength donor and acceptor pairs in the red chameleons [90], and (iii) improvement of the restricted dynamic range of chameleons through redesign or optimization of the CaM/M13 interaction [91, 92].

Since the initial development of FRET technology, the use of FRET in cell biological experiments has exploded over the past decade. Since we cannot review all of the papers, we give examples of several outstanding areas of research to address the advantages of FRET measurements.

13.3.2.1 Probes for Determining Protein Phosphorylation

Protein phosphorylation plays one of the most pivotal roles in intracellular signaling pathways. Hundreds of different protein kinases that catalyze protein phosphorylation rule complex networks of signaling pathways in the cells and control diverse cellular functions from survival to apoptosis. Protein phosphorylation is also involved in a variety of pathophysiologic states, including cancer, inflammatory disorders and cardiac diseases.

FRET-based fluorescent indicators have been developed for recognizing protein phosphorylation. A substrate sequence of a protein kinase of interest is first fused with a corresponding phosphorylation recognition domain via a flexible linker sequence consisting of glycine (G). The tandem fusion unit is further sandwiched with a pair of GFP variants like CFP and YFP, each of which serves as the donor and acceptor fluorophores for FRET. Phosphorylation of the substrate sequence prompts the subsequent binding with the adjacent phosphorylation recognition domain (Figure 13.4a) [93].

13.3.2.2 Probes for Determining Steroid-Activated Protein–Protein Interactions

NRs are one of the most abundant classes of transcriptional regulators in living subjects. They are known to regulate diverse functions, such as homeostasis, reproduction, development and metabolism. NRs function as ligand-activated transcription factors, and thus provide a direct link between signaling molecules that control these processes and transcriptional responses.

FRET can be utilized for constructing a fluorescent indicator to visualize the estrogen-dependent interaction between the estrogen receptor-α ligand-binding domain and an LXXLL motif (^{687}HKILHRLLQEG^{697}) of steroid receptor coactivator-1 [98]. This fusion protein is sandwiched between two different colored fluorescent proteins, CFP and YFP, in such a way that the excitation and emission spectra for these GFP variants are suitable for FRET in single living cells (Figure 13.4b).

13.3.2.3 A "Flip-Flop"-Type Indicator

As a new measure for protein phosphorylation, a FRET-based fluorescent indicator for tracing the dynamics of lipid second messengers was developed (Figure 13.4c) [95]. Phosphatidylinositol 3,4,5-trisphosphate (PIP$_3$) is a lipid second messenger, which regulates diverse cellular functions, including cell proliferation and apoptosis, and is also related to diabetes, cancer and other disease [100]. The indicator,

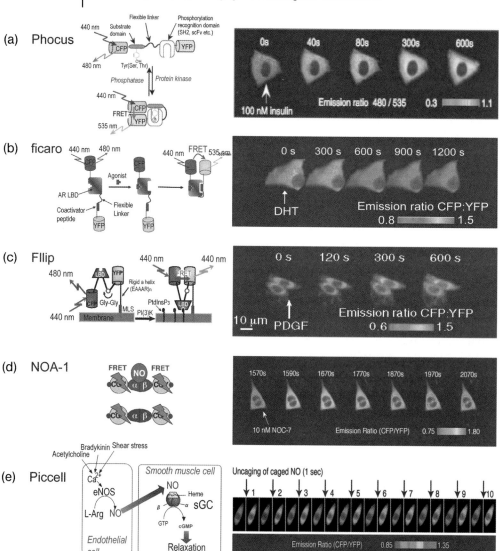

Figure 13.4 Fluorescent indicators for illuminating protein phosphorylation. (a) A fluorescent indicator for protein *phos*phorylation that can be *cus*tom-made (Phocus) [93]. (b) A *f*luorescent *i*ndicator to probe ligand-induced *co*nformational changes in the *a*ndrogen *r*eceptor (ficaro) [94]. (c) A *f*luorescent indicator for a *lip*id second messenger (Fllip) [95]. (d) A fluorescent indicator for *NO* with a signal *a*mplifier (NOA-1) [96]. (e) A fluorescent clone permanently expressing NOA-1 (Piccell) [97]. (All photographs were provided by the authors). Abbreviations: AR LBD, androgen receptor ligand binding domain; LBD, ligand binding domain; DHT, 5α-dihydrotestosterone; PDGF, platelet-derived growth factor; NO, nitric oxide; sGC, soluble gyanylyl cyclase; eNOS, endothelial nitric oxide synthase.

named "Fllip", comprises two GFP variants and a pleckstrin homology domain of "general receptor for phosphoinositides-1" as a PIP_3-binding domain. This flip-flop-type conformational change of the reporter protein is expected to result in intramolecular FRET from CFP to YFP, which makes it possible to detect the dynamics of PIP_3 at the cellular membrane. Some other types of FRET probes are exemplified in Figure 13.4.

In contrast to the great success of FRET-based assays, many practical issues still complicate FRET measurements and can lead to misleading or even meaningless results [101]. The practical concerns of FRET may be summarized as follows. (i) The biased brightness (quantum yield and extinction coefficient) between the donor and acceptor proteins may hamper the efficient measurement. For an efficient RET, it is best to use a donor and acceptor that are of comparable brightness. (ii) For efficient FRET measurements, the stoichiometry between donor and acceptor should be balanced. The FRET system for determining intermolecular protein interactions can be problematic when the donor and acceptor are not present in an exact 1:1 ratio. Even if the ratio of protein expression is 1:1, each fusion protein may be differently localized in the cellular compartments. (iii) Spectral cross-over between the donor and acceptor can disturb correct measurements. For instance, the acceptor can be excited directly with light that is chosen to excite the donor. In addition, fluorescence from the donor can similarly leak into the detection channel for the acceptor fluorescence. (iv) It is possible that the signaling fluorescence does not reflect protein–protein interactions. Although two labeled proteins may interact, the fluorescent labels are sterically hindered and there might not be a FRET signal. (v) The intrinsic drawbacks of fluorescent proteins can be problematic. For instance, fluorescent proteins generally have broad excitation and emission spectra causing significant cross-talk. The large size of fluorescent proteins can disturb the intended interactions between fused proteins (or peptides). In addition, many fluorescent proteins make weak dimers even in a basal condition, which inevitably cause a false-positive fluorescence. Poor photostability of the donor and/or acceptor may hamper efficient FRET measurements. GFP variants need from hours to days for maturation. This late maturation time can limit their broad usage in time-resolved studies. (vi) The last considerable limitation of FRET strategies is the requirement of a sophisticated light-filtering system and mathematical algorithms. (vii) Fluorescence microscope is limited in the number of detectable cells.

13.3.3
Bioluminescence Resonance Energy Transfer

Bioluminescence resonance energy transfer (BRET) is a technique for determining protein–protein interactions on the basis of the nonradiative transfer of energy between a bioluminescent donor protein, such as FLuc or RLuc, and a fluorescent acceptor protein [102–104]. BRET shares a similar working mechanism with FRET except that the donor molecule is a luciferase, not a fluorescent protein. BRET offers many distinct advantages over FRET owing to its extremely high quantum yield and high sensitivity for determining protein–protein interactions.

BRET is a useful technique to access protein–protein interactions because the BRET-permissive distance of less than 10 nm is very similar to the dimensions of biological macromolecular protein complexes. BRET was pioneered by Xu et al. in 1999, being used to investigate the dimerization of proteins in bacterial culture [105]. This first-generation BRET assay, called "BRET^1", consisted of RLuc and enhanced YFP (EYFP), and yielded a spectral separation of approximately 50 nm. A new generation BRET, called "BRET^2", was developed to increase the spectral separation between RLuc and GFP up to about 100 nm by utilizing the proprietary substrate coelenterazine DeepBlueC with an emission maximum between 390 and 400 nm with a λ_{max} of 395 nm [106].

In spite of the distinctive merits for determining protein–protein interactions, BRET still has some intrinsic drawbacks to be overcome. (i) BRET consumes a specific substrate for imaging protein–protein interactions. Owing to the substrate shortage, assays should be carried out in a relatively short time (typically, minutes). (ii) BRET is generally carried out using coelenterazine h or DeepBlueC as the RLuc substrate. However, these are unstable in aqueous solutions, particularly at 37 °C. (iii) Enzymes such as RLuc can catalyze the substrate independently in media containing serum (autoluminescence). (iv) Although an advanced BRET pairs RLuc (475 nm) and EYFP (530 nm), the emission of these BRET combinations is still not red-shifted sufficiently for optimal utility *in vivo* [105]. (v) The enzyme reaction emitting luminescence may change the pH and accumulate byproducts such as oxycoelenterazine monoanions. The byproducts can absorb light, causing modification of the emission spectra of donor and/or acceptor in the BRET system.

To partly address the drawbacks, Pfleger et al. developed a new BRET system, termed extended BRET (eBRET). eBRET utilizes a protected form of coelenterazine h (EnduRen), which is metabolized to coelenterazine h by endogenous esterases within cells [107]. Although the use of EnduRen prolongs the monitoring time and provides a high signal:background ratio, the absolute light intensity from eBRET is very weak compared to conventional BRET assays, which hampers the sensitive detection of protein–protein interactions.

Hoshino et al. recently presented a BRET-based probe with an excellent efficiency in energy transfer, named BRET-based Autoilluminated Fluorescent protein on EYFP (BAF-Y), emitting largely enhanced fluorescence [108]. Their successful probe was achieved with a RLuc variant (RLuc8) fused with EYFP at the N-terminal end. The probe not only induces a red shift of the emission peak, but largely enhances the integrated luminescence intensity of luciferase.

Future studies on BRET should be directed (i) to find an optimal structure of probes allowing efficient RET, (ii) to find an efficient pair of luciferase and red-shifted fluorophore that enables us to monitor molecular interactions *in vivo*, and (iii) to find a luciferase minimally influenced by pH and cations. Considering the rapid pace of progress of this line of studies, these goals should be accomplished in the near future. Further, finding an efficient luciferase and fluorescent protein pair will expand the utility of BRET in many directions.

13.3.4
Protein-Fragment Complementation Assay

Intracellular protein–protein interactions are a central event for the generation of biological regulatory specificity. The majority of bioanalytical methods have been developed for the investigation of protein–protein interactions. The protein-fragment complementation assay (PCA) is one of the major strategies for determining protein–protein interactions in cell lines.

The basic mechanism is as follows. Monomeric photoproteins such as GFP and luciferase can be split into two portions with resulting temporally inactive fragments. Any proteins of interest can be fused genetically to the split N- and C-terminal fragments. The interaction between the fused proteins triggers an approximation of the adjacent photoprotein fragments and subsequent recovery of the function. This combination of two fragments of a photoprotein to restore activity has been termed protein-fragment complementation. The known optimal dissection sites are summarized in Table 13.4.

Protein-fragment complementation between protein fragments was originally observed in 1958 by Richards using subtilisin-cleaved bovine pancreatic ribonuclease [121]. The first conditional association of protein fragments that requires fusion of the fragments to specific interaction partners was demonstrated in 1994 with fragments of ubiquitin [122]. Complementation between fragments of a fluorescent protein was first detected in *Escherichia coli* by Ghosh et al. in 2000 [123]. Fragments of YFP were also shown to produce fluorescent complexes in mammalian cells when fused to CaM and M13 [124]. Gambhir et al. first reported an approach for optical imaging of protein–protein interactions in living subjects by using a split bioluminescent reporter [51].

PCAs provide distinctive merits in determining protein–protein interactions compared to other methods. (i) In principle, PCA can be more sensitive than other assays because complementation of functional proteins creates new functions. (ii) PCA is technically straightforward, needs only simple instrumentation and requires little data processing for interpretation. (iii) The rapid association and moderate dissociation time enables us to detect reversible associations and kinetics between model proteins within minutes. (iv) PCA can provide general applicability in protein interactions with considerable spatial and temporal resolution. Any protein interactions can be potentially examined with PCA. (v) PCA can be intrinsically thought of as a high-throughput measurement, because protein complementation itself is completed in a few minutes.

On the other hand, PCA has some drawbacks that may be encountered in real use. (i) PCA can suffer from a time lag upon fragment reconstitution. It potentially hampers real-time determination of molecular events. (ii) Many complementation methods have been constructed on the basis of split luciferases, which require addition of exogenous substrates and/or cofactors for detection of complementation. (iii) Complementation methods using fluorescent proteins require time for fluorophore maturation. In addition, the fluorescent complex formation is irreversible

Table 13.4 Optimal dissection sites of luciferases for molecular imaging.

Luciferase	Optimal dissection sites (amino acids)	Interaction protein pair	Method	Reference
firefly luciferase	1–415/416–550	AR LBD and FQNLF motif	intramolecular complementation	[57]
	1–437/438–454	IRS-1 and SH2 domain of phosphatidylinositol 3-kinase	intein-mediated protein splicing	[109]
	2–416/398–550	FRB and FKBP12	intermolecular complementation	[110]
	1–437/438–550	FRB and FKBP12	intermolecular complementation	[111]
	1–445/446–550			
	1–475/245–550			
	1–475/265–550			
	1–475/300–550			
	1–416/417–550	AR and Src	intermolecular complementation	[112]
click beetle luciferase	1–439/440–542	AR LBD and LXXLL motif	intramolecular complementation	[113]
	1–439/443–542			
	1–439/437–542			

in vitro. These characteristics limit the applicability of the assays to studies of the kinetics of molecular events and average efficiencies of complex formation over relatively long times (hours). (iv) PCA, like other methods, requires genetic modification for generating protein-fused fragments. This tagging can modify the intrinsic properties and activities of the host proteins. (v) Most of the PCAs are based on an intermolecular complementation between two independent fragments. These types of PCAs are validated on the premise that the two component proteins of the analyzed system should be equally expressed beforehand. Biased expression of the two component proteins inevitably results in an inefficiency of the probing systems.

The general strategy of protein complementation has great potential for determining various intracellular molecular events. The basic concept of split proteins

Table 13.4 (*Continued*)

Luciferase	Optimal dissection sites (amino acids)	Interaction protein pair	Method	Reference
Renilla luciferase	1–229/230–311	full-length AR	intein-mediated protein splicing	[114]
	1–91/92–311	dimerization between ERK2 and ERK2	intramolecular complementation	[115]
	1–229/230–311	MyoD and Id	intermolecular complementation	[116]
Gaussia luciferase	18–109/110–185	FRB and FKBP12	intermolecular complementation	[117]
	18–105/106–185	CaM and M13; AR LBD and LXXLL motif; ER LBD and Src SH2 domain	intramolecular complementation	[143]
β-lactamase	26–196/198–290	FKBP and FRB	intermolecular complementation	[118]
β-galacto-sidase	α-subunit and ω-subunit	EGF receptors	intermolecular complementation	[119]
ubiquitin	1–34/35–76	TOM1 and TOM2	intermolecular complementation	[120]

The slash means the dissected position.

EGF, epidermal growth factor; IRS-1, phosphorylated insulin receptor substrate-1; FKBP12, mammalian target of rapamycin and FK506-binding protein 12; FRB, rapamycin-binding domain; ER LBD, ligand-binding domain of estrogen receptor; AR LBD, ligand-binding domain of androgen receptor; CaM, calmodulin; TOM1, tobamovirus multiplication 1; TOM2, tobamovirus multiplication 2.

and subsequent reconstitution can be applicable to a broad range of research fields. The advantages of this strategy have been successfully proved with hundreds of protein pairs in eukaryotic and prokaryotic cells, and/or living animals. In principle, all of the functional proteins including enzymes and fluorescent proteins can be dissected in two fragments, which may temporally delete the activities. Technical hurdles of this strategy are relatively low compared to others such as protein splicing and FRET. A protein-splicing scheme requires consideration of a precise matching of fusion proteins for high efficiency, while FRET requires a large fluorescence microscope with a sophisticated filter system and a trained technician. On the other hand, a split luciferase system does not need any large instrumentations and costs. Owing to the distinctive advantages, PCA will enable a wide range of novel applications to drug discovery, molecular imaging, toxicity evaluation and proteomics research.

13.3.5
Intein-Mediated Protein-Splicing Assay

Specific interactions between proteins in mammalian cells play key roles in many essential biological processes. The protein-splicing assay (PSA) was previously introduced for determining protein–protein interactions. Protein splicing is a naturally occurring, posttranslational processing event involving precise excision of an *in*ternal pro*tein* segment, the intein, from a primary translation product with concomitant ligation of the flanking sequences, the exteins (*ex*ternal pro*tein*) [125]. Since the discovery of protein splicing, more than 170 putative inteins have been identified in eubacteria, archeal and eukaryotic unicellular organisms [126]. A typical intein segment consists of 400–500 amino acid residues and contains four conserved protein splicing motifs, A, B, F and G, as well as a homing endonuclease sequence embedded between motifs B and F (Figure 13.5). At intein–extein junctions, conserved amino acid residues are directly involved in the protein-splicing reaction. While VDE (a homing endonuclease originally discovered in *Saccharomyces cerevisiae*) and most inteins are composed of a single polypeptide chain, a pair of functional and naturally split intein-coding sequences have been found from the split *dnaE* genes in the genome of *Synechocystis* sp. PCC6803 [129]. In this case, the DnaE intein can mediate a *trans*-splicing reaction with a higher efficiency when fused to foreign proteins. An important general feature of protein splicing is a self-catalyzed excision of the intein and ligation of the flanking exteins without any exogenous cofactor or energy source such as ATP or GTP. This feature encourages great ideas of reconstitution of split reporter proteins and their applications in molecular imaging and bioanalysis. The features of functional proteins for protein splicing are listed in Table 13.5.

An initial bioanalytical application of protein splicing was accomplished with VDE intein (N-VDE: 1–184 amino acids, C-VDE: 389–454 amino acids) and split GFP (N-GFP: 1–128 amino acids, C-GFP: 129–239 amino acids) [130]. As a proof of this principle, CaM and its target peptide (M13) were, respectively, labeled with N- and C-terminal fragments of the split VDE intein. The interaction of CaM and M13 in *E. coli* resulted in the formation of fluorescent GFP. Ozawa *et al.* found that enhanced GFP (EGFP) dissected at the position between 157 and 158 from the N-terminus showed strong fluorescence intensity upon protein interactions [131]. They further applied the split GFP reconstitution system to the screening of endocrine disrupting chemicals using the interaction between AR and c-Src in living eukaryotic cells [135].

The basic concept was extended to split FLuc as an optical probe for detecting protein–protein interactions [109].

Table 13.5 Representative fragments of functional proteins for protein splicing strategies.

Split reporter	Applied intein	Determined binding pairs	Organism	References
GFP (N-GFP: 1–128 amino acids; C-GFP: 129–239 amino acids)	VDE (N-VDE: 1–184 amino acids, C-VDE: 389–454 amino acids)	CaM and M13	E. coli	[130]
GFP (N-GFP: 1–157 amino acids; C-GFP: 158–239 amino acids)	DnaE (N-DnaE: 1–123 amino acids, C-DnaE: 1–36 amino acids)	IRS-1 and SH2 domain of phosphatidylinositol 3-kinase	E. coli	[131]
none	S. cerevisiae VMA (N-VMA: 1–184, C-VMA: 390–454)	FKBP and FRB	E. coli	[132]
firefly (N-FLuc: 1–437; C-FLuc: 438–544 amino acids)	DnaE	IRS-1 and SH2 domain of phosphatidylinositol 3-kinase	CHO cells	[109]
firefly (N-FLuc: 1–437; C-FLuc: 438–544 amino acids)	DnaE	MyoD and Id	293T cells and nude mice	[131]
firefly [N-FLuc: 1–490 (H489K) amino acids; C-FLuc: 438–544 (K491C) amino acids]	S. cerevisiae VMA (N-VMA: 1–182, C-VMA: 390-454)	FKBP and FRB	E. coli	[133]
ubiquitin [N: 1–37 (I13A or I13G) amino acids; C: 35–76 amino acids]	used a natural hydrolysis of ubiquitin instead of a splicing intein	FKBP and FRB	HeLa cells and SYF cells	[134]

CaM, calmodulin; M13, a CaM-binding peptide; IRS-1, insulin receptor substrate-1; MyoD, a myogenic regulatory protein; Id, a negative regulator of myogenic differentiation; 293T cell, a human embryonic kidney cancer cell line; CHO, Chinese hamster ovary cell line; HeLa, cervical cancer cell line.

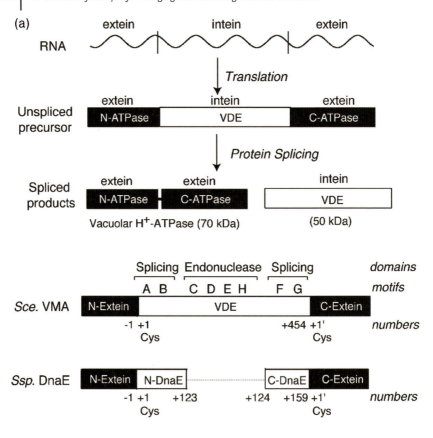

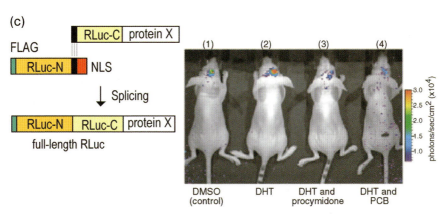

Figure 13.5 Intein-mediated PSA and subsequent molecular imaging. (a) Mechanism of intein-mediated protein splicing. (b) Cartoon diagram of protein-splicing steps in the mitochondrial matrix. The green fluorescence indicates the reconstitution of GFP in the mitochondrial matrix (OM = outer membrane; IM = inner membrane; MTS = mitochondria targeting signal; TMRE = tetramethylrhodamine ethyl ester) [127]. (c) Intein-mediated reconstitution of RLuc as a result of nuclear trafficking of a target protein. Hormonal

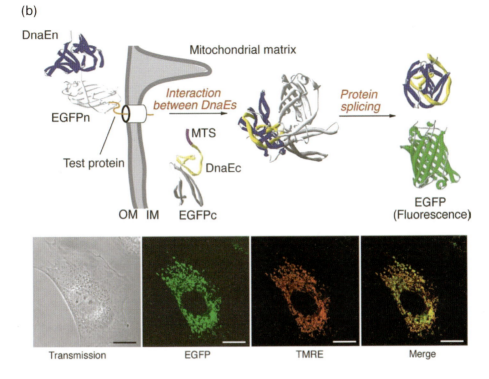

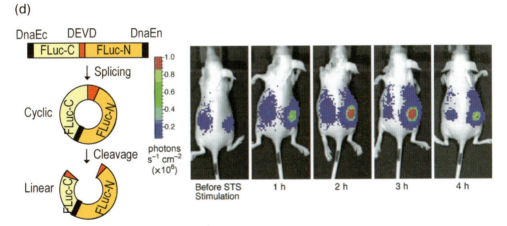

activities of 5α-dihydrotestosterone (DHT) and synthetic chemicals were determined in the brain of living mice (NLS = nuclear localization signal; DMSO = dimethyl sulfoxide; PCB = polychlorinated biphenyls) [114]. (d) Determination of caspase-3 activities based on the reconstitution of firefly luciferase through a protein splicing and cleavage of a four amino acid sequence (DEVD) by caspase-3 [128]. (All photographs are from the authors' previous papers).

In addition to phosphorylation-mediated interactions between IRS-1 and SH2, the method of split luciferase reconstitution has been used for noninvasively imaging MyoD–Id interaction in living mice [51].

One of the most distinct features of eukaryotic cells is the compartmentalization of each protein. Preferential localization of a protein is often an essential step determining its function. Therefore, functional assays aimed at characterizing the cellular localization of proteins are very important for understanding complicated protein networks. Ozawa et al. developed a method for identifying mitochondrial proteins from large-scale cDNA libraries [127].

Conditional protein splicing was well demonstrated by Muir et al., who very recently invented a protein-splicing system activated by red (660 nm) and far-red (750 nm) light in yeast [136]. The wavelengths are well tolerated by biological systems and excellently penetrate tissues.

Dynamics of the protein movement inside living cells is an important event in eukaryotic cells. A typical example is nuclear transport of a particular protein regulating gene expression in response to extracellular signals. For high-throughput analysis of the protein movement inside the cells, Kim et al. developed a bioluminescent reporter for detecting transport into the nucleus of AR using a RLuc reconstitution system [137].

One of the advantages of the use of a split GFP reporter is that fluorescence is an attractive readout for rapid and high-throughput approaches because of the availability of technologies such as fluorescent activated cell sorting and automated fluorescence microscopy. Meanwhile, split luciferases offer distinct advantages when compared with the split GFP reporter: the split luciferase reporters have the potential to noninvasively image protein–protein interactions or protein localization in living subjects owing to the background-free and highly sensitive detection. In addition, a bioluminescence-based detection of intracellular molecular events is quantitative rather than qualitative and its dynamic range is generally wider that one of fluorescence-based schemes.

There are a number of limitations of the intein-mediated splicing assays using split GFP and luciferase reporters. (i) It takes several hours to obtain fluorescent and bioluminescent signals after the protein-splicing reaction. During this time, the full-length reporter may be in part degenerated. Thus, temporal dynamics of proteins occurring over a short time (e.g., less than 1 h) is difficult to trace. (ii) Owing to the self-catalytic nature of the DnaE intein, the background luminescence can be unexpectedly high according to some experimental setups. According to the authors' consideration, the background is additionally influenced by the nature of a target protein fused to the DnaE intein. (iii) Although PSAs provide a unique and attractive readout for protein interactions, they are rather sophisticated compared to others, and require consideration for a precise match between the fusion proteins for efficient splicing reactions.

In spite of the drawbacks, further understanding of protein-splicing mechanisms, discovery of new inteins and development of new split reporters will certainly expand the basic tools of protein-splicing technology. Development of a variety of attractive applications with protein-splicing technology will increase in the near future.

13.3.6
Circular Permutation of Fluorescent and Luminescent Proteins

Another potential approach to molecular imaging is to exploit rearrangement within a single fluorophore, an approach known as circular permutation (CP). Fusion protein probes with circular permutated functional proteins may be a successful example of the protein rearrangement.

Endogenous Ca^{2+} levels were conventionally determined with synthetic chemical probes. As an alternative to the classic scheme for Ca^{2+} imaging, CP of fluorescent proteins was originally conducted by Tsien et al. for tracing intracellular dynamics of Ca^{2+} [1]. GFP variants themselves are poor candidates to be engineered for CP because of their monolithic cylindrical symmetry and the complexity of their maturation process. Tsien et al. genuinely exploited the characteristic molecular structure of GFP, the fluorescence of which arises from a posttranslational modification of three amino acids within the core of 11 surrounding β-sheets, which form a "β-can" structure. In the pioneering CP GFP indicator, an insertion of peptides at β-sheet linkers temporally disrupts fluorescence because of solvent penetration within the protein core, which interferes with fluorophore–β-sheet interactions. The inhibitory action is relieved by condensive interaction between the inserted peptides [138].

In the indicator, the original N- and C-termini of GFP were linked with a GS linker, and new N- and C-termini were created at a hinge region. The new ends of GFP were, respectively, extended with CaM and M13. Upon stimulation of Ca^{2+}, an intramolecular interaction between CaM and M13 occurs, which enhances the fluorescence. This strategy markedly increased the fluorescence of the molecule, with a roughly 7-fold increase in fluorescence intensity compared to the background, and these molecules were termed "camgaroos" [138].

Kawai et al. presented a tandem fusion protein containing a pleckstrin homology domain and a circularly permutated fluorescent protein for determining phosphorylation of mitogen-activated protein kinase [139]. The substrate domain used in this study is a peptide sequence that is phosphorylated by insulin receptor. Phosphorylation of the substrate domain induces its interaction with the phosphorylation recognition domain, which causes a conformational change in the circularly permutated fluorescent protein and a change in its fluorescence. By tailoring the substrate domains and the phosphorylation recognition domains in these cyan and green indicators, the present approach should be applicable to the *in vivo* analysis of a broad range of protein phosphorylation processes, together with other intracellular signaling processes.

The most expected merit with CP is the decrease of the background intensities and the subsequent improvement of signal:background ratios. Both CP-based, fluorescent and luminescent approaches for Ca^{2+} imaging exhibited improved signal:background ratios [140].

Although the basic concept of CP was originally introduced with GFP variants, it can be also fabricated in luciferases. Probes with circularly permutated enzymes may be generally utilized for tracing molecular dynamics of target proteins and intracellular protein signaling. The studies offer a new strategy for luciferase-aided probing systems that overcomes the limitations of conventional luminescent probes.

13.4
Conclusions and Perspectives

13.4.1
Frontier Research in Analysis

The history of analysis proves that a new breakthrough in assays is directly connected to expanding the scope of our knowledge. Although recent assays based on advanced knowledge have contributed greatly to better insights into the molecular mechanisms, many frontier regions of natural science still remain unexplored. This is because we do not even know how to determine the phenomenon in the frontier regions, owing to the lack of an appropriate method to access them.

Seeing is believing. However, "seeing" a life phenomenon is a tough mission. Many practical hurdles still exist in the "seeing" technology. (i) The fundamental unit of living organisms is the cell. However, the cells consist of complex compartments, in which tremendous numbers of unknown proteins and lipids work. Even though we determine a protein to play a key role in life phenomenon, the key protein is generally expressed in an extremely small amount and does not emit any optical signal for spectrometric analysis. (ii) The proteins abundant in the cells have nano- or subnanometer-scale dimensions. This means that analysis for proteins and lipids should be accessed at the nanometer scale. (iii) The ideal assays should trace molecular events, noninvasively. In practice, determination of the molecular events without any help of genetic modification or spectrometry is nearly impossible. Any invasive strategies for probing a target protein inevitably modify the readout of molecular events of interest. Researchers should note the relevance of the outcomes, which are potentially influenced by nonphysiological circumstances causing molecular abnormalities.

13.4.2
Analytical Achievements Inspired by Nature

Many modern advances in analysis have been achieved by ideas inspired by nature. PCA, PSA and two-hybrid assays are appropriate examples of this. As reviewed above, protein splicing is a naturally occurring, self-catalytic, posttranslational processing event of prematured proteins. Frontier researchers in this field took advantage of the self-catalytic maturation process of proteins in conditional reassembly of functional proteins. Protein complementation is one of the most widely occurring, natural molecular events in mammalian cells for activating diverse lineages of signaling cascades. A great number of proteins, including NRs, produce a dimer or tetramer for initiating signaling cascades. NRs are first dimerized to trafficking into the nucleus upon ligand activation. Epidermal growth factor receptors on the plasma membrane make dimers for transferring the external stimulation into the inner signaling cascades. Many fluorescent proteins emit red light after dimerization or tetramerization. Reporter gene assays and two-hybrid assays mimic the generally occurring expression events of proteins in the cells. Readers may find many examples related to this consideration. Likewise, we are certain that various molecular mechanisms of living organisms will be excellent prototypes for future analytical measures.

13.4.3
Future Directions

In the past decade, many new imaging technologies that have been developed have been greatly supported by the advanced properties of "lighting" proteins. FRET, BRET and PCA will prevail even in the next decade as powerful tools. The path of the technological advances is clearly directed to noninvasive, quantitative, specific, signal-enhanced and real-time investigation of intracellular molecular events. In addition, technological progress in the development of genetic probes should be directed to mimic the sense organs of the living subjects.

Genetically encoded fluorescent and bioluminescent probes fundamentally depend on the lighting property of functional proteins. Improvement of the properties of functional proteins is directly connected to advances in fusion protein probes. Many functional proteins emitting fluorescence have been discovered during the last decade. Great efforts have been concentrated on the improved optical properties of the functional proteins in addition to less sensitivity to the environment (e.g., pH and ions). Although such efforts have produced a great number of choices and improvements in the properties of functional proteins, an additional exploration should be directed to allow for more intensive imaging experiments and efficiently folding monomeric proteins in the case of fluorescent proteins. The long-wavelength end of the functional protein spectrum allows more sensitive and efficient imaging in thick tissue and whole animals.

The main concern for functional proteins emitting bioluminescence relates to the problems with the substrates. Such drawbacks of luciferases will be overcome by finding mutational variants and new luciferases with improved properties.

Another pivotal ingredient in the advance of imaging technologies is the improvement of the instrumentation. The development should be directed to be truly quantitative, highly sensitive and very comprehensive.

References

1 Tsien, R.Y. (1998) *Annual Review of Biochemistry*, **67**, 509.
2 Shimomura, O., Johnson, F.H. and Saiga, Y. (1962) *Journal of Cellular Comparative Physiology*, **59**, 223.
3 Prasher, D.C., Eckenrode, V.K., Ward, W.W., Prendergast, F.G. and Cormier, M.J. (1992) *Gene*, **111**, 229.
4 Chalfie, M., Tu, Y., Euskirchen, G., Ward, W.W. and Prasher, D.C. (1994) *Science*, **263**, 802.
5 Inouye, S. and Tsuji, F.I. (1994) *FEBS Letters*, **341**, 277.
6 Patterson, G.H., Knobel, S.M., Sharif, W.D., Kain, S.R. and Piston, D.W. (1997) *Biophysical Journal*, **73**, 2782.
7 Siegel, R.M., Chan, F.K.M., Zacharias, D.A., Swofford, R., Holmes, K.L., Tsien, R.Y. and Lenardo, M.J. (2000) *Science STKE*, **2000**, pl1.
8 Heim, R. and Tsien, R.Y. (1996) *Current Biology*, **6**, 178.
9 Cubitt, A.B., Woollenweber, L.A. and Heim, R. (1999) *Methods in Cell Biology*, **58**, 19.
10 Tsien, R.Y. (2005) *FEBS Letters*, **579**, 927.
11 Miyawaki, A., Nagai, T. and Mizuno, H. (2003) *Current Opinion in Chemical Biology*, **7**, 557.
12 Verkhusha, V.V. and Lukyanov, K.A. (2004) *Nature Biotechnology*, **22**, 289.
13 Yang, F., Moss, L.G. and Phillips, G.N., Jr (1996) *Nature Biotechnology*, **14**, 1246.
14 Stauber, R.H., Horie, K., Carney, P., Hudson, E.A., Tarasova, N.I., Gaitanaris, G.A. and Pavlakis, G.N. (1998) *Biotechniques*, **24**, 462.
15 Patterson, B., Domanik, R., Wernke, P. and Gombrich, M. (2001) *Acta Cytologica*, **45**, 36.
16 Rizzo, M.A., Springer, G.H., Granada, B. and Piston, D.W. (2004) *Nature Biotechnology*, **22**, 445.
17 Chattoraj, M., King, B.A., Bublitz, G.U. and Boxer, S.G. (1996) *Proceedings of the National Academy of Sciences of the United States of America*, **93**, 8362.
18 Nagai, T., Ibata, K., Park, E.S., Kubota, M., Mikoshiba, K. and Miyawaki, A. (2002) *Nature Biotechnology*, **20**, 87.
19 Shaner, N.C., Campbell, R.E., Steinbach, P.A., Giepmans, B.N., Palmer, A.E. and Tsien, R.Y. (2004) *Nature Biotechnology*, **22**, 1567.
20 Matz, M.V., Fradkov, A.F., Labas, Y.A., Savitsky, A.P., Zaraisky, A.G., Markelov, M.L. and Lukyanov, S.A. (1999) *Nature Biotechnology*, **17**, 969.
21 Campbell, R.E., Tour, O., Palmer, A.E., Steinbach, P.A., Baird, G.S., Zacharias, D.A. and Tsien, R.Y. (2002) *Proceedings of the National Academy of Sciences of the United States of America*, **99**, 7877.
22 Yurchenko, E., Friedman, H., Hay, V., Peterson, A. and Piccirillo, C.A. (2007) *Transgenic Research*, **16**, 29.
23 Kogure, T., Karasawa, S., Araki, T., Saito, K., Kinjo, M. and Miyawaki, A. (2006) *Nature Biotechnology*, **24**, 577.
24 Gurskaya, N.G., Fradkov, A.F., Terskikh, A., Matz, M.V., Labas, Y.A., Martynov, V.I., Yanushevich, Y.G., Lukyanov, K.A. and Lukyanov, S.A. (2001) *FEBS Letters*, **507**, 16.
25 Wang, L., Jackson, W.C., Steinbach, P.A. and Tsien, R.Y. (2004) *Proceedings of the National Academy of Sciences of the United States of America*, **101**, 16745.
26 Patterson, G.H. and Lippincott-Schwartz, J. (2002) *Science*, **297**, 1873.
27 Verkhusha, V.V. and Sorkin, A. (2005) *Chemistry and Biology*, **12**, 279.
28 Chudakov, D.M., Verkhusha, V.V., Staroverov, D.B., Souslova, E.A., Lukyanov, S. and Lukyanov, K.A. (2004) *Nature Biotechnology*, **22**, 1435.
29 Bevis, B.J. and Glick, B.S. (2002) *Nature Biotechnology*, **20**, 83.
30 Chudakov, D.M., Belousov, V.V., Zaraisky, A.G., Novoselov, V.V., Staroverov, D.B., Zorov, D.B., Lukyanov, S. and Lukyanov, K.A. (2003) *Nature Biotechnology*, **21**, 191.
31 Anderson, K.I., Sanderson, J., Gerwig, S. and Peychl, J. (2006) *Cytometry. Part A: Journal of the International Society for Analytical Cytology*, **69**, 920.
32 Lukyanov, K.A., Fradkov, A.F., Gurskaya, N.G., Matz, M.V., Labas, Y.A., Savitsky, A.P., Markelov, M.L., Zaraisky, A.G., Zhao, X., Fang, Y., Tan, W. and

Lukyanov, S.A. (2000) *Journal of Biological Chemistry*, **275**, 25879.

33 Ando, R., Mizuno, H. and Miyawaki, A. (2004) *Science*, **306**, 1370.

34 Muller-Taubenberger, A. and Anderson, K.I. (2007) *Applied Microbiology and Biotechnology*, **77**, 1.

35 Heim, R., Prasher, D.C. and Tsien, R.Y. (1994) *Proceedings of the National Academy of Sciences of the United States of America*, **91**, 12501.

36 Duisit, G., Conrath, H., Saleun, S., Folliot, S., Provost, N., Cosset, F.L., Sandrin, V., Moullier, P. and Rolling, F. (2002) *Molecular Therapy*, **6**, 446.

37 Bi, J.X., Wirth, M., Beer, C., Kim, E.J., Gu, M.B. and Zeng, A.P. (2002) *Journal of Biotechnology*, **93**, 231.

38 Gross, L.A., Baird, G.S., Hoffman, R.C., Baldridge, K.K. and Tsien, R.Y. (2000) *Proceedings of the National Academy of Sciences of the United States of America*, **97**, 11990.

39 Mena, M.A., Treynor, T.P., Mayo, S.L. and Daugherty, P.S. (2006) *Nature Biotechnology*, **24**, 1569.

40 Dickson, R.M., Cubitt, A.B., Tsien, R.Y. and Moerner, W.E. (1997) *Nature*, **388**, 355.

41 Shaner, N.C., Steinbach, P.A. and Tsien, R.Y. (2005) *Nature Methods*, **2**, 905.

42 Maruvada, P., Baumann, C.T., Hager, G.L. and Yen, P.M. (2003) *Journal of Biological Chemistry*, **278**, 12425.

43 Lippincott-Schwartz, J. and Patterson, G.H. (2003) *Science*, **300**, 87.

44 Jares-Erijman, E.A. and Jovin, T.M. (2006) *Current Opinion in Chemical Biology*, **10**, 409.

45 Schultz, M., Watzl, S., Oelschlaeger, T.A., Rath, H.C., Gottl, C., Lehn, N., Scholmerich, J. and Linde, H.J. (2005) *Journal of Microbiological Methods*, **61**, 389.

46 Cabantous, S., Terwilliger, T.C. and Waldo, G.S. (2005) *Nature Biotechnology*, **23**, 102.

47 Fischer, T., Gemeinhardt, I., Wagner, S., Stieglitz, D.V., Schnorr, J., Hermann, K.G., Ebert, B., Petzelt, D., Macdonald, R., Licha, K., Schirner, M., Krenn, V., Kamradt, T. and Taupitz, M. (2006) *Academic Radiology*, **13**, 4.

48 Terskikh, A., Fradkov, A., Ermakova, G., Zaraisky, A., Tan, P., Kajava, A.V., Zhao, X., Lukyanov, S., Matz, M., Kim, S., Weissman, I. and Siebert, P. (2000) *Science*, **290**, 1585.

49 Brannigan, J.A. and Wilkinson, A.J. (2002) *Nature Reviews Molecular Cell Biology*, **3**, 964.

50 Weissleder, R. and Ntziachristos, V. (2003) *Nature Medicine*, **9**, 123.

51 Paulmurugan, R., Umezawa, Y. and Gambhir, S.S. (2002) *Proceedings of the National Academy of Sciences of the United States of America*, **99**, 15608.

52 Paulmurugan, R. and Gambhir, S.S. (2006) *Proceedings of the National Academy of Sciences of the United States of America*, **103**, 15883.

53 Viviani, V.R., Uchida, A., Viviani, W. and Ohmiya, Y. (2002) *Photochemistry and Photobiology*, **76**, 538.

54 Hastings, J.W. (1996) *Gene*, **173**, 5.

55 Chalfie, M. and Kain, S. (1998) *Green Fluorescent Protein: Properties, Applications, and Protocols*, Wiley-Liss, New York.

56 Loening, A.M., Wu, A.M. and Gambhir, S.S. (2007) *Nature Methods*, **4**, 641.

57 Kim, S.B., Awais, M., Sato, M., Umezawa, Y. and Tao, H. (2007) *Analytical Chemistry*, **79**, 1874.

58 Markova, S.V., Golz, S., Frank, L.A., Kalthof, B. and Vysotski, E.S. (2004) *Journal of Biological Chemistry*, **279**, 3212.

59 de Wet, J.R., Wood, K.V., Helinski, D.R. and DeLuca, M. (1985) *Proceedings of the National Academy of Sciences of the United States of America*, **82**, 7870.

60 Lorenz, W.W., McCann, R.O., Longiaru, M. and Cormier, M.J. (1991) *Proceedings of the National Academy of Sciences of the United States of America*, **88**, 4438.

61 Verhaegent, M. and Christopoulos, T.K. (2002) *Analytical Chemistry*, **74**, 4378.

62 Wood, K.V., Lam, Y.A., Seliger, H.H. and McElroy, W.D. (1989) *Science*, **244**, 700.

63 Viviani, V.R., Arnoldi, F.G.C., Venkatesh, B., Neto, A.J.S., Ogawa, F.G.T., Oehlmeyer, A.T.L. and Ohmiya, Y. (2006) *Journal of Biochemistry*, **140**, 467.

64 Thompson, E.M., Nagata, S. and Tsuji, F.I. (1989) *Proceedings of the National*

Academy of Sciences of the United States of America, **86**, 6567.
65. Inouye, S., Watanabe, K., Nakamura, H. and Shimomura, O. (2000) *FEBS Letters*, **481**, 19.
66. Faust, M.A. and Gulledge, R.A. (2002) *Identifying Harmful Marine Dinoflagellates*, National Museum of Natural History, Washington, DC.
67. Hastings, J.W. (1983) *Journal of Molecular Evolution*, **19**, 309.
68. Rees, J.F., de Wergifosse, B., Noiset, O., Dubuisson, M., Janssens, B. and Thompson, E.M. (1998) *Journal of Experimental Biology*, **201**, 1211.
69. McElroy, W.D., Seliger, H.H. and White, E.H. (1969) *Photochemistry and Photobiology*, **10**, 153.
70. de Wet, J.R., Wood, K.V., DeLuca, M., Helinski, D.R. and Subramani, S. (1987) *Molecular and Cellular Biology*, **7**, 725.
71. Gandelman, O., Allue, I., Bowers, K. and Cobbold, P. (1994) *Journal of Bioluminescence and Chemiluminescence*, **9**, 363.
72. Lembert, N. and Idahl, L.A. (1995) *Biochemical Journal*, **305**, 929.
73. Lippard, S.J. and Berg, J.M. (1994) *Principles of Bioinorganic Chemistry*, University Science Books, Mill Valley, CA.
74. Domaille, D.W., Que, E.L. and Chang, C.J. (2008) *Nature Chemical Biology*, **4**, 168.
75. Lichtman, J.W. and Conchello, J.A. (2005) *Nature Methods*, **2**, 910.
76. Georget, V., Lobaccaro, J.M., Terouanne, B., Mangeat, P., Nicolas, J.C. and Sultan, C. (1997) *Molecular and Cellular Endocrinology*, **129**, 17.
77. Htun, H., Barsony, J., Renyi, I., Gould, D.L. and Hager, G.L. (1996) *Proceedings of the National Academy of Sciences of the United States of America*, **93**, 4845.
78. Mattheyses, A.L., Hoppe, A.D. and Axelrod, D. (2004) *Biophysical Journal*, **87**, 2787.
79. Lidke, D.S., Nagy, P., Barisas, B.G., Heintzmann, R., Post, J.N., Lidke, K.A., Clayton, A.H., Arndt-Jovin, D.J. and Jovin, T.M. (2003) *Biochemical Society Transactions*, **31**, 1020.
80. Rizzo, M.A. and Piston, D.W. (2005) *Biophysical Journal*, **88**, L14.
81. Clayton, A.H., Hanley, Q.S., Arndt-Jovin, D.J., Subramaniam, V. and Jovin, T.M. (2002) *Biophysical Journal*, **83**, 1631.
82. Kau, T.R., Way, J.C. and Silver, P.A. (2004) *Nature Reviews Cancer*, **4**, 106.
83. Villalobos, V., Naik, S. and Piwnica-Worms, D. (2007) *Annals Reviews of Biomedical Engineering*, **9**, 321.
84. Parrish, A.R., Wang, W. and Wang, L. (2006) *Current Opinion in Neurobiology*, **16**, 585.
85. Piston, D.W. and Kremers, G.J. (2007) *Trends in Biochemical Sciences*, **32**, 407.
86. Haugland, R.P., Spence, M.T.Z. and Johnson, I.D. (1996) *Handbook of Fluorescent Probes and Research Chemicals*, 6th edn, Molecular Probes, Eugene, OR.
87. Miyawaki, A., Llopis, J., Heim, R., McCaffery, J.M., Adams, J.A., Ikura, M. and Tsien, R.Y. (1997) *Nature*, **388**, 882.
88. Souslova, E.A. and Chudakov, D.M. (2007) *Biochemistry (Russia)*, **72**, 683.
89. Griesbeck, O., Baird, G.S., Campbell, R.E., Zacharias, D.A. and Tsien, R.Y. (2001) *Journal of Biological Chemistry*, **276**, 29188.
90. Mizuno, H., Sawano, A., Eli, P., Hama, H. and Miyawaki, A. (2001) *Biochemistry*, **40**, 2502.
91. Nguyen, A.W. and Daugherty, P.S. (2005) *Nature Biotechnology*, **23**, 355.
92. Palmer, A.E., Giacomello, M., Kortemme, T., Hires, S.A., Lev-Ram, V., Baker, D. and Tsien, R.Y. (2006) *Chemistry and Biology*, **13**, 521.
93. Sato, M., Ozawa, T., Inukai, K., Asano, T. and Umezawa, Y. (2002) *Nature Biotechnology*, **20**, 287.
94. Awais, M., Sato, M., Lee, X.F. and Umezawa, Y. (2006) *Angewandte Chemie (International Edition in English)*, **45**, 2707.
95. Sato, M., Ueda, Y., Takagi, T. and Umezawa, Y. (2003) *Nature Cell Biology*, **5**, 1016.
96. Sato, M., Hida, N. and Umezawa, Y. (2005) *Proceedings of the National Academy of Sciences of the United States of America*, **102**, 14515.
97. Sato, M., Nakajima, T., Goto, M. and Umezawa, Y. (2006) *Analytical Chemistry*, **78**, 8175.

98 Awais, M., Sato, M., Sasaki, K. and Umezawa, Y. (2004) *Analytical Chemistry*, **76**, 2181.
99 Kaihara, A., Kawai, Y., Sato, M., Ozawa, T. and Umezawa, Y. (2003) *Analytical Chemistry*, **75**, 4176.
100 Czech, M.P. (2000) *Cell*, **100**, 603.
101 Vogel, S.S., Thaler, C. and Koushik, S.V. (2006) *Science STKE*, **2006**, re2.
102 Prinz, A., Diskar, M. and Herberg, F.W. (2006) *Chembiochem*, **7**, 1007.
103 Pfleger, K.D. and Eidne, K.A. (2006) *Nature Methods*, **3**, 165.
104 De, A. and Gambhir, S.S. (2005) *FASEB Journal*, **19**, 2017.
105 Xu, Y., Piston, D.W. and Johnson, C.H. (1999) *Proceedings of the National Academy of Sciences of the United States of America*, **96**, 151.
106 Milligan, G. and Bouvier, M. (2005) *FASEB Journal*, **272**, 2914.
107 Pfleger, K.D., Dromey, J.R., Dalrymple, M.B., Lim, E.M., Thomas, W.G. and Eidne, K.A. (2006) *Cellular Signaling*, **18**, 1664.
108 Hoshino, H., Nakajima, Y. and Ohmiya, Y. (2007) *Nature Methods*, **4**, 637.
109 Ozawa, T., Kaihara, A., Sato, M., Tachihara, K. and Umezawa, Y. (2001) *Analytical Chemistry*, **73**, 2516.
110 Luker, K.E., Smith, M.C., Luker, G.D., Gammon, S.T., Piwnica-Worms, H. and Piwnica-Worms, D. (2004) *Proceedings of the National Academy of Sciences of the United States of America*, **101**, 12288.
111 Paulmurugan, R. and Gambhir, S.S. (2005) *Analytical Chemistry*, **77**, 1295.
112 Kim, S.B., Kanno, A., Ozawa, T., Tao, H. and Umezawa, Y. (2007) *ACS Chemistry and Biology*, **2**, 484.
113 Kim, S.B., Otani, Y., Umezawa, Y. and Tao, H. (2007) *Analytical Chemistry*, **79**, 4820.
114 Kim, S.B., Ozawa, T., Watanabe, S. and Umezawa, Y. (2004) *Proceedings of the National Academy of Sciences of the United States of America*, **101**, 11542.
115 Kaihara, A. and Umezawa, Y. (2008) *Chemistry: An Asian Journal*, **3**, 38.
116 Paulmurugan, R. and Gambhir, S.S. (2003) *Analytical Chemistry*, **75**, 1584.
117 Remy, I. and Michnick, S.W. (2006) *Nature Methods*, **3**, 977.
118 Galarneau, A., Primeau, M., Trudeau, L.E. and Michnick, S.W. (2002) *Nature Biotechnology*, **20**, 619.
119 Blakely, B.T., Rossi, F.M., Tillotson, B., Palmer, M., Estelles, A. and Blau, H.M. (2000) *Nature Biotechnology*, **18**, 218.
120 Fetchko, M. and Stagljar, I. (2004) *Methods (Duluth)*, **32**, 349.
121 Richards, F. (1958) *Proceedings of the National Academy of Sciences of the United States of America*, **44**, 162.
122 Johnsson, N. and Varshavsky, A. (1994) *Proceedings of the National Academy of Sciences of the United States of America*, **91**, 10340.
123 Ghosh, I., Hamilton, A.D. and Regan, L. (2000) *Journal of the American Chemical Society*, **122**, 5658.
124 Nagai, T., Sawano, A., Park, E.S. and Miyawaki, A. (2001) *Proceedings of the National Academy of Sciences of the United States of America*, **98**, 3197.
125 Gimble, F.S. (1998) *Chemistry and Biology*, **5**, R251.
126 Perler, F.B. (2002) *Nucleic Acids Research*, **30**, 383.
127 Ozawa, T., Sako, Y., Sato, M., Kitamura, T. and Umezawa, Y. (2003) *Nature Biotechnology*, **21**, 287.
128 Kanno, A., Yamanaka, Y., Hirano, H., Umezawa, Y. and Ozawa, T. (2007) *Angewandte Chemie (International Edition in English)*, **46**, 7595.
129 Wu, H., Hu, Z. and Liu, X.Q. (1998) *Proceedings of the National Academy of Sciences of the United States of America*, **95**, 9226.
130 Ozawa, T., Nogami, S., Sato, M., Ohya, Y. and Umezawa, Y. (2000) *Analytical Chemistry*, **72**, 5151.
131 Ozawa, T., Takeuchi, T.M., Kaihara, A., Sato, M. and Umezawa, Y. (2001) *Analytical Chemistry*, **73**, 5866.
132 Mootz, H.D. and Muir, T.W. (2002) *Journal of the American Chemical Society*, **124**, 9044.
133 Schwartz, E.C., Saez, L., Young, M.W. and Muir, T.W. (2007) *Nature Chemical Biology*, **3**, 50.
134 Pratt, M.R., Schwartz, E.C. and Muir, T.W. (2007) *Proceedings of the National Academy of Sciences of the United States of America*, **104**, 11209.

135 Umezawa, Y., Ozawa, T. and Sato, M. (2002) *Bulletin of the Chemical Society of Japan*, **75**, 1423.
136 Tyszkiewicz, A.B. and Muir, T.W. (2008) *Nature Methods*, **5**, 303.
137 Kim, S.B., Ozawa, T. and Umezawa, Y. (2005) *Analytical Chemistry*, **77**, 6588.
138 Baird, G.S., Zacharias, D.A. and Tsien, R.Y. (1999) *Proceedings of the National Academy of Sciences of the United States of America*, **96**, 11241.
139 Kawai, Y., Sato, M. and Umezawa, Y. (2004) *Analytical Chemistry*, **76**, 6144.
140 Kotlikoff, M.I. (2007) *Journal of Physiology*, **578**, 55.
141 Kahru, A., Dubourguier, H.C., Blinova, I., Ivask, A., Kasemets, K. (2008) *Sensors*, **8**, 5153.
142 Scofield, S.R., Tobias, C.M., Rathjen, J.P., Chan, J.H., Lavelle, D.T., Michelmore, R.W. and Staskawicz, B.J. (1996) *Science*, **274**, 2063.
143 Kim, S.B., Sato, M. and Tao, H. (2009) *Analytical Chemistry*, **81**, 67.

Index

a

absorption coefficient 103
Aβ peptide, *see* peptide
activation agents 9
adhesion
– adhesion force(s) 72, 73, 80
– adhesion phenomenon 81
– cell adhesion 80
adjuvant, effect on immune response 144
alcohol dehydrogenase (ADH) inhibition 206–209
Aldrich 111
Alzheimer, Alois 193
Alzheimer's disease 101
– role of β sheets 193
– symptoms 193, 194
– treatment 194, 195
amino acids, role in membrane rupture process 252
aminopyrazoles 193, 195–201
amine 5, 126
angiotensin-converting enzyme (ACE) inhibitors 202
antibody
– as biorecognition elements 223
– bifunctional 115, 116–118
– bispecific 116, 117
– catalytic, *see also* cancer therapy
– chemically programmed (cpAb) 128, 131
– in cancer therapy 114–128
– in chiral molecular imprinting 97
– in prodrug therapy 116
– monoclonal 113
antibody-directed enzyme prodrug therapy (ADEPT) 116
antigen
– effect of adjuvant 144
– glycosphingolipids (GSLs) 159–161
– in cancer therapy 116
– in vaccine development 53, 54, 144
antimicrobial peptides (AMPs) 247, 251, 253, 254, 256, 258
assays
– chloramphenicol transacetylase (CAT) 310, 311
– cyclic luciferase 315
– for analysis of biological interactions 230, 232
– growth inhibition 114, 125, 127
– protein-fragment complementation (PCA) 323–326
– protein-splicing assay (PSA) 315, 326–331
– two-hybrid 314, 317
atomic force microscopy (AFM)
– cantilever 73, 74, 76, 89
– in study of composite fibers 14, 15
– in study of silica– cell interaction 70
– probe 73–76, 78–80, 82
– scanners 73
– sensitivity 80
– software 72
– system 70
– technique 71, 72
ATP binding cassette (ABC) transporters 281, 287
Avanti Polar Lipids 144

b

bacterial display 11
β-sheet capping 193–201
B_6 ester derivatives 113
binding
– coordinative 44
– covalent 44
– ephedrine 103
– ionic 44

– nonspecific 103, 105, 225
– of D-histidine to L-histidine-imprinted films 107
– of nicotine 102
– of template to film 104
– template/analyte rebinding 105
bioconjugation, see conjugation
biocompatibility in drug delivery systems 40
biodegradability in drug delivery systems 40
biofunctionalization 42, see also nanoparticles
bioluminescence, in luciferases
– functions 310
– origins and characteristics 310
bioluminescence resonance energy transfer (BRET) 315, 321, 322, 333
biomaterials for drug delivery 33, 52
biomolecule(s)
– conjugation techniques 42
– immobilization 97, 226, 227
– purified, examples 49
biomolecular sensing, see biosensors
biopanning technique 8
biosensors, see also CPs; PDAs; SLO
– development 177–190
– surface plasmon resonance (SPR)-based 219–238
– using the protein-binding family 281–294
birecognition elements
– immobilization 224
– surface chemistry 224
blue fluorescent protein (BFP) 300, 303, 318
Bristol-Myers Squibb 130
bucky balls, see fullerenes

c

camptothecin (CPT) 125
cancer, detection
– of acute myeloid leukemia 125
– of breast 127, 129, 132, 272
– of colon 127
– of erythroleukemia 125
– of Kaposi's sarcoma 127
– of MOLT-3 T cell leukemia 122
– of murine NXS2 neuroblastoma 118, 119, 122
– of prostate 131
cancer research 272
cancer therapy 114–128
CD1d 158, 163–167

cell membrane, composition 247, 248
cell separation 69
cell–silica particle interaction 70
– silica–cell contact 81
– silica–cell interaction 85
cells
– cancer cells 78, 80–83, 87, 89
– cervical cancer cells 72
– epithelial cells 72
– human epithelial cells 73
– human epithelial cervical cells 70, 72, 78
– normal cells 78, 80–83, 85, 87, 88
– normal cervical cells 72, 77
– red blood cells 69
cell surface 70, 73–75, 78, 80–83, 89
chemical adaptor 123, 124
chemical ligation 5
chemical modification
– of silicon oxide 35
– of viruses, in combination with genetic modification 8
chemical nanotechnology 31
chemical sensors for molecular imaging 313, 316
chemotherapeutic agents, concentration 111
chemotherapy 116–118, see also cancer therapy
chiral discrimination 97, 104
chiral drugs, chemical structure 100
chiral molecular imprinting, see MI
chiral polymer 98
chiral specificity 100, 103
chiral templates 97
chromatographic separations 97, 98, see also MI
circular dichroism (CD) 283, 286, 291, 292
circular permutation 331, 332
critical micelle concentration (CMC) 39
coat protein(s)
– gene encoding 7
– M13 bacteriophage 3, 12
– tobacco mosaic virus (TMV) 21
colloidal dispersion 87
colorimetric signals 178, signaling 179–189
conductance switching 21
conductivity, anisotropic 19
conjugated polymers (CPs)
– applications 177
– as sensing materials 177
– examples 177
– properties 177

conjugation
– capture molecule 56
– conventional bioconjugation strategies 5
controlled 1D assembly of nanostructures 14
controlled 3D assembly of M13 bacteriophage 19
controlled 3D assembly of TMV 19
convective alignment 18
convective assembly process 19
core/shell particle 31, 50, 51, 54
coupling
– between recognition layers and sensor surfaces 98, 99
– carbodiimide 5
– diazonium 5
– of proteins 43
– polymer–sensor 101
critical particle size D_c 37
crystallographic defects 34
cyan fluorescent protein (CFP) 318, 319, 321
cyclic voltammetry 103
cyclization 118, 127
cysteine 7–9, 20
cytokine(s)
– proinflammatory 139
– stimulation activity of monophosphoryl lipid A 144
cytoplastic membrane 80

d

D-dopa, **1** 101
Debye length 78
deflection signal 70
Dennis, Edward A. 143
deposition, metals 9
Derjaguin approximation 79
1,3-diketone **10** 114
differential scanning calorimetry (DSC) 283
diffusional path 99
dimerization 84
dip-pen nanolithography (DPN) 17, 18
direct dissolution method 40
direct-write lithographic method 17
discoids 85
drug release 112–115, 123, 125
drug sensors
– development 97, 99
– electrochemical 101, 102
– enantioselectivity 101, 103, 107
– label-free 99

– mass sensor 106, 107
– – quartz crystal microbalance sensor 106
– optical 103–105
– redox characteristics 101
drugs, anticancer
– camptothecin (CPT) 125
– doxorubicin 115, 126, 133
– dynemicin 126–128
– 5-fluorodeoxyuridine **2** 112
– nitrogen mustard anticancer drug **5** 112

e

electrical double-layer interaction 78
electrochemical impedance technique 107
electromagnetic radiation 31
electrophilic substitution reaction 5
electrospinning method 16, 18
electrospray ionization (ESI) 199
electrospraying 39
electrostatic 82
emission spectrographs 87
enantiomer 97–99, 107
enantioselectivity, *see* drug sensors
endocytosis 53, 54
endohedral materials 42
endotoxin, bacterial
– discovery 137, 138
– inflammatory potential 138
energy transfer 21
enhanced permeability and retention (EPR) 120
enzyme inhibition 202–210
epitope approach 33, 46, 47, 49
Escherichia coli
– D-galactose/D-glucose-binding protein (GGBP) 282–285
– glutamine-binding protein (GlnBP) 285–287
– inhibition 112
– lipid A, synthesis 138, 140
ester(s)
– activated 43
– derivatives, B_6 113, 114
1-ethyl-3-(3-dimethylaminopropyl) carbodiimide hydrochloride (EDC) 43
etoposide, prodrug actrivation, 38C2-mediated 118–120
exponential force dependencies 78, 79
– fatty acid 138
– functions 42
– in prodrug activation 112–114

f

fd phage 20
fibers
– composite 14–17
– conductive 16
– continuous 16
– 1D 12
– fiber-like structures 14, 15
– long 15, 16
– M13, applications 16
– microfibers 16
– nanofibers 15, 16
– nonconjugated 18
– polyvinylpyrolidone (PVP), virus-blended 16
films
– acrylic 103
– A7–ZnS 21
– casting 40
– in drug sensors 106, 107
– Langmuir-Blodgett (LB) 224
– molecularly imprinted polymers (MIPs), electropolymerized 103
– monolayer 19
– mutant M13–ZnS 21
– sol-gel
– – imprinted with L-histidine 106
– – imprinted with nafcillin 104
– – imprinted with propranolol 103
– thin
– – applications 99
– – fabrication by 2D assembly 17
"flip-flop" type probe 314, 315, 321
Fllip, see "flip-flop" type probe
fluorescence, see also fluorescent proteins
– resonance energy transfer 105, 178
– sensors, "turn-on" 178, 181–185, "mix-and-detect" 186–189
– use in biomolecular sensing 177
fluorescence anisotropy 314
fluorescence correlation spectroscopy (FCS) 199
fluorescence emissions 86, 87
fluorescence polarization 316
fluorescence resonance energy transfer (FRET)
– in determining protein phosphorylation 319–321
– in determining protein–protein interactions 319
– in DNA detection 178
– in glucose binding 284, 285
– in molecular imaging 317–321
– in nanoparticle detection 105
– in probe preparation 265, 269–274
fluorescence signal 86, 87, 89
fluorescence spectroscopy 283, 284, 286, 287, 289, 291
fluorescent dyes 32
fluorescent microscopy 87
fluorescent particles 83, 86, 87
– fluorescent silica particles 84, 85
– ultra-bright fluorescent silica particles 70, 85, 87
fluorescent proteins, see also BFP; CFP; GFP; RFP; YFP
– applications 307, 308
– factors influencing activities of 317
– general requirements 307
– with changing with changing intensity or color 306, 307
fluorescent tagging 84
5-fluorodeoxyuridine 2 112
Förster distance 268
Förster resonance energy transfer, see FRET
Force curves 72, 73, 78, 80, 89
– normal cell force curves 80
– retraction force curves 80
– retracting curves 72
Fourier transform infrared (FTIR) spectroscopy 283, 286–289, 291
fullerenes 37, 42
functional peptides
– in development of fusion protein probes 312
– properties required 312, 313
functional shells 42–49
functional silanes 35
functional surfmers 43
fusion proteins
– A7 13
– immunogenic peptides 13
– in production of nanostructures 7
– J140– VIII 13
– probes for molecular imaging 307, 313, 316
– variable lymphocyte receptor (VLR) 152

g

D-galactose/D-glucose-binding protein (GGBP) 282–285
genetic engineering 7
genetic modification 1
genetic mutation 8
glucose-6-phosphate dehydrogenase (G6PD) inhibition 207–209

glutamine-binding protein (GlnBP) 285–287
green fluorescent protein (GFP)
– as fusion protein probes 307, 316
– as markers for gene expression 299, 300
– compared with RFP 301, 302
– in determining molecular dynamics 316, 317
– in determining protein–protein interactions 317–319
– use in FRET 317, 318
– variants, attributes 299–307
– wild-type, limitations 302, 306
grinding 32
glycocalyx molecules 82
glycosphingolipids (GSLs), see sphingolipids

h

Hank's balanced salt solution (HBSS) 73, 74, 83, 87
hemolysis 69
Hertz–Sneddon model 75
hexa-acylation 144
hybrid materials 12
hybrid viruses 7
hybrid virus–inorganic nanostructures 20
hybrid virus–silicon nanotube 20
hydrogen 82
hydrogen bonding 72
hysteresis 72

i

impedance spectroscopy 103
immune response 137
– effect of adjuvant 144
– role of MD-2/lipid IVa complex 149, 150
immune response
– innate natural immunity 137
– T cell immunity 53, 54
immunization, reactive 114
immunosensors, surface plasmon-based 219, 223, see also biosensors
inorganic core materials
– crystalline nanoparticles 34
– metal and metal oxides 36, 37
– silica nanoparticles 35, 36
inorganic nanowires 8, 10, 11, 14
inorganic nanoparticles 32
inorganic oxides, preparation of 20
interface, assembly 14

interfacial energy 12, 19
interfering compounds 103
intramolecular folding sensor 315
isothermal titration calorimetry 208

k

Kdo_2-lipid A, see LIPID MAPS
keratinocyte serum-free medium 72
Kirin Pharmaceuticals 159

l

Langmuir-Blodgett (LB)
– deposition 224
– films 224
– self-assembly 224
leucine-rich repeats (LRRs) 139
L-dopa 104
ligands, interaction with receptors 159
light-emitting devices 34
light-harvesting systems 9, 20
limit of detection (LOD) 101
– of surface plasmon resonance 233–235, 237, 238
lipids
– bilayer 247, 248
– composition 248
– Eritoran 151
– function 249
– identification, see LIPID MAPS
– in drug delivery 41
– lipid A, see LPS
– – chemical structure 138
– – derivatives, monosaccharide 146
– – hexa-acyl 144
– – in development of antibiotics 140
– – intermediates 153
– – minimal, modified 140–143
– – monophosphoryl lipid A 144
– – synthesis 140
– mycobacterial 160, 161
– physiological 41
– structure 249
– synthetic 41
lipid-binding proteins (LPBs) 138, 148
– structure 148–152
Lipid Metabolites And Pathways Strategy (LIPID MAPS) 143, 144
lipocalins and odorant-binding protein 289–293
lipopolysaccharide (LPS) 137, 138, see also endotoxin
– biosynthesis 140, 144
– discovery 137, 138

– effect of lipid IVa   149
– in detection of endotoxin in water   147
– neutralization   147
– recognition   138, 139
– relationship with Toll-like receptor   139, 151
liposomes
– mannosylated   53
– neutral   54
– poloaxmer F127   41
– stealth   41
lithium-ion batteries, storage capacity   22
localized surface plasmon resonance (LSPR)
– compared with SPR   235
– Mie theory   236
long-chain base (LCB)   153, 156
LpXF   142
luciferases
– as functional proteins   299, 308
– as reporters for bioanalysis and molecular imaging   310, 311
– attributes   299
– drawbacks   312
– from insects, marine organisms and prokaryotes   308–311
luminescent proteins, factors influencing activities   317
luminex   55, 56
lysomotropic carrier systems   40

m
macropinocytosis   53
magnetic beads/tweezers   70
magnetic moment   31
magnetic nanoparticles, *see also* nanoparticles
– applications   32, 37, 49, 50
– attributes
– – diameter, *see* critical particle size D_c
– – superparamagnetic property   37, 38
– biopurification   49, 50
– examples   37
– production   37
mass spectroscopy   199
M13 bacteriophage
– as template for inorganic materials synthesis   11
– in nanowire synthesis   12
– mutants, streptavidin-binding   20
– mutated   18
– native M13   11, 16
membrane (cell) disruption   251

metal oxide(s)   9, 32, 34
microemulsation   41
microscopy   71
microvilli   82, 89
microridges   82, 89
mirror image binding profile   103
moiety
– aminopyrazole   196
– β-diketone   131, 132
– carboxylic acid   114
– cysteine   43
– Kdo   146
– *N*-isopropyl acrylamide   40
– molecular recognition   32
– simple phosphate   142
– targeting   120
molecular clips   202, 205, 210
molecular dynamics (MD)   283, 284, 287, 291
molecular fluorophores
– as donors and acceptors in FRET-based assays and sensors   265
– in development of luminescent quantum dots   265
– limitations   265, 269
molecular imaging
– classical methods   313, 316, 317
– new methods   317–332
molecular imprinting (MI)
– chiral   97–107
– challenges   45, 46
– development   97
– of peptides   32, 45, 46
– of polymers   31, 56, *see also* MIPs
– of proteins   32, 45
– noncovalent   45
– reaction   33
– understanding the mechanisms
– – by isothermal titration chemistry (ITC)   45, 48
– – in chromatographic separations   97
– using protein surface recognition   210, 214
molecular recognition
– biomimetic feature   32
– in molecularly imprinted polymer (MIP) generation   47
– of target enantiomer   97
– recognition sites   97
– role in nature   44
molecular tweezers   202, 205, 206, 210
molecularly imprinted polymers (MIPs)
– applications

– – creation of recognition elements for chiral drugs 99
– – high-performance liquid chromatography 55, 98
– – immunoassays 55, 56
– – protein purification 55, 56
– – sensor technology 55, 56
– – solid-phase extraction 55
– – waste-water treatment 55
– coupling to sensor surface 99
– ephedrine imprinted 103
– nanoparticulate 32
– nanospheric 47
– recognition layers 101
– preparation 33
– quartz crystal microbalance (MIP–QCM) chiral sensor 106
– tools for developing 44–49
– – design of experiments (DoE) 44, 45
– – experimental high-throughput screening 44
– – nuclear magnetic resonance (NMR) spectroscopy 45
molecule(s)
– analyte 103
– – L-dopa 104
– amino-functionalized 43
– immunostimulatory 137
– polymer 120
– membrane-active, see peptide, membrane-active
monomers
– achiral 98
– functional 45
– in protein surface recognition 210–214
motifs
– functional binding 8
– materials-specific 12
– substrate-specific 12
multivalency 210
murine NXS2 neuroblastoma 118, 119
mutagenesis
– polymerase chain-reaction based site directed 7
– single amino acid substitution 7
M13 viral systems 20
M13 virus, genetically engineered
– as template in fabrication of composite nanowires 22
– conjugated with quantum dots 18

n

molecularly imprinted polymers (nanoMIPs)
– applications 44
– L-boc-nanoMIPs 49
– techniques for generating 44–49
nanocapsules, poly(alkylcyanoacrylate) 40
Nanocytes technology 54, 55
nanodevices 34
nanoparticles
– benefits 33, 51–54
– biodegradable 40
– biofunctionalization 42, 56
– biomimetic 31, 33, 42, 56
– controlling the size and morphology 34–36
– core materials 31
– gold 31, 34, 39, 54, 234, 235, 237
– limitations 52
– mannose-coated 53
– metal oxide 9
– poly(cyanoacrylate) 41, 54
– properties required 52
– solid phospholipid 41
– surface functionalization 35, 36, 54, 56, see also biofunctionalization
nanocircuits 20
nanoprecipitation 39, 40
nanorings
– formation 17
– structure 16
nanotechnology 1, 31, 70
nanotemplate hybridization 18
nanowires 12, 13, 22, 23
natural killer T (NKT) cells 158–165
NF-κB 139
nucleation
– by engineered virus 13
– of wurtzite ZnS 12, 13
– of ZnS nanocrystals 13
neurotransmitter 101

o

odorant-binding proteins (OBPs) 289–293
optical microscopy 87
optical tweezers 70
orbital shakers 83
organic core materials
– fullerenes 37, 42
– lipids 37, 41
– polymers 37–39
organic shell 31, 32, 42
Oswald ripening 47

p

cytolytic pore-forming toxins (PFTs)   179, see also PDAs
Parkinson's disease   101
peptides
– A7   12, 13
– Aβ
– – aggregation kinetics, methods of determining   199
– – inhibition   195, see also aminopyrazoles
– – role in Alzheimer's disease   194
– aminopyrazoles   195
– antimicrobial   247, 251, 253, 254, 256, 258
– anti-streptavidin   17
– hexahistidine   17
– membrane-active
– – applications   247
– – mode of action   249–252
– – properties   249
– natural   247, 252–254
– peptidomimetics   256
– supplementation fusiogenic   54
– synthetic   247, 254
Percoll   69
Pfeiffer, Richard   138
phage display   7
phospholipid membrane   70
photostability   84
pig odorant-binding protein (pOBP)
– functional characteristics and applications   292, 293
– structural characteristics   291, 292
PMB, effect on lypopolysaccharide toxicity   144
polyaniline (PANI)   14, 16
polycrystalline   12
polydiacetylene (PDA)
– advantages   178
– in development of biosensors for toxin detection   179–181
– in fabrication of "turn-on" fluorescence sensors   181–185, see also fluorescence
poly(lactic acid)   40
poly(lactic-co-glycolide)   40, 41
polymer(s)
– amphiphilic   40, in protein surface recognition   210
– conjugated   177, see also CPs
– copolymer   40, 122, 124, 134, in protein surface recognition   210, 214
– polymer-directed enzyme prodrug therapy (PDEPT)   120–122
– polymer-drug   121
– sol-gel   101
– synthetic   44
polymerization
– copolymerization, radical   32
– emulsion   39
– interfacial   40
– miniemulsion   32, 44, 47–49
polysaccharides   69, 82
polyvinylpyrrolidone polymer   69
prodrugs   203, see also prodrug activation
prodrug activation   111–114
– conversion of a drug to a prodrug   116
– etoposide, antibody-mediated   118
– in antitumor therapy, see cancer therapy
– prodrug **1**   112, 115
– prodrug **4**   112, 113
– prodrug **7**   113, 114
– prodrug **11**   115, 116
– prodrug **14**, see etoposide
– prodrug **19**   126
– single triggered trimeric prodrug   125
– with antibody 38C2   114–116, 118–120
primary cervical carcinomas   72
probe–cell contact   74
probe–cell interaction   78
proinflammatory cytokines   139
protease inhibitors   203
protease probes, QD FRET-based   269–274
proteoglycans   82
protein-fragment complementation assay (PCA)   323–326
proteins, see also coat proteins; fusion proteins; peptides
– abnormal folding   210
– A7 engineered P8   13
– artificial protein binders   210
– β-sheet, role in Alzheimer's disease   193
– CD14   138, 139, 144, 148
– D-galactose/D-glucose-binding protein (GGBP)   282–285
– D-trehalose/D-maltose-binding protein (TMBP)   287–289
– FhuA   148
– fluorescent, see BFP; CFP; GFP; RFP; YFP
– glutamine-binding protein (GlnBP)   285–287
– gold staining   32
– immobilization   42, 43, 54
– lipid-binding proteins (LPBs)   138, 148
– – structure   148–152
– lipocalins and odorant-binding protein (OBP)   289–293
– misfolded components   194
– myeloid differentiation-2 (MD-2)   139, 149
– – MD-2/lipid IVa complex   149

– – MD-2/TLR4 complex   149–152
– odorant-binding proteins (OBPs)   289–293
– P3   3, 7, 12, 17
– P6   7
– P7   8
– P8   3, 7, 12
– P9   7, 8, 17
– photoproteins, see luciferases
– purification   33
– solute-binding, functions and examples   281
– surface recognition   210–214
– transport system   281–282
– viral synthesis   7
protein phosphorylation, probes for determining   319
protein–protein interactions
– methods for determining
– – by BRET   321, 322
– – by FRET   318
– – by PCA   323–326
– – by two-hybrid systems   314
– probes for determining   319
protein tagging   314, 317
pulsed field-gradient longitudinal eddy-current delay techniques   211
pyridoxal-5'-phosphate (PLP)   155

q

quantum dots (QDs)
– as alternatives to traditional molecular fluorophores   269
– as reporting elements in chiral templates   105
– incorporated in molecularly imprinted polymers   105
– in imaging applications   34
– in development of bioanalytical probes   265
– in lypopolysaccharide synthesis   147
– luminescent QDs
– – characteristics and types   266–268
– – in development of protease probes   269–274
– – linked to molecular fluorophores   265
– – use in cancer detection   265
– surface functionalization   267
– water-soluble QDs, applications   267, 268

r

reactions
– alkaline hydrolysis   35
– carbodiimide coupling   5, 43
– chemical reaction vs transition state analog   114
– copper-catalyzed azide–alkyne cycloaddition   5
– covalent conjugation   43
– doxorubicin prodrug 11 activation   115
– molecular imprinting   33
– retro-aldol-retro-Michael   115, 116, 118, 125, 126, see also prodrug activation
– uncatalyzed   116
– sol-gel   32, 35
reactive immunization, see immunization
real-time biomolecular interaction analysis   219
receptors
– artificial   97
– biomolecular   32
– CD14   138, 139, 148
– CD1d   158, 163–167
– FhuA   148
– macrophage phagocytic   53
– natural killer T cell   158, 159–161, 166, 167
– ligand, interaction with   159
– T cell receptors (TCRs)   158–168
– TLR4   139, 145, 147, see also TLR4–MD-2 complex
– tumor-related   54
– variable lymphocyte receptor (VLR), hagfish   152
red fluorescent proteins (RFPs)   301, 302, 304, 306, 307
refractive index   220–222, 224, 230, 234, 237, 293
reflectrometric interference spectroscopy (RIfS)   212
region of constant compliance   72

s

"sea grass effect", mechanism   82, 83
selective enzymatic activation of a nontoxic drug to a toxic drug   112
selectivity, in drug delivery   123
self-assembled monolayers (SAMs)   224, 225, 227, 228
self-immolative dendrimer(s)   125
sensors, protein-based, see also biosensors; drug sensors
– for explosive compounds   284
– for glucose   284
– for glutamine   286
– for gliadin   286
– optical   286, 294

sepsis
- causes   139
- cost   139
- mortality   140
serine palmitoyltransferase (SPT) in sphingolipid biosynthesis   155–157
shells
- functional   42–49
- organic   31, 32, 42
sick cell anemia   210
signal transduction methods   103–107
silanes, functional   35
silanol groups   69, 82
silica   69
- colloidal   69
- crystalline   69
- fused   69
- organically modified   31
- pure   69
- silica ball   74, 82
- silica matrix   84
- silica nanoparticles   70
- silica particles   69, 70, 73, 74, 75, 80, 83–85, 87–89
- silica probe   80, 82
- silica sphere   73–75, 82, 83
- silica toxicity   69
- silica-coated polymeric beads   69
silicon acid   70
silicosis   69
single-stranded viral RNA   7
small angle X-ray scattering (SAXS)   15
solute-binding proteins, functions and examples   281
solvent displacement   39
specific capacity   22
specific optical absorbance   31
sphingolipids
- biosynthesis   154–157
- glycosphingolipids (GSLs)   154–168
- structure, function and metabolism   153
square wave voltammetry (SWV)   101
steric repulsion   78, 82
stimulators
- natural   137
- synthetic   137
Stöber process   35, 36
streptolysin O (SLO), detection of   179–181
sulfonamide Edonentan   130
superconducting quantum interference device (SQUID)   37, 38
superparamagnetism   34, 37, 38, 50
supramolecular assembly of polydiacetylenes   177, 179–181, *see also* PDAs; SLO

surface plasma resonance (SPR), *see also* biosensors
- bioassays for enhancing sensitivity and limit of detection (LOD)   233
- biorecognition elements   223
- – properties required   229
- compared with localized SPR (LSPR)   236
- in biomolecular interaction analysis   219, 220, 227–230
- in mass sensors   107
- working principle   220–222
surfactant molecules   84
surfmers, functional   43

t
T cells
- immunity   53, 54
- natural killer   158, *see also* receptors
templates
- biological   5
- chiral   97
- molecular   33, 44, 97
- template/analyte rebinding   105
- template–monomer complex   44
tetraethyl orthosilicate (TEOS)   35
thiopene   15
Thudichum, Johann   153
thymidylate synthetase   112
time-resolved small angle X-ray scattering (TRSAXS)   15
TLR4–MD-2 complex   149–152
tobacco mosaic virus (TMV)
- capsid monomer   6
- coat proteins, cysteine-substituted   2, 7, 9
- composite fibers, TMV/PANI (polyaniline)   14
- C-terminus   7
- native TMV   9, 10
- head-to-tail assembly   14
- nanoarrays fabrication   17
- nanotemplates   18
Toll receptor   139, 146, *see also* receptors, TLR4
transmission electron microscopy (TEM), in composite fiber study   14, 15
D-trehalose/D-maltose-binding protein (TMPB)   287–289
tumor necrosis factor-α (TNF-α)   43
- in cancer therapy   54, 55
- in sepsis   139
type 3 library   8, 12
type 8 library   8, 12
tyrosine   5

u
"UAG" stop codon   7

v
vaccine development   1, 53
van der Waals attraction   82
van der Waals force   72
van der Waals interactions   89
virus, see also M13 bacteriophage; TMV
– A7–P8 engineered   13
– bifunctional   17
– capsid   4
– cowpea chlorotic mottle virus   5
– cowpea mosaic virus   5
– hybrid   7
– hybrid virus–inorganic nanostructures   20
– multifunctional   1
– pVIII engineered   12
– rod-like   19, 20, 22
– tomato mosaic virus
– virus-like particles (VLPs)   1
– wild-type   13, 18
variable lymphocyte receptor (VLR)   152
viscoelastic effects   73

w
wet-spinning process   16
wurtzite CdS and ZnS phases   13
wurtzite ZnS structure, hexagonal   13

x
xMAP technology, see luminex
X-ray crystallography   1, 308

y
yellow fluorescent protein (YFP)   300, 303, 304, 318, 319, 322, 323
Young's modulus   78

z
ZnS nanocrystals   20
ZnS systems, mineralization   12